MIT-Pappalardo Series in Mechanical Engineering

Series Editors: Rohan C. Abeyaratne and Nam P. Suh

Energy and the Environment
James A. Fay and Dan S. Golomb

Axiomatic Design: Advances and Applications
Nam P. Suh

ENERGY AND THE ENVIRONMENT

ENERGY AND THE ENVIRONMENT

James A. Fay

Department of Mechanical Engineering
Massachusetts Institute of Technology

Dan S. Golomb

Department of Environmental, Earth, and Atmospheric Sciences
University of Massachusetts Lowell

New York ◆ Oxford
OXFORD UNIVERSITY PRESS
2002

Oxford University Press

Oxford   New York
Athens   Auckland   Bangkok   Bogotá   Buenos Aires   Cape Town
Chennai   Dar es Salaam   Delhi   Florence   Hong Kong   Istanbul   Karachi
Kolkata   Kuala Lumpur   Madrid   Melbourne   Mexico City   Mumbai   Nairobi
Paris   São Paulo   Shanghai   Singapore   Taipei   Tokyo   Toronto   Warsaw

and associated companies in
Berlin   Ibadan

Published by Oxford University Press, Inc.
198 Madison Avenue, New York, New York, 10016
http://www.oup-usa.org

Oxford is a registered trademark of Oxford University Press

Library of Congress Cataloging-in-Publication Data

Fay, James A.
    Energy and the environment / James A. Fay, Dan S. Golomb.
        p. cm.—(The MIT-Pappalardo Series in Mechanical Engineering)
    Includes bibliographical references and index.
    ISBN 978-0-19-515092-6
    1. Power resources—Environmental aspects. I. Golomb, D. II. Title. III. Series.
TD195.E49 F39 2002
    333.79′14–dc21   2001036249

Printing number: 9 8 7 6 5 4

Printed in the United States of America
on acid-free paper

CONTENTS

LIST OF TABLES

FOREWORD

In 1996, the MIT Department of Mechanical Engineering adopted a new undergraduate curriculum to enhance the learning process of its students. In this new curriculum, key concepts of engineering are taught in four integrated sequences: the thermodynamics/heat transfer/fluid mechanics sequence, the mechanics/materials sequence, the design/manufacturing sequence, and the systems/dynamics/control sequence. In each one of the four sequences, the basic principles are presented in the context of real engineering problems that require simultaneous use of all basic principles to solve engineering tasks ranging from synthesis to analysis. Active learning, including hands-on experience, is a key element of this new curriculum.

To support new instructional paradigms of the curriculum, the faculty began the development of teaching materials such as books, software for web-based education, and laboratory experiments. This effort at MIT is partially funded by the Neil and Jane Pappalardo fund, a generous endowment created at MIT in support of this project by the Pappalardos. Mr. Neil Pappalardo, an alumnus of MIT, is the founder and CEO of Medical Technology Information, Inc., and Mrs. Jane Pappalardo is a graduate of Boston University, active in many civic functions of Massachusetts.

Oxford University Press and MIT have created the MIT-Pappalardo Series in Mechanical Engineering to publish books authored by its faculty under the sponsorship of the Pappalardo fund. All the textbooks written for the core sequences, as well as other professional books, will be published under this series.

This volume, *Energy and the Environment* by James A. Fay and Dan S. Golomb, differs from the others in that it is not itself a subject in the core curriculum. Instead, it is an upper-level subject that draws upon the dynamics, fluid mechanics, thermodynamics, heat transfer, and related sciences of the core curriculum. While exposing the student to a societal problem of great current concern—namely, the use of energy and the local, regional, and global environmental effects that use engenders—it utilizes core curriculum skills in describing and analyzing the modern technology being used to ameliorate these adverse environmental effects. It enables the student to integrate this understanding into an appreciation of both the technology and science that must be employed by nations to maintain a livable environment while providing improved economic circumstances for their populations.

Energy and the Environment provides many provocative examples of advancing a student's skills in engineering fundamentals. Calculating how much power is needed to propel an automobile, how mechanical power can be extracted from the dynamical motion of the wind or ocean waves and the pull of gravity on river flows and tidal motions, how fuel cells and batteries generate electric power from chemical reactions, how power can be generated by the combustion of fossil fuels in conventional power plants, and how gaseous atmospheric contaminants can change the earth's temperature requires integration of the understanding achieved in core studies. Equally important is the quantitative understanding of the contamination of the atmosphere and surface waters by the toxic byproducts of energy use, their effects upon human health and natural ecological systems, and how these effects can be ameliorated by improvements in the technology of energy use.

We expect that the addition of this volume to the others of this series will expand the student's understanding of the role of mechanical engineering in modern societies.

Rohan C. Abeyaratne
Nam P. Suh
Editors
MIT-Pappalardo Series

PREFACE

The impetus for creating this book was provoked by one of us (DSG) as a consequence of lecturing on the subject of energy and the environment for the past 10 years at the University of Massachusetts Lowell to students in the Colleges of Engineering and Arts and Sciences. In all those years a diligent search did not unearth a suitable textbook to match the syllabus of that course. To be sure, numerous texts exist on the subjects of energy, energy systems, energy conversion, energy resources, and fossil, nuclear, and renewable energy. Also, there are texts on air pollution and its control, effluents and solid waste from energy mining and usage, the greenhouse effect, and so on. However, we were unable to find a contemporary text that discusses on a deeper technical level the relationship between energy usage and environmental degradation or that discusses the means and ways that efficiency improvements, conservation, and shifts to less polluting energy sources could lead to a healthier and safer environment.

Our book is intended for upper-level undergraduate and graduate students and for informed readers who have had a solid dose of science and mathematics. While we do try to refresh the student's and reader's memory on some fundamental aspects of physics, chemistry, engineering and geophysical sciences, we are not bashful about using some advanced concepts, the appropriate mathematical language, and chemical equations. Each chapter is accompanied by a set of numerical and conceptual problems designed to stimulate creative thinking and problem solving.

Chapter 1 is a general introduction to the subject of energy, its use, and its environmental effects. It is a preview of the subsequent chapters and sets the context of their development.

In Chapter 2 we survey the world's energy reserves and resources. We review historic trends of energy usage and estimates of future supply and demand. This is done globally, by continent and country, by energy use sector, and by proportion to population and gross domestic product. The inequalities of global energy supply and consumption are discussed.

Chapter 3 is a refresher of thermodynamics. It reviews the laws that govern the conversion of energy from one form to another—that is, the first and second laws of thermodynamics and the concepts of work, heat, internal energy, free energy, and entropy. Special attention is given to the combustion of fossil fuels. Various ideal thermodynamic cycles that involve heat or combustion engines are discussed—for example, the Carnot, Rankine, Brayton, and Otto cycles. Also, advanced and combined cycles are described, as well as nonheat engines such as the fuel cell. The principles of the production of synthetic fuels from fossil fuels are treated.

The generation and transmission of electrical power, as well as the storage of mechanical and electrical energy, are covered in Chapter 4. Electrostatic, magnetic, and electrochemical storage of electrical energy is treated, along with various mechanical energy storage systems.

The generation of electricity in fossil-fueled power plants is thoroughly discussed in Chapter 5. The complete workings of a fossil-fueled power plant are described, including fuel storage and preparation, burners, boilers, turbines, condensers, and generators. Special emphasis is placed on emission control techniques, such as particulate matter control with electrostatic precipitators, sulfur oxide control with scrubbers, and nitric oxide control with low-NO_x burners and flue gas denitrification. Alternative coal-fired power plants are discussed, such as fluidized bed combustion and coal gasification combined cycle.

In Chapter 6 we describe electricity generation in nuclear-fueled power plants. Here we review the fundamentals of nuclear energy: atoms, isotopes, the nucleus and electrons, protons and neutrons, radioactivity, nuclear stability, fission, and fusion. The nuclear fuel cycle is described, including mining, purification, enrichment, fuel rod preparation, and spent fuel (radioactive waste) disposal. The workings of nuclear reactors are discussed, including control rods, moderators, and neutron economy, as well as the different reactor types: boiling water, pressurized water, and breeder reactors.

The principles of renewable energy utilization are explained in Chapter 7. This includes hydropower, biomass, geothermal, solar thermal and photovoltaic, wind, tidal, ocean wave, and ocean thermal power production. Attention is given to the capacity factor and capital cost of these systems.

Chapter 8 is devoted principally to the automobile, because road vehicles consume roughly one-third of all primary energy and also because the transportation of people and goods is so dependent upon them. The characteristics of the internal combustion engine are described, for both gasoline and diesel engines. The importance of vehicle characteristics for vehicle fuel efficiency is stressed. Electric drive vehicles are described, including battery-powered and hybrid vehicles. Vehicle emissions are explained, and the technology for reducing them is described.

A survey of the environmental effects of fossil fuel usage begins in Chapter 9. In this chapter we discuss urban and regional air pollution, the transport and dispersion of particulate matter, sulfur oxides, nitrogen oxides, carbon monoxide, and other toxic pollutants from fossil fuel combustion, and the effects of these pollutants on human health, biota, materials, and aesthetics. The phenomena of photochemical smog, acid deposition and regional haze are also described. Also treated are the impacts of energy usage on water and land.

Chapter 10 continues the survey of environmental effects of fossil fuel combustion with particular reference to global climate change resulting from anthropogenic enhancement of the greenhouse effect. Here we discuss the carbon dioxide emission trends and forecast, the global carbon cycle, and the uptake of CO_2 by the oceans and biota. The physics of the greenhouse effect is described in some detail, as well as the predicted consequences to the planet and its inhabitants if CO_2 emissions continue unabated.

We conclude with Chapter 11, a reemphasis of the important relationships among the science, technology, and economics of energy usage and its environmental effects. We note the limited success of regulation of urban and regional air pollution in industrialized nations, and we also note the great challenge that lies ahead in dealing with global climate change.

Finally, we include Appendix A, an explanation of the scientific and engineering units that are commonly used in energy studies, easing the translation from one set to another.

The authors wish to express their appreciation to colleagues who aided in the review of the manuscript: John Heywood, Wai K. Cheng, and Jason Mark. Of course, the authors bear complete responsibility for the accuracy of this text. We also thank George Fisher for preparing many of the tables and figures.

James A. Fay
Massachusetts Institute of Technology

Dan S. Golomb
University of Massachusetts Lowell

Energy and the Environment

1.1 INTRODUCTION

Modern societies are characterized by a substantial consumption of fossil and nuclear fuels needed to provide for the operation of the physical infrastructure upon which these societies depend: the production of food and water, clothing, shelter, transportation, communication, and other essential human services. The amount of this energy use and its concentration in the urban areas of industrialized nations has caused the environmental degradation of air-, water- and land-dependent ecosystems on a local and regional scale, as well as adverse health effects in human populations. Recent scientific studies have forecast potentially adverse global climate changes that would result from the accumulation of gaseous emissions to the atmosphere, principally carbon dioxide from energy related sources. This accumulation is aggravated by an expected expanding consumption of energy both by industrialized nations and by developing nations seeking to improve the living standards of their growing populations. The nations of the world, individually and collectively, are undertaking to limit the damage to human health and natural ecosystems that attend these current problems and to forestall the development of even more severe ones in the future. But because the source of the problem, energy usage, is so intimately involved in nations' and the world's economies, it will be difficult to ameliorate this environmental degradation without some adverse effects on the social and economic circumstances of national populations.

To comprehend the magnitude of intensity of human use of energy in current nations, we might compare it with the minimum energy needed to sustain an individual human life, that of the caloric value of food needed for a healthy diet. In the United States, which is among the most intensive users of energy, the average daily fossil fuel use per capita amounts to 56 times the necessary daily food energy intake. On the other hand, in India, a developing nation, the energy used is only 3 times the daily food calorie intake. U.S. nationals expend 20 times the energy used by Indian nationals, and their per capita share of the national gross domestic product is 50 times greater. Evidently, the economic well-being of populations is closely tied to their energy consumption.

When agricultural technology began to displace that of the hunter–gatherer societies about 10,000 years ago, activities other than acquiring food became possible. Eventually other sources of mechanical energy—that of animals, wind, and water streams—were developed, augmenting human labor and further enhancing both agricultural and nonagricultural pursuits. As world population increased, the amount of crop and pasture land increased in proportion, permanently replacing natural forest and grassland ecosystems by less diverse ones. Until the beginning of the industrial revolution several centuries ago, this was the major environmental impact of human activities. Today, we are approaching the limit of available land for agricultural purposes, and only more intensive use of it can provide food for future increases of world population.

The industrial revolution drastically changed the conditions of human societies by making available large amounts of energy from coal (and later oil, gas, and nuclear fuel) far exceeding that available from the biofuel, wood. Some of this energy was directed to increasing the productivity of agriculture, freeing up a large segment of the population for other beneficial activities. Urban populations grew rapidly as energy-using activities, such as manufacturing and commerce, concentrated themselves in urban areas. Urban population and population density increased, while those of rural areas decreased.

By the middle of the twentieth century, nearly all major cities of the industrialized world experienced health-threatening episodes of air pollution, and today this type of degradation has spread to the urban areas of developing countries as a consequence of the growing industrialization of their economies. Predominantly, urban air pollution is a consequence of the burning of fossil fuels within and beyond the urban region itself. This pollution can extend in significant concentrations to rural areas at some distance from the pollutant sources so that polluted regions of continental dimensions even include locations where there is an absence of local energy use.

Despite the severity of urban pollution, it is technically possible to reduce it to harmless levels by limiting the emission of those chemical species that cause the atmospheric degradation. The principal pollutants comprise only a very small fraction of the materials processed and can be made even smaller, albeit at some economic cost. In industrialized countries, the cost of abating urban air pollution is but a minor slice of a nation's economic pie.

While the industrialized nations grapple with urban and regional air pollution, with some success, and developing nations lose ground to the intensifying levels of harmful urban air contamination, the global atmosphere experiences an untempered increase in greenhouse gases, those pollutants that are thought to cause the average surface air temperature to rise and climate to be modified. Unlike the urban pollutants, most of which are precipitated from the atmosphere within a few days of their emission, greenhouse gases accumulate in the atmosphere for years, even centuries. The most common greenhouse gas is carbon dioxide, which is released when fossil fuels are burned. As it is not possible to utilize the full energy of fossil fuels without forming carbon dioxide, it will be very difficult to reduce the global emissions of carbon dioxide while still providing enough energy to the world's nations for the improvement of their economies. While there is technology available or being developed that would make possible substantial reductions in global carbon dioxide emissions, the cost of implementation of such control programs will be much larger than that for curbing urban air pollution.

1.1.1   An Overview of This Text

This book describes the technology and scientific understanding by which the world's nations could ameliorate the growing urban, regional, and global environmental problems associated with energy use while still providing sufficient energy to meet the needs of populations for a humane existence. It focuses on the technology and science, the base on which any effective environmental control program must be built. It does not prescribe control programs, because they must include social, economic, and political factors that lie outside the scope of this book. We do not delve deeply into the science and technology, but do provide an adequate description of the fundamental principles and their consequences to the topic at hand. We present a bibliography in each chapter for the use of the reader who wants to pursue some aspects at greater depth.

The major sources of energy for modern nations are fossil fuels, nuclear fuels, and hydropower. Non-hydro renewable energy sources, such as biomass, wind, geothermal, solar thermal, and photovoltaic power, account for only a small portion of current energy production. Like other mineral deposits, fossil fuels are not distributed uniformly around the globe, but are found on continents and their margins that were once locations of great biomass production. They need to be discovered and removed, and often processed, before they can be available for energy production. Current and expected deposits would appear to last for a few centuries at current consumption rates. Within the time horizon of most national planning, there is no impending shortage of fossil fuel despite the continual depletion of what is a finite resource. In contrast, renewable energy sources are not depletable, being supplied ultimately by the flux of solar insolation that impinges on the earth.

Like food, energy needs to be stored and transported from the time and place where it becomes available to that where it is to be used. Fossil and nuclear fuels, which store their energy in chemical or nuclear form indefinitely, are overwhelmingly the preferred form for storing and transporting energy. Electrical energy is easily transmitted from source to user, but there is no electrical storage capability in this system. Hydropower systems store energy for periods of days to years in their reservoirs. For most renewable energy sources, there is no inherent storage capability so they must be integrated into the electrical network. Many forms of mechanical and electrical energy storage are being developed to provide for special applications where storage in chemical form is not suitable. Efficient transformation of energy from mechanical to electrical form is an essential factor of modern energy systems.

Although fossil fuels may be readily burned to provide heat for space heating, cooking, or industrial and commercial use, producing mechanical or electrical power from burning fuels required the invention of power producing machines, beginning with the steam engine and subsequently expanding to the gasoline engine, diesel engine, gas turbine, and fuel cell. The science of thermodynamics prescribes the physicochemical rules that govern how much of a fuel's energy can be transformed to mechanical power. While perfect machines can convert much of the fuel's energy to work, practical and economic ones only return between a quarter and a half of the fuel energy. Nevertheless, the technology is rich and capable of being improved through further research and development, but large increases in fuel efficiency are not likely to be reached without a considerable cost penalty.

Initially, steam engines were used to pump water from mines, to power knitting mills, and to propel trains and ships. Starting in the late nineteenth century, electrical power produced by steam engines became the preferred method for distributing machine power to distant end-users. By the time electricity distribution had become universal (supplying mechanical power, light, and communication signals), the generation of electrical power in steam power plants had become the largest segment of energy use. Currently, 55% of world fossil fuel is consumed in electric power plants.

The modern fossil-fueled steam power plant is quite complex (see Figure 1.1). Its principal components—the boiler, the turbine, and the condenser—are designed to achieve maximum thermal efficiency. But the combustion of the fuel produces gaseous and solid pollutants, among which are the following: oxides of carbon, sulfur, and nitrogen; soot; toxic metal vapors; and ash. Removing these pollutants from the flue gases requires complex machinery, such as scrubbers and electrostatic precipitators, that increases the operating and capital cost of the power plant and consumes a small percentage of its electrical output. The removed material must be disposed safely in a landfill. But because of the size and technical sophistication of these plants, they provide a more certain avenue of improvement in control than would many thousands of small power plants of equal total power.

Nuclear power plants utilize a steam cycle to produce mechanical power, but steam for the turbine is generated by heat transfer from a hot fluid that passes through the nuclear reactor, or by

Figure 1.1 A large coal-fired steam–electric power plant whose electrical power output is nearly 3000 megawatts. In the center is the power house and tall stacks that disperse the flue gas. To the left, a cooling tower provides cool water for condensing the steam from the turbines. To the right, high-voltage transmission lines send the electric power to consumers. (By permission of Brian Hayes.)

direct contact with the reactor fuel elements. The main disadvantage of a nuclear power plant, which does not release any ordinary pollutants to the air, is the difficulty of assuring that the immense radioactivity of its fuel is never allowed to escape by accident. Nuclear power plant technology is technically quite complex and expensive. The environmental problems associated with preparing the nuclear fuel and sequestering the spent fuel have become very difficult and expensive to manage. In the United States, all of these problems make new nuclear power plants more expensive than new fossil fuel power plants.

Renewable energy sources are of several kinds. Wind turbines and ocean wave energy systems convert the energy of the wind and ocean waves that stream past the power plant to electrical power. Hydropower and ocean tidal power plants convert the gravitational energy of dammed up water to electrical power (see Figure 1.2). Geothermal and ocean thermal power plants make use of a stream of hot or cold fluid to generate electric power in a steam power plant. A solar thermal power plant absorbs sunlight to heat steam in a power cycle. Photovoltaic systems create electricity by direct absorption of solar radiation on a semiconductor surface. Biomass-fueled power plants directly burn biomass in a steam boiler or utilize a synthetic fuel made from biomass. Most of these energy systems experience low-energy flux intensity, so that large structures are required per unit of power

Figure 1.2 A run-of-the-river hydropower plant on the Androscoggin River in Brunswick, Maine (United States). In the center is the power house, on the right is the dam/spillway, and on the left is a fish ladder to allow anadromous species to move upriver around the dam. Except when occasional springtime excessive flows are diverted to the spillway, the entire river flow passes through the power house.

output, compared to fossil-fueled plants. On the other hand, they emit no or few pollutants, while contributing no net carbon dioxide emissions to the atmosphere. Their capital cost per unit of power output is higher than that of fossil plants, so that renewable plants may not become economical until fossil fuel prices rise.

Transportation energy is a major sector of the energy market in both industrialized and developing nations. Automobiles are a major consumer of transportation energy and emitter of urban air pollutants. The technology of automobiles has advanced considerably in the last several decades under regulation by governments to reduce pollutant emissions and improve energy efficiency. Current automobiles emit much smaller amounts of pollutants than their uncontrolled predecessors as a consequence of complex control systems. Considerable gains in energy efficiency seem possible by introduction of lightweight body designs and electric drive systems powered by electric storage systems or onboard engine-driven electric generators, or combinations of these.

Air pollutants emitted into the urban atmosphere by fossil fuel users and other sources can reach levels harmful to public health. Some of these pollutants can react in the atmosphere by absorbing sunlight so as to form even more harmful toxic products. This soup of direct and indirect pollutants is termed *smog*. One component of smog is the toxic oxidant ozone, which is not directly emitted by any source. Because of the chemical complexity of these photochemical atmospheric reactions, great effort is required to limit all the precursors of photochemical smog if it is to be reduced to low levels.

Carbon dioxide and other greenhouse gases warm the lower atmosphere by impeding the radiative transfer of heat from the earth to outer space. Limiting the growth rate of atmospheric carbon dioxide requires either (a) reducing the amount of fossil fuel burned or (b) sequestering the carbon dioxide below the earth's or ocean's surface. To maintain or increase the availability of energy while fossil fuel consumption is lowered, renewable or nuclear energy must be used. Of course, improving the use efficiency of energy can result in the lowering of fossil fuel use while not reducing the social utility of energy availability. By combination of all these methods, the rate of rise of atmospheric carbon dioxide can be ameliorated at an economic and social cost that may be acceptable.

The amelioration of environmental degradation caused by energy use is a responsibility of national governments. By regulation and by providing economic incentives, governments induce energy users to reduce pollutant emissions by changes in technology or use practices. Bilateral or global treaties can bring about coordinated multinational actions to reduce regional or global environmental problems, such as acid deposition, ozone destruction, and climate change. The role of technology is to provide the necessary reduction in emissions while still making available energy at the minimum increase in cost needed to attain that end.

1.2 ENERGY

There is a minimum amount of energy needed to sustain human life. The energy value of food is the major component, but fuel energy is needed for cooking and, in some climates, for heating human shelter. In an agricultural society, additional energy is expended in growing, reaping, and storing food, making clothing, and constructing shelters. In modern industrial societies, much more energy than this minimum is consumed in providing food, clothing, shelter, transportation, communication, lighting, materials, and numerous services for the entire population.

It is a basic principle of physics that energy cannot be destroyed, but can be transformed from one form to another. When a fuel is burned in air, the chemical energy released by the rearrangement of fuel and oxygen atoms to form combustion products is transformed to the random energy of the hot combustion product molecules. When food is digested in the human digestive tract, some of the food energy is converted to energy of nutrient molecules and some warms the body. When human societies "consume" energy, they transform it from one useful form to a less useful form, in the process providing a good or service that is needed to maintain human life and societies.

A quantitative measure of the ongoing good that energy "consumption" provides to society is the time rate of transformation of the useful energy content of energy-rich materials, such as fossil and nuclear fuels. In 1995, this worldwide consumption rate amounted to 12.1 terawatts (TW)[1], or about 2 kilowatts (kW) per capita.[2] Of this world total, the United States consumed 2.9 TW, or about 13 kW per capita, which is the largest of any nation. However unevenly distributed among the world's population, the world energy consumption rate far exceeds the minimum required to sustain human life.

The capacity to consume energy at this rate is a consequence of the technology developed in industrialized nations to permit the efficient extraction and utilization of these fuels by only a small fraction of the population.[3] But the earth's fossil and nuclear fuel resources are being depleted at a rate that will render them very scarce in future centuries, even if they are used more efficiently than in the past. The current cost of these fuels, however, has remained low for decades as recovery technology has improved enough to offset the distant threat of scarcity.

---

[1] One terawatt = 1E(12) watts. See Appendix A for a specification of scientific notation for physical units.

[2] This rate is 16 times the per capita food energy consumption of 120 W.

[3] This situation is analogous to the industrialization of agriculture in advanced economies, whereby a few percent of the population provides food for all.

There are other, less energy-rich sources of energy which are not depletable. These are the so-called renewable energies, such as those of solar insolation, wind, flowing river currents, tidal flows, and biomass fuels. In fact, these are the energies that were developed on a small scale in preindustrial societies, providing for ocean transportation, cooking, sawing of lumber, and milling of grain. Industrial age technologies have made it possible to develop these sources today on a much larger scale, yet in aggregate they constitute less than 8% of current world energy consumption. Renewable energy is currently more costly than fossil energy, but not greatly so, and may yet become more economical when pollution abatement costs of fossil and nuclear energy are factored in.

How is energy used? It is customary to divide energy usage among four sectors of economic activity: industrial (manufacturing, material production, agriculture, resource recovery), transportation (cars, trucks, trains, airplanes, pipelines and ships), commercial (services), and residential (homes). In the United States in 1996, these categories consumed, respectively, 36%, 27%, 16%, and 21% of the total energy. Considered all together, energy is consumed in a myriad of individual ways, each of which is an important contributor to the functioning of these sectors of the national economy.

One prominent use of energy, principally within the industrial and commercial sectors, is the generation of electric power. This use of energy now constitutes 36% of the total energy use worldwide, but 44% in the United States. Combined with the transportation sector, these two uses comprise 70% of the total U.S. energy use. For this reason, electric power production and transportation form the core energy uses discussed in this text.

How is energy supplied? Except for renewable energy sources (including hydropower), the main sources of energy are fossil and nuclear fuels, which are depletable minerals that must be extracted from the earth, refined as necessary, and transported to the end-user in amounts needed for the particular uses. Given the structure of modern industrialized economies, supplying energy is a year-round activity in which the energy is consumed within months of being extracted from its source.[4] While there are reserves of fossil and nuclear fuels that will last decades to centuries at current consumption rates, these are not extracted until they are needed for current consumption.[5] Because fossil and nuclear fuel reserves are not uniformly distributed within or among the continents, some nations are fuel poor and others fuel rich. The quantities of fuel traded among nations is a significant fraction of overall energy production.

1.2.1 Electric Power

One hallmark of industrialization in the twentieth century has been the growth of the electric power sector, which today consumes about 36% of the world's energy in the production of an annual average of 1.4 TW of electric power. In the United States, 44% of total energy is used to generate an annual average of 0.4 TW of electrical power. Nearly all of this electric power is produced in large utility plants, each generating in the range of 100 to 1000 megawatts (MW). Fossil and nuclear fuels supply 63% and 17%, respectively, of the total electric power, the remainder being generated

---

[4]The United States has established a crude oil reserve for emergency use, to replace a sudden cutoff of foreign oil supplies. The reserve contains only several months' supply of imported oil.

[5]In contrast, food crops are produced mostly on an annual basis, requiring storage of food available for marketing for the better part of a year.

in renewable energy plants, of which hydropower (19%) is the overwhelming contributor.[6] The generation and distribution of electric power to numerous industrial, commercial, and residential consumers is considered today to be a requirement for both advanced and developing economies.

The electric energy produced in power plants is very quickly transmitted to the customer, where it is instantaneously consumed for a multitude of purposes: providing light, generating mechanical power in electric motors, heating space and materials, powering communication equipment, and so on. There is practically no accumulation of energy in this system, in contrast with the storage of fuels (or water in hydrosystems) at power plants, so that electric energy is produced and consumed nearly simultaneously.[7] Electric power plants must be operated so as to maintain the flow of electric power in response to the instantaneous aggregate demand of consumers. This is accomplished by networking together the electric power produced by many plants so that a sudden interruption in the output of one plant can be replaced by the others.

1.2.2 Transportation Energy

Transportation of goods and people among homes, factories, offices, and stores is a staple ingredient of industrialized economies. Ground, air, and marine vehicles powered by fossil-fueled combustion engines are the principal means for providing this transportation function.[8] Transportation systems require both vehicle and infrastructure: car, truck, and highway; train and railway; airplane and airport; ship and marine terminal. Ownership, financing, and construction of the infrastructure is often distinct from that of the vehicle, with public ownership of the infrastructure and private ownership of the vehicle being most common.

In economic terms, the largest transportation sector is that of highways and highway vehicles. Worldwide, highway vehicles now number about 600 million, 200 million of them in the United States. In the United States, 96% of the road vehicles are passenger automobiles and light-duty trucks. The world and U.S. vehicle populations are growing at annual rates of 2.2% and 1.7%, respectively. On average, U.S. vehicles are replaced every 13 years or so, providing an opportunity to implement relatively quickly improvements in vehicle technology.[9]

Transportation fuels are nearly all petroleum-derived. In the United States, transportation consumes 70% of the petroleum supply, or 32% of all fossil fuel energy. Highway vehicles account for 46% of petroleum consumption, or 21% of all fossil fuel energy. Transportation systems are especially vulnerable to interruptions in the supply of imported oil, which now exceeds the supply from domestic production. Unlike some stationary users of oil, transportation vehicles cannot substitute coal or gas for oil in times of scarcity.

While substantial reduction in highway vehicle air pollutant emissions has been achieved in the United States since 1970, and more reductions are scheduled for the first decade of the twenty-first century, the focus of vehicle technology has shifted to improving vehicle fuel economy. Doubling current fuel efficiencies without penalizing vehicle performance is technically possible, at a vehicle

---

[6]When comparing the amount of hydropower energy with that of fossil and nuclear, the former is evaluated on the basis of fuel energy needed to generate the hydroelectric power output of these plants.

[7]In some renewable energy electric power systems, such as wind and photovoltaic power systems, there is usually no energy storage; these systems can comprise only a part of a reliable electric energy system.

[8]In developing countries, human-powered bicycles may be important components of ground transportation.

[9]In contrast, fossil and nuclear power plants have a useful life of 40 years or more.

manufacturing cost penalty that will be offset in part by fuel cost savings. Automobiles promise to be one of the more cost-effective ways for reducing oil consumption and carbon emissions.

1.2.3 Energy as a Commodity

Because of the ubiquitous need for energy, combined with the ability to store and utilize it in many forms, energy is marketed as a commodity and traded internationally at more or less well established prices. For example, in recent years the world crude oil price has ranged from about 15–35 \$/barrel, or about 2–5 \$/GJ.[10] Coal is generally cheaper than oil, whereas natural gas is more expensive. The difference in price reflects the different costs of recovery, storage, and transport. Nuclear fuel refined for use in nuclear electric power plants is less expensive than fossil fuels, per unit of heating value.

Coal is the cheapest fuel to extract, especially when mined near the earth's surface. It is also inexpensive to store and transport, both within and between continents. But it is difficult to use efficiently and cleanly, and in the United States it is used mainly as an electric utility fuel. Oil is more expensive than coal to recover, being more dispersed within geologic structures, but is more easily transported by pipeline and intercontinentally by supertanker. It is almost exclusively the fuel of transportation vehicles, and it is also the fuel of choice for industrial, commercial, and residential use in place of coal. Like oil, natural gas is recovered from wells, but is not easily stored or shipped across oceans. It commands the highest price because of the greater cost of recovery, but is widely used in industry, commerce, and residences because of its ease, efficiency, and cleanliness of combustion.

In contrast with fossil and nuclear fuels, renewable energy is not transportable (except in the form of electric power) or storable (except in hydropower and biomass systems). Renewable hydropower electricity is a significant part of the world electric power supply and is sold as a commodity intra- and internationally.

Synthetic fuels, such as hydrogen, ethanol, and producer gas, are manufactured from other fossil fuels. By transforming the molecular structure of a natural fossil fuel to a synthetic form while preserving most of the heating value, the secondary fuel may be stored or utilized more easily, or provide superior combustion characteristics, but is inevitably more expensive than its parent fuel.[11]

On the time scale of centuries, the supplies of fossil and nuclear fuels will be severely depleted, leaving only deposits that are difficult and expensive to extract. The only sources that could supply energy indefinitely beyond that time horizon are nuclear fusion and renewable energy. These are both capital-intensive technologies. Their energy costs will inevitably be competitive with fossil and fission fuels when the latter become scarce enough.[12]

---

[10]A barrel of crude oil contains about 6 GJ (6 MBtu) of fuel heating value.

[11]Plutonium-239, a fissionable nuclear fuel, is formed from uranium-238, a nonfissionable natural mineral, in nuclear reactors. In this sense it is a synthetic nuclear fuel, which can produce more energy than is consumed in its formation, unlike fossil fuel-based synthetic fuels.

[12]If fusion power plants will be no more expensive than current fission plants, at about 0.3–1 dollar per thermal watt of heat input, then the capital cost of supplying the current U.S. energy consumption of about 3 TW would be 1–3 trillion dollars (T\$). The cost of this energy would be several times current costs.

1.3 THE ENVIRONMENT

The twentieth century, during which industrialization proceeded even faster than population growth, marked the beginning of an understanding, both popular and scientific, that human activity was having deleterious effects upon the natural world, including human health and welfare. These effects included increasing pollution of air, water, and land by the byproducts of industrial activity, permanent loss of natural species of plants and animals by changes in land and water usage and human predation, and, more recently, growing indications that the global climate was changing because of the anthropogenic emissions of so-called greenhouse gases.

At first, attention was focused on recurring episodes of high levels of air pollution in areas surrounding industrial facilities, such as coal burning power plants, steel mills, and mineral refineries. These pollution episodes were accompanied by acute human sickness and the exacerbation of chronic illnesses. After mid-century, when industrialized nations' economies recovered rapidly from World War II and expanded greatly above their prewar levels, many urban regions without heavy industrial facilities began to experience persistent, chronic, and harmful levels of photochemical smog, a secondary pollutant created in the atmosphere from invisible volatile organic compounds and nitrogen oxides produced by burning fuels and the widespread use of manufactured organic materials. Concurrently, the overloading of rivers, lakes, and estuaries with industrial and municipal wastes threatened both human health and the ecological integrity of these natural systems. The careless disposal on land of mining, industrial, and municipal solid wastes despoiled the purity of surface and subsurface water supplies.

As the level of environmental damage grew in proportion to the rate of emission of air and water pollutants, which themselves reflected the increasing level of industrial activity, national governments undertook to limit the rate of these emissions by requiring technological improvements to pollutant sources. As a consequence, by the century's end ambient air and water pollution levels were decreasing gradually in the most advanced industrialized nations, even though energy and material consumption was increasing. Nevertheless, troubling evidence of the cumulative effects of industrial waste disposal became evident, such as acidification of forest soils, contamination of marine sediments with municipal waste sludge, and poisoning of aquifers with drainage from toxic waste dumps. Not the least of the impending cumulative waste problems is the disposal of used nuclear power plant fuel and its reprocessing wastes.

Environmental degradation is not confined to urban regions. In preindustrial times, large areas of forest and grassland ecosystems were replaced by much less diverse crop land. Subsequently, industrialized agriculture has expanded the predominance of monocultured crops and intensified production by copious applications of pesticides, herbicides, and inorganic fertilizers. Valuable topsoil has eroded at rates above replacement levels. Forests managed for pulp and lumber production are less diverse than their natural predecessors, the tree crop being optimized by use of herbicides and pesticides. In the United States, factory production of poultry and pork have created severe local animal waste control problems.

The most threatened, and most diverse, natural ecosystems on earth are the tropical rain forests. Tropical forest destruction for agricultural or silvicultural uses destroys ecosystems of great complexity and diversity, extinguishing irreversibly an evolutionary natural treasure. It also adds to the burden of atmospheric carbon dioxide in excess of what can be recovered by reforestation.

The most sobering environmental changes are global ones. The recent appearance of stratospheric ozone depletion in polar regions, which could increase harmful ultraviolet radiation at the earth's surface in mid-latitudes should it increase in intensity, was clearly shown by scientific research to be a consequence of the industrial production of chlorofluorocarbons. (By international treaty, these chemicals are being replaced by less harmful ones, and the stratospheric ozone destruction will eventually be reduced.) But the more ominous global pollutants are infrared-absorbing molecules, principally carbon dioxide but including nitrous oxide and methane, that are inexorably accumulating in the atmosphere and promising to disturb the earth's thermal radiation equilibrium with the sun and outer space. It is currently believed by most scientists that this disequilibrium will cause the average atmospheric surface temperature to rise, with probable adverse climatic consequences. Because carbon dioxide is formed ineluctibly in the combustion of fossil fuels that produce much of current and expected future energy use, and is known to accumulate in the atmosphere for centuries, its continued emission into the atmosphere presents a problem that cannot be managed except on a global scale. It is a problem whose control would greatly affect the future course of energy use for centuries to come.

1.3.1 Managing Industrial Pollution

To address the problem of a deteriorating environment, industrialized nations have undertaken to regulate the emission of pollutants into the natural environment, whether it be air, water, or land. The concept that underlies government control is that the concentration of pollutants in the environment must be kept below a level that will assure no harmful effects in humans or ecological systems. This can be achieved by limiting the mass rate of pollutant emissions from a particular source so that, when mixed with surrounding clean air or water, the concentration is sufficiently low to meet the criterion of harmlessness.[13]

In the case of multiple sources located near to each other, such as automobiles on a highway or many factories crowded together in an urban area, the additive effects require greater reduction per source than would be needed if only one isolated source existed. In industrialized countries and regions, the cumulative effects of emissions into limited volumes of air or water result in widespread contamination, with both local and distant sources contributing to local levels.

The ultimate example of cumulative effects is the gradual increase in the global annual average atmospheric carbon dioxide concentration caused by the worldwide emissions from burning of fossil fuels and forests. Because the residence time of carbon dioxide in the atmosphere is of the order of a century, this rise in atmospheric concentration reflects the cumulative emissions over many prior decades. Unlike urban or regional air pollutant emission reduction, reducing carbon dioxide emissions will not reduce the ambient carbon dioxide level, only slow its inexorable rise.

The scientific and technological basis for national and international management of environmental pollution is the cumulative understanding of the natural environment, the technology of industrial processes that release harmful agents into the environment, and the deleterious effects upon humans and ecological systems from exposure to them. By itself, this knowledge cannot secure a solution to environmental degradation, but it is a requisite to fashioning governmental programs for attaining that purpose.

---

[13]In regulatory procedures, it is usually not necessary to prove absolute harmlessness, but only the absence of detectable harm.

Global Energy Use and Supply

2.1 INTRODUCTION

The industrial revolution has been characterized by very large increases in the amount of energy available to human societies compared to their predecessors. In preindustrial economies, only very limited amounts of nonhuman mechanical power were available, such as that of domesticated animals, the use of wind power to propel boats and pump water, and the use of water power to grind grain. Wood was the principal fuel to cook food, to heat dwellings, and to smelt and refine metals. Today, in industrial nations, or in the urban-industrial areas of developing nations, the availability of fossil and nuclear fuels has vastly increased the amount of energy that can be expended on economic production and personal consumption, helping to make possible a standard of living that greatly exceeds the subsistence level of preindustrial times. Furthermore, the population of the world increased severalfold since the preindustrial era, thus requiring the recovery of ever-increasing amounts of energy resources. However, these resources are not evenly distributed among the countries of the world, and they are finite.

The principal sources of energy in present societies are fossil energy (coal, petroleum, and natural gas), nuclear energy, and hydroenergy. Other energy sources, the so-called renewables, are presently supplying a very small fraction of the total energy consumption of the world. The renewables include solar, wind, geothermal, biomass, ocean-thermal, and ocean-mechanical energy. In fact, hydroenergy may also be called a renewable energy source, although usually it is not classified together with solar, wind, or biomass. Increased use of renewable energy sources is desirable because they are deemed to cause less environmental damage, and their use would extend the available resources of fossil and nuclear energy.

In this chapter we describe the supply and consumption patterns of energy in the world today, along with the historical trends, with emphasis on available resources and their rate of depletion. In recent years the effects of the global consumption of fossil fuels on the increase of atmospheric concentration of CO_2 has become an international concern. In examining the global energy use, it is useful to include in our accounting the concomitant CO_2 emissions to provide a perspective on the problem of managing the potential threat of global climate change due to these emissions.

2.2 GLOBAL ENERGY CONSUMPTION

The trend of world energy consumption from 1970 to 1997 and projections to 2020 is depicted in Figure 2.1. The worldwide energy consumption in 1997 was 380 Quads.[1] In 1997, the industrialized

[1]1 Quad (Q) = 1 quadrillion (1E(15)) British thermal units (Btu) = 1.005 E(18) joules (J) = 1.005 exajoules (EJ) = 2.9307 E(11) kilowatt hours (kWh). See Tables A.1 and A.2.

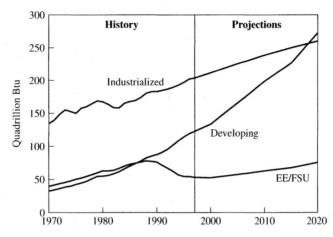

Figure 2.1 Trend of world's energy consumption for 1970–1997 and a projection to 2020. (Data from U.S. Department of Energy, Energy Information Agency, 2000. *International Energy Outlook 2000.*)

countries, also called "developed" countries, consumed 54% of the world's energy, the "less developed" countries consumed 31.5%, and the eastern European and former Soviet Union countries consumed 14.5%. It is interesting to note that in 2020, the projection is that the less developed countries will consume a greater percentage of the world's energy than the industrialized countries.

Table 2.1 lists the 1996 population, total energy use, Gross Domestic Product (GDP), energy use per capita, and energy use per GDP of several developed and less developed countries. The United States is the largest consumer of energy (88.2 Q), followed by China (35.7 Q) and India (30.6 Q). The United States consumes 23.2% of the world's energy with 4.6% of the world's population; western Europe consumes 16.7% of the world's energy with 6.5% of the world's population. China consumes about 10% of the world's energy with 21% of the world's population, whereas India consumes 3% of the energy with 16.3% of the population.

Among the listed countries, Canada, Norway, and the United States are the world's highest users of energy per capita: 395, 390, and 335 million Btu per capita per year, respectively. Russia consumes 181 MBtu/cap y, Japan 171, United Kingdom 169, Germany 168, and France 162. The less developed countries consume much less energy per capita. For example, Mexico consumes 59 MBtu/cap y, Brazil 43, China 29.4, Indonesia 54.1, and India 32.6. The world average consumption is 63 MBtu/cap y.

If we compare the energy consumption per GDP, a different picture emerges. Among developed countries, Canada uses 24.5 kBtu/$ GDP (reckoned in constant 1987 dollars), Norway 16.7, United States 16.2, United Kingdom 12.5, Germany 9.1, France 9, Italy 8.4, and Japan 7.1. Canada, Norway, and the United States use more energy per GDP than the other western European countries and Japan, in part because of the colder climate, larger living spaces, longer driving distances, and larger automobiles. On the other hand, Russia and the less developed countries (with the exception of Brazil) spend a higher rate of energy per dollar GDP than do Canada, United States, Japan, and the European countries: Russia (108.3 kBtu/$ GDP), Indonesia (81), China (67), and Mexico (36). This is an indication that much of the population in these countries does not (yet) contribute significantly to the GDP. Furthermore, their industrial facilities, power generation, and heating

TABLE 2.1 Population, Energy Use, GDP, Energy Use per Capita, and Energy Use per $ GDP in Several Countries, 1996[a]

Country	Population (million)	Energy Use (Quad)	GDP[b] ($ billion)	Energy Use per Capita (MBtu/cap/y)	Energy Use per $ GDP (kBtu/$ GDP)
Developed					
United States	263	88.2	5452	335	16.2
Canada	30	11.7	477	395	24.5
Russia	148	26.8	247	181	108.3
Japan	125	21.4	3007	171	7.1
Germany	82	13.7	1501	168	9.1
France	58	9.4	1402	162	9.0
Italy	58	7.4	884	127	8.4
United Kingdom	58	9.8	786	169	12.5
Norway	4	1.6	102	390	16.7
Less Developed					
China	1212	35.7	533	29.4	67
India	936	30.6	379	32.6	24
Indonesia	194	10.5	129	54.1	81
Brazil	156	6.7	333	43	20
Mexico	95	5.6	155	59	36
World Total	5724	380	NA[c]	66.4	NA

[a]Data from U.S. Department of Energy, Energy Information Agency, 1997. *International Energy Outlook 1997.*
[b]Gross Domestic Product in constant 1987 U.S. dollars.
[c]NA, not applicable.

(or cooling) systems apparently are less efficient or in other ways more wasteful of energy than in Canada, United States, western Europe, and Japan.

2.3 GLOBAL ENERGY SOURCES

The primary energy sources supplying the world's energy consumption in 1997 were petroleum (39%), coal (25%), natural gas (21.5%), nuclear-electric (6.3%), hydroelectric (7.5%), and geothermal and other renewables (0.7%) (see Figure 2.2).[2,3] The trend of the growth of energy sources from 1970 to 1997 and the prediction to 2020 is given in Figure 2.3. The projection for the next two decades is that nuclear's share will decline and the share of renewables will increase,

---

[2]Primary energy is energy produced from energy resources such as fossil or nuclear fuels, or renewable energy. It is distinguished from secondary energy, such as electric power or synthetic fuel, which is derived from primary energy sources.

[3]In converting nuclear and renewable (e.g., hydro) energy to primary energy in Quads, the U.S. Energy Information Agency (EIA) uses the thermal energy that would be used in an equivalent steam power plant with a thermal efficiency of about 31%.

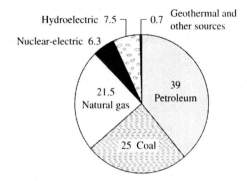

Figure 2.2 Proportions (%) of world's energy consumption supplied by primary energy sources, 1997. (*Source:* Same as in Figure 2.1.)

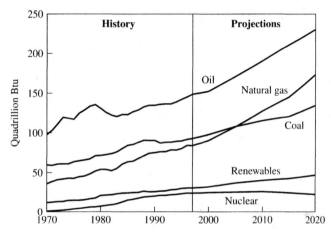

Figure 2.3 The trend of the growth of energy sources from 1970 to 1997 and the prediction to 2020. (*Source:* Same as in Figure 2.1.)

presumably with increase of the use of solar, wind, and biomass energy. The consumption of all fossil fuels will also increase in the next decades, with the rise of natural gas use exceeding that of coal by the year 2020.

Taken as a linear growth rate over the 10 years 1987–1997, the worldwide energy consumption was increasing at approximately 1.55% per year. Coal consumption grew by 0.8%/y on the average, natural gas 2.45%/y, petroleum 1.1%/y, nuclear-electric 2.2%/y, hydroelectric 2.1%/y, and geothermal and other energy sources 13%/y. However, as mentioned above, the latter constitute only a small fraction of the current energy consumption. In the United States, energy consumption increased 1.7%/y on the average over the 10 years. China's energy consumption grew 5.3%/y on the average, whereas India's energy use increased about 6.6%/y. Most of the growth is due to increased fossil fuel consumption.

In 1996, the total energy consumption in the United States was close to 90 Q. The distribution of the U.S. energy consumption by energy source is presented in Figure 2.4. Petroleum contributed 39.7%, natural gas 25.1%, coal 22.8%, nuclear-electricity 8%, hydroelectricity 4%,

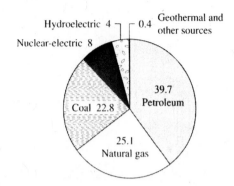

Figure 2.4 Proportions (%) of U.S. energy consumption supplied by primary energy sources, 1996. (Data from U.S. Department of Energy, Energy Information Agency, 1997. *Monthly Energy Review,* April 1997.)

and geothermal and other renewables 0.4%. These proportions are not greatly different from those of the world as a whole. In 1996, about 50.5% of the U.S. petroleum and 12% of natural gas consumption was supplied by foreign sources.

2.4 GLOBAL ELECTRICITY CONSUMPTION

Electricity is a secondary form of energy, because primary energy (fossil, nuclear, hydropower, geothermal, and other renewable sources of energy) is necessary to generate it. The trend of the world's electricity production from 1990 to 1997 and the prediction to 2020 is depicted in Figure 2.5. In 1997, the world's total electricity production was close to 12 trillion kilowatt hours. By 2020, the production is predicted to increase to over 21 trillion kWh.

Of the 1997 electricity production, 63% was from fossil energy, 19% was from hydroenergy, 17% was from nuclear energy, and less than 1% was from geothermal and other renewable sources (see Figure 2.6). Because the worldwide thermal efficiency of power plants is about 33.3%, in 1997 these plants consumed about 32.6% of the world's primary energy and about 55.5% of the world's fossil energy. The majority of the latter (over 80%) was in the form of coal. In the United States, Europe, Japan, and some other countries, in the past decades, natural gas became a preferred fuel for electricity generation, and many new power plants were built that employ the method of Gas Turbine Combined Cycle (GTCC), which is described in Section 5.3.1.

The reliance on energy sources for electricity production varies from country to country. For example, in the 1996, U.S. electricity production amounted to 3079 billion kWh. Of this, coal contributed 56.4%, nuclear power plants 21.9%, hydroelectric power plants 10.7%, natural gas 8.6%, petroleum 2.2%, and geothermal and other sources less than 0.3% (see Figure 2.7).

Hydropower is a significant contributor to electricity generation in many countries. For example, in Norway practically all electricity is produced by hydropower, in Brazil 93.5%, New Zealand 74%, Austria 70%, and Switzerland 61%. China and India produce about 19% of their electricity from hydropower. While hydroelectricity is a relatively clean source of energy and there is still a potential for its greater use worldwide, most of the accessible and high "head" hydrostatic dams are already in place. Building dams in remote, inhospitable areas will be expensive and hazardous. Furthermore, there is a growing public opposition to damming up more rivers and streams for

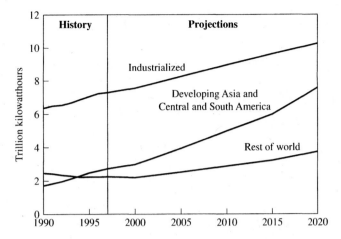

Figure 2.5 Trend of world's electricity consumption from 1990 to 1997 and the prediction to 2020. (*Source:* Same as in Figure 2.1.)

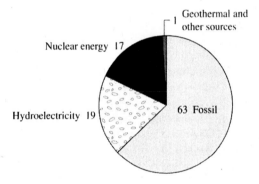

Figure 2.6 Proportions (%) of the the world's electricity generation supplied by primary energy sources, 1997. (Data from U.S. Department of Energy, Energy Information Agency, 1997. *International Energy Outlook 1997.*)

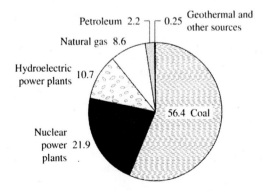

Figure 2.7 Proportions (%) of primary energy sources supplying U.S. electricity generation, 1996. (*Source:* Same as in Figure 2.4.)

environmental reasons and also because of the dislocation of population that is often involved in creating upstream reservoirs.

Geothermal energy supplies a significant portion of electricity in the following countries: El Salvador (28.5%), Nicaragua (18.5%), Costa Rica (10.3%), New Zealand (5.7%), Iceland (5.3%), Mexico (3.8%), Brazil (2.6%), Indonesia (1.8%), and Italy (1.6%). Geothermal energy has a great potential for supplying heat and electricity to many areas of the world. However, at present, geothermal energy is only competitive with fossil energy where the geothermal sources are on, or near, the surface of the earth.

In contrast to fossil-fueled power plants, nuclear power plants do not emit any CO_2 into the atmosphere, nor do they emit the other fossil-energy-related pollutants (SO_2, NO_x, particulate matter). However, the fear of nuclear accidents and the unresolved problem of nuclear waste disposal has brought the construction of additional nuclear power plants to a halt in many countries. In the United States, several nuclear power plants are presently being decommissioned even before their normal retirement date. On the other hand, in some countries, new nuclear power plants are being constructed, and nuclear energy does provide a significant portion of the total electricity production. For example, in France 76% of the electricity is nuclear-electric, South Korea 36%, Germany 29%, Taiwan 27%, and Japan 26%.

2.5 GLOBAL CARBON EMISSIONS

Table 2.2 lists the total carbon emissions, carbon emissions per capita and per GDP of several countries in the world in 1996. (Emissions are reckoned in mass of carbon, not that of CO_2.) In terms of absolute quantities, the United States and China are the largest emitters of carbon [1407 and 871 million metric tons per year (Mt/y), respectively], followed by Russia (496 Mt/y). In terms of per capita emissions, the United States and Canada were the largest emitters, 5270 and 4040 kilogram per capita per year, followed by Russia, 3340 kg/cap y. In countries where nonfossil energy is used for electricity generation and other purposes, the per capita carbon emissions are lower. Thus, while the energy consumption per capita in Germany and France are similar (168 and 162 MBtu/cap y, respectively), the carbon emissions are quite different (2790 and 1600 kg/cap y, respectively). This reflects the greater use of nuclear energy for electricity generation in France. Similarly, Switzerland and New Zealand have lower carbon emissions per capita (1470 and 2100 kg/cap y, respectively) than other industrial countries, because of their use of hydroenergy and geothermal energy. The world average is 1100 kg/cap y. The United States emits about five times as much carbon per capita as the world's average.

In terms of carbon emissions per dollar GDP, an interesting picture emerges. The ratio in the United States and Canada is 0.26 and 0.25 kg carbon per dollar GDP (reckoned in 1987 U.S. dollars), respectively, whereas in Japan, Germany, France, Italy, and the United Kingdom, it ranges from 0.1 to 0.2 kg/$. In part this stems from the higher consumption of energy per unit of GDP in the United States and Canada, but also from the fact that the United States and Canada use more fossil fuel per capita for space heating, space cooling, and transportation than do the European countries and Japan. In Russia, the ratio of carbon emission per dollar GDP is 2.01, China 1.62, India 0.65, Indonesia 0.62, and Mexico 0.63 kg/$. In these countries, fossil fuel is not used as efficiently in the production of GDP. The exception is Brazil, where the ratio is 0.2 kg/$, probably on account of Brazil's greater use of hydroenergy and biomass energy. (Note that emissions from forest burning are not included in these estimates.)

TABLE 2.2 Carbon Emissions, Carbon Emissions per Capita, and per $ GDP in Several Countries of the World, 1996.[a]

Country	Carbon Emissions (Mt C/y)[b]	Carbon Emissions per Capita (kg/cap y)	Carbon Emissions per $ GDP (kg/$ GDP)[c]
United States	1407	5270	0.26
Canada	119	4040	0.25
Russia	496	3340	2.01
Japan	307.5	2460	0.1
Germany	228	2790	0.15
France	93	1600	0.09
Italy	112	1960	0.13
United Kingdom	148	2530	0.19
Norway	20	4560	0.19
Switzerland	11	1470	0.06
New Zealand	7.5	2100	0.17
China	871	730	1.62
India	248	270	0.65
Indonesia	81	410	0.62
Brazil	68	430	0.2
Mexico	98	1070	0.63
World Total	6250	1090	NA[d]

[a] Data from U.S. Department of Energy, Carbon Dioxide Information Analysis Center, Oak Ridge National Laboratory, Oak Ridge.
[b] Million metric tons of carbon per year
[c] Gross Domestic Product in constant 1987 U.S. dollars.
[d] NA, not applicable.

2.6 END-USE ENERGY CONSUMPTION IN THE UNITED STATES

In order to gain some insight as to where lie the greatest potentials in energy savings, it is useful to consider the consumption of energy in each end use sector. The major sectors are residential–commercial, industrial, and transportation. We shall use the United States as an example. In other countries the end-use pattern may differ somewhat, depending on the industrial output of the country (heavy vs. light industry), climate (heating vs. air conditioning, or neither), and automobile usage (personal vs. freight; distances traveled). In 1996, the U.S. total primary energy consumption was close to 90 Q. Of this the industrial sector consumed 36.2%, residential–commercial 36.6%, and transportation 27.2%. In 1973, the shares of the three sectors were: industry 43%, residential–commercial 32%, and transportation 25%. This reflects the trend of (a) population growth and (b) a shift from an industrial to a service oriented economy, and within the industrial sector, a shift from heavy ("low-tech") to light ("high tech") industry. In the United States, over the years 1973–1996 the energy consumption per GDP also declined. In 1973, it was 19,000 Btu per constant 1992 dollar; in 1996, 13,000 Btu per 1992 dollar. This shift reflects the increasing share of the service industry to the total economy.

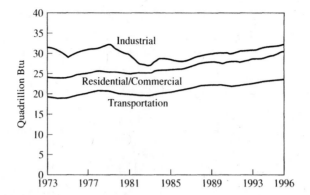

Figure 2.8 Trend of U.S. energy consumption for industrial, residential–commercial, and transportation sectors in 1973–1996. (*Source:* Same as in Figure 2.4.)

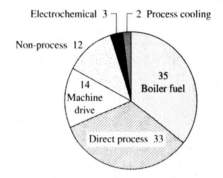

Figure 2.9 Proportions (%) of primary energy use in the U.S. industrial sector. (Data from U.S. Department of Energy, Energy Information Agency, 1994. *Manufacturing Energy Consumption Survey, 1994.*)

The trend of annual consumption of energy in the three sectors over the years 1973–1996 is shown in Figure 2.8. In those years the energy consumption for industrial production barely increased, whereas the consumption for the residential–commercial and transportation sectors increased significantly. Let us consider the pattern of energy consumption within each sector.

2.6.1 Industrial Sector

Of the total energy used by the industrial sector, about 35% is used for boiler fuel. The major part of this boiler fuel is used for direct industrial processes, including electricity generation; a smaller part is used for space heating. Direct process heat consumes 33%, machine drive 14%, nonprocess uses 12%, electrochemical processes 3%, and process cooling and other uses 2% of the total energy consumed by the industrial sector (see Figure 2.9). Even though the industrial sector has become more energy efficient over the past years, there is still room for improvement. Industry could save energy by process modification, better heat exchangers, more efficient drive mechanisms, and load-matched, variable-speed motors. In some cases, cogeneration can save energy. In cogeneration, heat (steam) and electricity required by the industrial process and the other energy needs for the facility (e.g., space heating) are supplied from the same power plant. The fuel savings are especially pronounced when low-quality heat is adequate for the industrial process or space heating, such as the heat rejected by the steam to a condenser after driving a turbine.

2.6.2 Residential Sector

Of the total energy used by the residential sector, about 40% is used for appliances and lighting (mainly electricity), 34% for space heating (mainly fossil fuel as petroleum and natural gas), 16% for water heating (mainly electricity and natural gas), and 10% for air conditioning (mainly electricity). This is depicted in Figure 2.10. Significant savings in space heating could be realized by conservation (e.g., lowering the thermostat in winter and raising it in summer) and by better insulation. Solar heating could be more widely utilized both for space and water heating. Appliances can be made more energy efficient and smaller. Lighting could be converted from incandescent to fluorescent bulbs.

2.6.3 Commercial Sector

Of the total energy used in the commercial sector, about 26% is used for lighting, 19% for space heating, 9% for office equipment, 8.5% for water heat, 7.5% for air conditioning, 7% for ventilation, 5% for refrigeration, 3.5% for cooking, and 14.5% for other uses (see Figure 2.11). As

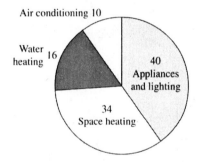

Figure 2.10 Proportions (%) of primary energy use in the U.S. residential sector. (Data from U.S. Department of Energy, Energy Information Agency, 1990. *Residential End Use Energy Consumption, 1990.*)

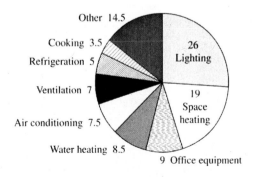

Figure 2.11 Proportions (%) of primary energy use in the U.S. commercial sector. (Data from U.S. Department of Energy, Energy Information Agency, 1990. *Energy Markets and End Use, 1990.*)

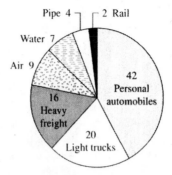

Figure 2.12 Proportions (%) of primary energy use in the U.S. transportation sector. (Data from U.S. Department of Transportation, 1993. *Transportation Energy Data Book,* Edition 15.)

in the residential sector, this sector could also realize large energy savings, especially in lighting, space heating, cooling, and ventilation.

2.6.4   Transportation Sector

Of the total energy used by the transportation sector, about 42% is used for personal automobile transport, 20% for light trucks (often just used for personal transport), 16% for heavy freight, 9% for air, 7% for water, 4% for pipe, and 2% for rail transport (see Figure 2.12). Considering that in the United States the transportation fuel is practically all derived from petroleum, that over 50% of petroleum is imported, that automobiles are responsible for about 50% of ground-level air pollution, and that the transportation sector contributes about one-third of U.S. CO_2 emissions, the transportation sector could realize significant savings in energy consumption and reduction in air pollutant and carbon emissions. This could be accomplished by (a) reducing the weight of automobiles, (b) increased engine fuel efficiency, (c) using fuel cell and battery powered electric drive cars, (d) using hybrid internal combustion engine-electric powered cars, and (e) increased use of public transportation. Some of these alternatives are further discussed in Chapter 8.

2.7   GLOBAL ENERGY SUPPLY

In this section we shall address only the supply of fossil energy—that is, coal, petroleum, natural gas, and unconventional sources of fossil energy, such as oil shale, tar sand, geopressurized methane, and coal seam methane.

2.7.1   Coal Reserves

Coal is found practically on every continent and subcontinent. It is found buried deeply in the ground or under the seabed, or close to the surface. Coal characteristics vary widely according to its biological origin (forests, low growing vegetation, swamps, animals) and geological history (age, overburden, temperature, pressure). Thus, the chemical and physical characteristics of coal are also highly variable, such as the content of moisture, minerals (ash), sulfur, nitrogen, and oxygen; heat

TABLE 2.3 Composition and Characteristics of Coal, Percent by Weight[a]

	Pennsylvania Anthracite	Pittsburgh Seam Bituminous	Illinois No. 6 Bituminous	Montana Subbituminous	North Dakota Lignite
Moisture	4.5	2.0	13.2	10.5	33.7
Volatile matter	1.7	30.5	36.0	34.7	26.6
Fixed carbon	84.1	58.2	41.8	43.6	32.5
Sulfur	0.7	2.2	3.4	1.2	0.9
Ash	0.7	9.3	9.0	11.2	7.2
HHV, Btu/lb	12,750	13,620	11,080	10,550	7,070

[a]Percentages may not add up to 100% because of other elements present in coal.

value; hardness; porosity; and so on. Table 2.3 lists the characteristics and composition of several U.S. coals. The variability from coal to coal is clearly evident. For example, the carbon content varies from 62.9% for North Dakota lignite (a relatively young coal) to 93.9% for Pennsylvania anthracite (a relatively old coal). The sulfur content varies from 0.7% to 3.4% by weight, and the higher heating value (HHV) varies from 7070 Btu/lb (16,430 kJ/kg) for North Dakota lignite to 13,620 Btu/lb (31,650 kJ/kg) for Pittsburgh seam bituminous.[4]

Because of the widely varying characteristics of coals, it is difficult to estimate the precise energy reserves residing in the world's coal deposits. The world total coal reserves are estimated at 1.037E(12) metric tons.[5] About one-half is bituminous and anthracite coal, whereas the other half is subbituminous and lignite coal. Assuming that the average HHV of bituminous and anthracite coal is 12,500 Btu/lb (29,050 kJ/kg), and that of subbituminous and lignite is 8200 Btu/lb (19,055 kJ/kg), the world's coal reserves have a total heating value of about 24,000 Q. The 1995 world coal consumption amounted to a little more than 93 Q/y. If the present consumption level were to continue into the distant future, the world coal reserves would last about 250–300 years. However, if coal consumption keeps increasing at a rate of 0.8%/y (see Section 2.3), the lifetime of the world's coal reserves would be only about 140 years.

The countries where the world's major coal reserves are found are (in percent of the total) United States (26), former Soviet Union countries (25), China (12), Australia (10), Germany (7), South Africa (7), Poland (4), and other countries (9) (see Figure 2.13).

In addition to the above *reserves,* coal may be found in yet unproven reservoirs. Unproven reservoirs are called *resources.*[6] Some estimates place the coal resources at about 140,000 Q.[7] The

---

[4]The higher heating value (HHV) includes the latent heat of condensation of the moisture content of the coal and the water vapor formed in combustion, whereas the lower heating value (LHV) excludes it.

[5]Energy Information Agency, 1997. *International Energy Outlook,* DOE/EIA-0484(97).

[6]Reserves of a given fossil fuel are those quantities that geological and engineering information indicate with reasonable certainty to be extractable under existing economic and operating conditions. Resources are those quantities that from geological and engineering information may exist, but their extraction will require different economic and operating conditions.

[7]Anonymous, 1978. *World Energy Conference.* Guildford: IPC Science and Technology Press.

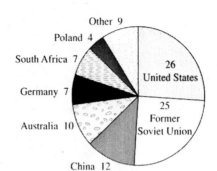

Figure 2.13 World's coal reserves, percent. (*Source:* Same as in Figure 2.6.)

resources could supply the present consumption rate for about 1500 years. The resources are mainly located in China, the former Soviet Union countries, the United States, and Australia. However, the resources may be located at great depth under the ground or under the continental shelves. The cost of exploiting these resources will certainly be much greater than that of the reserves.

2.7.2   Petroleum Reserves

The terms petroleum and mineral oil, or oil for short, are synonymous. The crude oil that is found in various parts of the world differs in quality and composition, depending on the biological origin and geological history. It is found in geological reservoirs under the ground or under the seabed at depths up to several thousand meters.

A recent survey of the U.S. Geological Survey (USGS) estimated the world's oil reserves as 1.6 E(12) barrels.[8, 9] The distribution of the oil reserves among the major oil reservoirs of the world is as follows (in percent): Middle East (42), North America, including US, Canada and Mexico (15), Russia, including Siberia (14), North and West Africa (7.2), South and Central America (6.1), Asia and Pacific (5.5), Caspian Basin (3.5), Western Europe, including the North Sea (3.1), and others (3.6) (see Figure 2.14).

Taking an average heating value of crude oil as 5.8 E(6) Btu/bbl, the world's oil reserves amount to 9280 Q. The world's oil consumption in 1995 amounted to about 141 Q/y. If that consumption rate were to continue into the future, the world's oil reserves would last only for about 65–70 years. If oil consumption keeps increasing at a rate of 1.1%/y, the lifetime of the world's oil reserves would be only about 50 years.

In a 1998 *Science* article, Kerr predicted that the peak production rate of crude oil will occur sometime between 2005 and 2020.[10] After that, the production rate will decline, which means that the consumption rate must also decline. This also means that the world's energy appetite must be supplied by sources other than oil.

---

[8]U.S. Geological Survey, 1997. *Ranking of the World's Oil and Gas Reserves.* USGS Report 97-463.

[9]1 barrel (bbl) = 42 U.S. gallons = 159 liters. See Table A.2.

[10]Kerr, R.A., 1998. The Next Oil Crisis Looms Large—and Perhaps Close. *Science* **281**, 1128–1131.

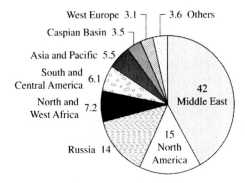

Figure 2.14 World's oil reserves, percent. (Data from U.S. Geological Survey, 1997. *Ranking of the World's Oil and Gas Provinces by Known Petroleum Volumes.* Report 97-463.)

2.7.3 Unconventional Petroleum Resources

In addition to conventional oil reserves, vast amounts of hydrocarbon fluids are distributed in various geological formations, such as oil shales and tar sands. Oil shale deposits are known to exist in the United States in the Colorado Basin (Colorado, Utah, and Wyoming) and in the Appalachian Basin (Pennsylvania, Virginia, and West Virginia). Tar sands are found in the Canadian Province of Alberta, as well as in Venezuela and Colombia. In the United States alone, it is estimated that deposits of oil shale contain perhaps close to 2000 Q of petroleum. With an estimated recovery factor of 60%, the U.S. oil shales may contain up to 1200 Q. This is about 10 times as much as the proven oil reserves in the United States. However, the exploitation of these unconventional petroleum resources may require greater financial and technological investments than those for the discovery and extraction of oil reserves. Furthermore, the extraction of petroleum from oil shale may impact the environment to a greater degree than that of pumping oil from on- or off-shore wells. On the average, oil shales contain between 60 and 120 liters of petroleum per ton of shale rock. The rock must be excavated and heated in retorts to drive out the liquid petroleum. Thus, a significant fraction of the derived petroleum must be burnt in order to heat the rock for further extraction of petroleum. The process will require complex and expensive control technology for the prevention of air emissions and liquid effluents. Also, the spent rock must be disposed of in an environmentally safe and aesthetic manner. The environmental controls alone will add greatly to the cost of extracting petroleum from the unconventional resources. After the OPEC oil embargoes in the 1970s, a consortium of oil companies started to produce pilot scale quantities of petroleum products from oil shale deposits in Colorado. However, after the prices of crude oil fell from a high of $35 per barrel in 1981 to the teens in the late 1980s, all oil shale activities in the United States ceased.

2.7.4 Natural Gas Reserves

The combustible part of natural gas (NG) consists mainly of methane (CH_4) with some admixture of heavier hydrocarbons (ethane, propane, and butane). However, frequently noncombustible gases are found mixed with NG, namely, N_2 and CO_2. For example, the recently discovered gas fields off the coast of the Indonesian archipelago contain up to 70% by volume CO_2. On the average, NG contains 74.4% by weight of carbon, 24.8% hydrogen, 0.6% nitrogen, and 0.2% oxygen.

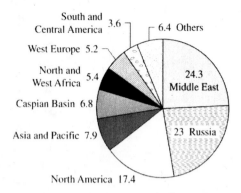

Figure 2.15 World's natural gas reserves, percent. (*Source:* Same as in Figure 2.14.)

On the average, the heating value of NG is 23,500 Btu/lb (55 MJ/kg), or 1032 Btu per cubic foot (38.5 MJ/m^3).

Natural gas is a very desirable fuel for several reasons. First, it is easy to combust because, being a gas, it readily mixes with air. Thus, the combustion is rapidly completed, and the boiler or furnace volume is smaller than that required for oil or coal combustion. Second, the combusted gas can directly drive a gas turbine with applications in power generation. Third, gas combustion does not produce particulate and sulfurous pollutants. Fourth, NG produces one-half the amount of CO_2 per unit heating value as does coal, and it produces three-quarters as much as oil.

The USGS estimate of the world's natural gas reserves is 6.75 E(15) cubic feet. Taking the heating value of NG as 1032 Btu/ft^3, the world's NG reserves amount to 6966 Q, comparable to those of oil. The world's rate of consumption of NG in 1995 amounted to 77.5 Q. If that rate were to continue in the future, the world's proven gas reserves would be depleted in about 85–90 years. If gas consumption keeps increasing at a rate of 2.45%/y, the lifetime of the world's gas reserves would be only about 50 years.

The gas reserves are distributed among the major reservoirs as follows (in percent): Middle East (24.3), Russia, including Siberia (23), North America, including the United States, Canada, and Mexico (17.4), Asia and Pacific (7.9), Caspian Basin (6.8), North and West Africa (5.4), West Europe, including the North Sea (5.2), South and Central America (3.6), and others (6.4) (see Figure 2.15).

2.7.5 Unconventional Gas Resources

Methane is known to exist also in unconventional reservoirs. These are (a) gas trapped in sandstone, (b) gas trapped in shale rock, (c) gas trapped in coal seams, (d) gas trapped in pressurized underground water reservoirs ("geopressurized methane"), and (e) methane clathrates, also called hydrates, found at some locations at the bottom of oceans and ice caps. Methane hydrates are basically ice crystals in which a methane "guest" molecule is trapped. Hydrates are formed under the high pressures and low temperatures prevailing at the ocean bottom or ice caps. The first four reservoirs in the United States alone may yield 600–700 Q, which is about 70% of the proven gas reserves in the United States and about 10% of the world's proven gas reserves.

TABLE 2.4 The World's Proven Fossil Fuel Reserves, Rates of Consumption, and Lifetimes

Fuel	Reserves (Q)	1995 Consumption (Q/y)	Rate of Growth (%/y) 1987–1997	Lifetime (y) No Growth	Lifetime (y) with Growth
Coal	24,000	93	0.8	258	140
Oil	9280	141	1.1	66	50
Gas	6966	78	2.5	90	50

The amount of methane hydrates at the ocean bottoms and ice caps may range from 1 to 2 E(16) kg.[11] Taking the heating value of methane as 4.76 E(4) Btu/kg, the heating value stored in methane hydrates could amount to 5–10 E(5) Q, two orders of magnitude larger than proven gas reserves. However, the amount of gas hydrates is speculative, and no technology exists yet to recover the methane from gas hydrates laying on the ocean bottom or under ice caps.

2.7.6 Summary of Fossil Reserves

The world's proven fossil fuel reserves are summarized in Table 2.4. Also listed are the lifetimes of the reserves for the case that 1995 consumption continues without growth into the future and for the case that the 1987–1997 growth rate continues into the future.

2.8 CONCLUSION

We reviewed the present and historic trends of energy consumption and supply patterns in the world as a whole, as well as in individual countries—by industrial sector, by end-use, and per capita. The so-called "developed" countries consume a much larger amount of energy and emit a much higher rate of CO_2 per capita than the "less developed" countries. However, the converse is true for energy use per Gross Domestic Product (GDP). The "less developed" countries have a higher ratio of energy consumption and a higher emission rate of CO_2 per dollar GDP than the "developed" countries.

Measured by the available proven fossil energy reserves, and present rate of consumption, coal may last 250–300 years, oil 65–70 years, and natural gas 85–90 years. Unconventional fossil energy resources, such as oil shale, tar sands, geopressurized methane, and methane hydrates, may extend the lifetime of fossil fuels severalfold, but their exploitation will require greatly increased capital investment and improved technology. The price of the delivered product will be much higher than is currently paid for these commodities.

The major conclusion is that for the sake of husbanding the fossil fuel reserves, as well as for the sake of mitigating air pollution and the CO_2-caused global warming, mankind ought to conserve these fuels, increase the efficiency of their uses, and shift to nonfossil energy sources.

---

[11]Kvenvolden, K. A., 1988. *Chem. Geol.* **71**, 41.

PROBLEMS

Problem 2.1

From Figure 2.1 determine the rate of growth r (%/y) of energy consumption for the years 1970–2000. For the industrial countries use a linear least-square fit through the data points. Use the intercept at 1970 as a base. For the developing countries use an exponential fit, $Q\{t\} = Q_0 \exp\{rt\}$.

Problem 2.2

From Table 2.1 produce a bar chart of energy use per capita (GJ/cap y) for the listed countries. Draw a dashed line through the bars at the level of the world's average per capita energy use. Do the same for energy use per dollar of GDP (MJ/$GDP).

Problem 2.3

From Figure 2.3 determine the rate of growth r (%/y) for coal, oil, and natural gas for the years 1970–2000. For coal use a linear least-square fit through the data points with the intercept at 1970 as a base; for oil and gas use an exponential fit.

Problem 2.4

From Table 2.2 produce a bar chart of carbon emissions (kg C/cap y) for the listed countries. Draw a dashed line through the bars at a level of the world's average carbon emissions. Do the same for carbon emissions in kg C/$GDP with a dashed line for the average of (a) the United States, Canada, Japan, Germany, France, Italy, United Kingdom, Norway, Sweden, and New Zealand and (b) Russia, China, India, Indonesia, and Mexico. Why is the average emission per $GDP so much lower for the group of countries (a) and so much higher for countries (b)?

Problem 2.5

Based on Figure 2.8, draw linear best fit lines through the data points of U.S. energy consumption (Q/y) for the years 1973–1996 for the three sectors: industry, residential–commercial and transportation. Determine the rate of growth or decline in %/y, using the intercept at 1973 as a base.

Problem 2.6

By consulting the relevant literature, try to give realistic estimates of energy savings, in percent, in the U.S. residential sector for appliances, lighting, space heat, water heat, and air conditioning over the next 20 years.

Problem 2.7

Do the same for U.S. personal automobiles including sport utility vehicles (SUV). Use available literature and Chapter 8 for your estimates.

Problem 2.8

From Table 2.3 calculate the higher heating value (HHV) for U.S. coals in kJ/kg.

Problem 2.9

The world's coal reserves are estimated to be 24,000 Q. How much is that in EJ? What will be the lifetime of coal reserves if the present consumption rate of 93 Q/y continues into the future and if the consumption increases by $r = 0.5, 0.8$, and $1\%/y$? (The lifetime T of a reserve is calculated from $T = r^{-1}\{\ln[r(Q_T/Q_0) + 1]\}$, where r is the rate of growth of consumption, Q_T is the total reserve, and Q_0 is the present consumption rate.)

Problem 2.10

The world's oil reserves are estimated at close to 10,000 Q. The present consumption rate is 140 Q/y. What will be the lifetime of oil reserves if the present consumption rate continues into the future and if the consumption rate increases by $1\%/y$, $1.5\%/y$, and $2\%/y$?

Problem 2.11

The world's natural gas reserves are estimated at close to 7000 Q. The present consumption rate is 80 Q/y. What will be the lifetime of NG reserves if the present consumption rate continues into the future and if the consumption rate increases by $2\%/y$, $2.5\%/y$, and $3\%/y$? Tabulate results of Problems 2.9–2.11 in the form of Table 2.4.

Problem 2.12

Estimate the proportions (%) of world's energy consumption in 2050 supplied by primary energy sources as shown in Figure 2.2 for 1997. Justify your proportions on the basis of what you read and know on the predicted availability, acceptability, and cost of the energy resources.

BIBLIOGRAPHY

Annual Review of Energy, 1975–1999. Volumes 1–24. Palo Alto: Annual Reviews.

Kraushaar, J. J., and R. A. Ristinen, 1993. *Energy and the Problems of a Technical Society,* 2nd edition. New York: John Wiley & Sons.

Matare, H. F., 1989. *Energy: Facts and Future.* Boca Raton: CRC Press.

Tester, J. F., D. O. Wood, and N. A. Ferrari, Ed., 1991. *Energy and the Environment in the 21st Century.* Cambridge: MIT Press.

U.S. Department of Energy, Energy Information Agency, 1997. *International Energy Outlook.* Washington, D.C.: DOE/IEA-048(97).

U.S. Department of Energy, Energy Information Agency, 2000. *International Energy Outlook.* Washington, D.C.: DOE/IEA-0484(2000).

Thermodynamic Principles
of Energy Conversion

3.1 INTRODUCTION

The development of the steam engine, an invention that powered the first two centuries of the industrial revolution, preceded the discovery of the scientific principle involved—namely, the production of mechanical work in a device that utilizes the combustion of fuel in air. The scientific understanding that explains the production of work in different kinds of combustion engines is derived from the laws of thermodynamics, developed in the nineteenth century. In the twentieth century these principles aided the development of engines other than the steam engine, such as the reciprocating gasoline and diesel engines, the gas turbine, and the fuel cell. With the aid of this scientific hindsight, in this chapter we will review how the laws of thermodynamics determine the functioning of these sources of mechanical energy, and especially how they limit the amount of mechanical work that can be generated from the burning of a given amount of fuel.

The source of mechanical power developed from the combustion of fossil fuels or the fission of nuclear fuel in an engine is the energy released by the change in molecular or nuclear composition. This released energy is never lost but is transformed into other forms, appearing as some combination of mechanical or electric energy, internal energy of the molecular or nuclear products of the reactions, or energy changes external to the engine caused by heat transfer from it. This conservation of energy is explicitly expressed by the first law of thermodynamics.

The principle of energy conservation, or first law, places an upper limit on the conversion of chemical or nuclear energy to mechanical work; that is, the work of an engine cannot exceed the energy available. Experience shows that the work is very significantly less than the energy released by fuel reactions in engines, a matter having important practical consequences. The scientific principle that explains why, and by how much, there is a work shortfall is called the second law of thermodynamics. In combination with the first law, it enables us to understand the limits to producing work from fuel and to improve the engines that have been invented to accomplish this task.

The laws of thermodynamics cannot substitute for invention, but they do inform us of the performance limits of a perfected invention. In this chapter we review the principles of operation of the major inventions that transform chemical to mechanical energy: the steam engine, the gasoline and diesel engines, the gas turbine, and the fuel cell. In each case we show that the laws of thermodynamics provide limits on how much of the fuel's energy can be converted to work and describe how various debilitating factors reduce the work output of practical devices to values below this limit. Thermodynamic analyses of this type provide guidance for improving the energy efficiency of mechanical power production.

In the next five sections of this chapter we summarize the relevant principles of thermodynamics as embodied in its two laws, including the concepts of energy, work, and heat, the definition of useful thermodynamic functions, and their application to the steady flow of working fluids, such as air and combustion gases. We then proceed to the special application of the combustion of fuels and the various thermodynamic cycles that explain how common heat and combustion engines function. A subsequent section treats separately the fuel cell, a more recent development that operates on a different principle than heat engines, producing work directly in electrical form. The fuel efficiencies of various power-producing cycles are then summarized. The chapter concludes with a short discussion of the energy efficiency of synthetic fuel production.

3.2   THE FORMS OF ENERGY

The concept of energy, which originated with Aristotle, has a long history both in science and as a colloquial term. It is a central concept in classical and quantum mechanics, where it appears as a constant of the motion of mechanical systems. In the science of thermodynamics, energy has a distinct definition that distinguishes it from heat, work, or power. In this section we define thermodynamic energy as a quantity that is derived from an understanding of the physical and chemical properties of matter.

3.2.1   The Mechanical Energy of Macroscopic Bodies

Newtonian mechanics identifies two forms of energy, the *kinetic energy* of a moving body and the *potential energy* of the field of force to which the body is subject. The kinetic energy KE is equal to the product of the mass M of the body times one-half of the square of its velocity V,

$$KE \equiv \frac{1}{2}MV^2$$

The potential energy PE of a body subject to a force $\mathbf{F}\{\mathbf{r}\}$ at a location $\mathbf{r}$ in space is equal to the work done in moving the body to the location $\mathbf{r}$ from a reference position $\mathbf{r}_{ref}$,

$$PE \equiv -\int_{\mathbf{r}_{ref}}^{\mathbf{r}} \mathbf{F}\{\mathbf{r}\} \cdot d\mathbf{r}$$

While the kinetic energy has always a positive value with a zero minimum, the potential energy's value is measured with respect to the reference value and may be positive or negative.

One consequence of Newton's laws of motion of a body in a force field is that the sum of the kinetic and potential energies is a constant of the motion; that is, it is not a function of time. Calling this sum the total energy E, we have

$$E \equiv KE + PE = \text{constant}$$

Absent any other force applied to the body, its energy E is unchanged despite its movement within the region of space available to it. We may call this the *principle of conservation of energy*.

A simple example of the motion of a body that possesses kinetic and potential energy is that of a satellite moving in orbit around the earth. The potential energy is that of the satellite mass in the earth's gravitational field, which increases inversely in proportion to the distance of the satellite from the center of the earth. If the satellite is in an eccentric orbit about the earth, the conservation of energy requires that the satellite speed (and kinetic energy) is a maximum where it dips closest to the earth's surface.

It is possible to change the energy E of a body by acting upon it with an external force. In the case of the satellite, a quick impulse from its rocket engine can change its velocity and kinetic energy, thereby changing its total energy E. The amount by which the total energy E is changed is directly related to the amount of the rocket impulse. It is thereby possible to extend the statement of the conservation of energy by taking into account the changes in E brought about by external impulses applied to the moving body. This is the most general form of the principle of conservation of energy.

3.2.2 The Energy of Atoms and Molecules

The matter of a macroscopic body is composed of microscopic atoms and/or molecules (themselves aggregates of atoms). Sometimes, as for gases, these molecules are so widely separated in space that they may be considered to be moving independently of each other, each possessing a distinct total energy. Otherwise, in the case of liquids or solids, each molecule is under the influence of forces exerted by nearby molecules, and we can only distinguish the aggregate energy of all the molecules of the body. We call this energy the *internal energy* and give it the symbol U. Even though the motion of microscopic molecules is not describable by Newtonian mechanics, it is still possible to consider their total energy to be the sum of the kinetic energies of their motion and the potential energies of their intermolecular forces.

It is not possible to observe directly the energies of individual atoms of a thermodynamic substance, but changes in its internal energy are indirectly measurable by changes in temperature, pressure, and density. These observables, called thermodynamic state variables, are the surrogates for specifying internal energy.

3.2.3 Chemical and Nuclear Energy

Molecules are distinct stable arrangements of atomic species. Their atoms are held together by strong forces that resist rearrangement of the atoms. To disassemble a molecule into its component atoms usually requires the expenditure of energy, so that the molecules of a body may be considered to possess an energy of formation related to how much energy was involved in assembling them from their constituent atoms. If the internal energy U of a material body is changed, but the individual molecules remain intact, then their chemical energy of formation remains unchanged and contributes nothing to the change in U. On the other hand, if a chemical change occurs, so that new molecular species are formed from the atoms present in the original species, there will be a redistribution of energy among the components of the internal energy, of which the chemical energy of formation of the molecules must be taken into account.

A similar energy change accompanies the formation of new atomic nuclei in the fission of the nuclei of heavy elements or the fusion of light ones. Because the binding forces that hold nuclei together are so much larger then those that hold molecules together, nuclear reactions are much more energetic than molecular ones. Nevertheless, we can consider both molecules and atomic nuclei to possess energies of formation that must be taken into account in expressing the conservation of energy for material bodies that experience chemical or nuclear changes in composition.

3.2.4 Electric and Magnetic Energy

Molecules that possess a magnetic or electric dipole moment can store energy when they are in the presence of a magnetic or electric field, in the form of magnetic or electric polarization of the material. This energy is associated with the interaction of the molecular dipoles of the material body with the external electric charges and currents that give rise to the applied electric or magnetic field. Since capacitors and inductors are common components of electronic and electrical circuits, this form of energy is important to their functioning.

3.2.5 Total Energy

The various forms of energy that can be possessed by a material body, as described above, can be added together to define a total energy, to which we give the symbol E,

$$E \equiv KE + PE + U + E_{chem} + E_{nuc} + E_{el} + E_{mag} \tag{3.1}$$

It is very seldom that more than just a few of these forms are significant in any practical process for which there are changes in the total energy. There are many examples. In a gasoline engine, the combustion of the fuel–air mixture involves U and E_{chem}; in a steam and gas turbine, only KE and U change; in a nuclear power plant fuel rod, U and E_{nuc} are involved; and in a magnetic cryogenic refrigerator, U and E_{mag} are important. Nevertheless, the manner in which the various forms of energy enter into the laws of thermodynamics is expressed through the total energy E, a result of very great generality and consequence.

3.3 WORK AND HEAT INTERACTIONS

Thermodynamics deals with the interaction of a thermodynamic material system and its environment.[1] It is through such interactions that we are able to generate mechanical power or other useful effects in the environment. There are two quite different but important modes of interaction of a system with its environment, called the work interaction and the heat interaction. Each of these is a process in which, over time, the system and its environment undergo physical and/or chemical changes related to the kind of interaction taking place, either work or heat (or both simultaneously).

As we shall see below, work and heat interactions are distinguishable from each other by the character of the changes in the system and the environment. Both are quantifiable in terms of the interaction, being expressed in energy units. Neither is a form of energy, but only a transaction

---

[1]In thermodynamics, the environment of a thermodynamic system is that part of its material surroundings with which the system interacts.

quantity that accounts for the character of the exchange of energy between a thermodynamic system and its environment.

3.3.1 Work Interaction

Newtonian mechanics employs the concept of work as the exertion of a force acting through a displacement or a couple acting through an angular displacement.[2] We say that the amount of work required to lift a mass m through a vertical distance r in the earth's gravitational field is the product of the magnitude of the gravity force, mg, times the distance r, where g is the magnitude of the local acceleration of gravity. In thermodynamics, by convention, positive work is defined to be the product of the force exerted by a system on the environment times any displacement of the environment that occurs while the force is acting. (By Newton's principle, the force that the system exerts on the environment is equal in magnitude but opposite in direction to the force that the environment exerts on the system.) If the force F_{en} exerted on the environment by the system is accompanied by an incremental displacement dr_{en} of the environment in the direction of F_{en}, the increment of work may be expressed as

$$dW \equiv F_{en} \, dr_{en} \tag{3.2}$$

When the system does work on the environment, dW is positive; when the environment does work on the system, then dW is negative.

There are many simple examples of a work interaction. If a gas is contained in a circular cylinder capped at one end and fitted with a movable piston at the other, then the force exerted by the gas on that portion of the environment that is the movable piston is pA, where p is the gas pressure and A is the piston face area. If the piston is displaced an incremental distance dr_{en} in the direction of the pressure force pA, the positive work increment dW in this displacement is

$$dW = pA \, dr_{en} = p(A \, dr_{en}) = p \, dV \tag{3.3}$$

where $dV = A \, dr_{en}$ is the increment in the volume V of the gas in the cylinder. Or the work interaction with the environment may involve the movement of an increment of electric charge dQ_{en} through an increase in electric potential ϕ_{en}, such as when a current flows from the system to and from an electric motor in the environment, for which the work increment is

$$dW = \phi_{en} \, dQ_{en} \tag{3.4}$$

Another common example is the rotation of the shaft of a turbine (a material system) that applies a torque T_{en} (or couple) to an electric generator (the environment) attached to the turbine, rotating it through an increment of angle $d\theta_{en}$ in the same direction as the torque, giving rise to a positive work increment,

$$dW = T_{en} \, d\theta_{en} \tag{3.5}$$

---

[2]A couple is the product of the distance separating two equal but opposite forces times the magnitude of the force.

These are but a few specific examples of the many possible kinds of work interaction between a thermodynamic system and its environment.

3.3.2 Heat Interaction

We are very familiar with the processes whereby substances are warmed or cooled. Cooking or refrigerating food requires increasing or decreasing its temperature by bringing it into contact with a warmer or cooler environment. A temperature difference between a system and its environment is required for a heat interaction to transpire. If the environment undergoes a temperature increase after a system warmer than the environment is brought into contact with it, then a heat interaction has taken place. The incremental amount of the heat interaction dQ, which in this case equals the energy transfer from the system to the environment, is equal to the product of the heat capacity C_{en} of the environment times its temperature increase dT_{en}. But by convention the energy transferred *to* a system in a heat interaction is regarded as a positive quantity so that in this case the energy transfer is negative. Consequently,

$$dQ = -C_{en}\, dT_{en} \qquad (3.6)$$

We usually describe this interaction as *heat transfer,* although it is energy which is being exchanged in a process solely involving a heat interaction.

Both heat and work quantities involved in an interaction of a system and its environment are recognized by their effects in the environment, as described in equations (3.2)–(3.6) above. Furthermore, both heat and work interactions may occur simultaneously, being distinguishable by their different physical effects in the environment (e.g., $F_{en}\, dr_{en}$ vs. $C_{en}\, dT_{en}$).

3.4 THE FIRST LAW OF THERMODYNAMICS

The first law of thermodynamics is an energy conservation principle. It relates the incremental change in energy dE of a system with the increments of work dW and heat dQ recognizable in the environment during an interaction of the system with its environment. In words, it states that the increment in system energy dE equals the increment in heat dQ transferred to the system minus the work dW done by the system on the environment,

$$dE = dQ - dW \qquad (3.7)$$

It is an energy conservation principle in the sense that the sum of the system energy change dE, the work dW, and the heat $-dQ$ added to the environment is zero; that is, this sum is a conserved quantity in any interaction with the environment.

Equation (3.7) expresses the first law in differential form. If many successive incremental changes are added to accomplish a finite change in the system energy E from an initial state i to a final state f, the first law may be expressed in integral form as

$$E_f - E_i = \int_i^f dQ - \int_i^f dW \qquad (3.8)$$

In this form, the first law expresses the finite change in energy of the system as equal to the sum of the heat transferred to the system minus the work done by the system on the environment during the process that brought about the change from the initial to the final state.

The integrals of the heat and work quantities on the right-hand side of the equation (3.8) cannot be evaluated unless the details of the process that caused the change from the initial to the final state of the system is known. In fact, there may be many different processes that can bring about the same change in energy of the system, each distinguished by different amounts of heat and work, but all having in common that the sum of the heat and work quantities added to the environment are the same for all such processes that change the system from the same initial to final states.

In some power-producing and refrigeration systems, a working fluid undergoes a series of heating, cooling, and work processes that returns the fluid to its initial state. Because $E_f = E_i$ for such a cyclic process, the integral expression of the first law of thermodynamics, equation (3.8), has the form

$$\oint dQ = \oint dW \tag{3.9}$$

where the symbol $\oint$ identifies the cyclic process for which the heat and work integrals are evaluated. In other words, in a cyclic process the net heat and work quantities are equal.

3.5   THE SECOND LAW OF THERMODYNAMICS

The goal of engineers who design power plants is to devise a system to convert the energy of a fuel into useful work. If we consider the combustion of a fossil fuel to provide a source of heating, then the desirable objective is to convert all of the fuel energy to work, as the first law, equation (3.9), allows. However, the second law of thermodynamics states that it is not possible to devise a cyclic process in which heating supplied from a single source is converted entirely to work. Instead, only some of the heat may be converted to work; the remainder must be rejected to a heat sink at a lower temperature than the heat source. In that way the net of the heat added and subtracted in the cycle equals the work done, as the first law requires.

It is not possible to express directly this second law statement in the form of an equation. However, it is possible to deduce three important consequences of the second law. The first is that there exists an absolute temperature scale, denoted by T, which is independent of the physical properties of any substance and which has only positive values. The second is that there is a thermodynamic property called entropy, denoted by S, whose incremental change is equal to the heat interaction quantity dQ divided by the system temperature T for any incremental process in which the system temperature remains spatially uniform, called a reversible heat addition, or

$$dS \equiv \left(\frac{dQ}{T}\right)_{rev} \tag{3.10}$$

The third deduction is called the inequality of Clausius. It states that, in any process, dS is equal to or greater than the ratio dQ/T,

$$dS \geq \frac{dQ}{T} \tag{3.11}$$

As a consequence, in a process for which $d\mathcal{Q} = 0$, which is called an *adiabatic* process, the entropy may remain the same or increase but may never decrease. An adiabatic process for which the entropy increases is an irreversible process because the reverse of this process, for which the entropy would decrease ($dS < 0$), violates the second law as expressed in Clausius' inequality.

There are many important consequences of Clausius' inequality that we will not examine in detail in this chapter, but will be identified as such at the appropriate occasion of use. Among them are the conditions for thermodynamic equilibrium, including thermochemical equilibrium, and the limits on the production of useful work in cycles or processes.[3]

3.6 THERMODYNAMIC PROPERTIES

The most common methods for utilizing the energy of fossil or nuclear fuels require the use of fluids as the means to generate mechanical power or to transport energy to a desired location. The thermodynamic properties of fluids thereby assume a great importance in the systems that transform energy.

We know that in a steam power plant the working fluid, water, undergoes large changes in temperature and pressure as it moves through the boiler, turbine, and condenser. The thermodynamic properties pressure p and temperature T are called *intensive* properties because their values are not proportionate to the mass of a fluid sample but are the same at all points within the sample. On the other hand, the energy E, volume V, and entropy S are *extensive* properties in that their values are directly proportionate to the mass of a fluid sample.[4] But if we divide an extensive property, such as E, by the mass M of fluid whose energy is E, then the ratio E/M, called the *specific energy,* is independent of the amount M of fluid. Denoting specific extensive properties by a lowercase letter, we have for the specific energy e, volume v, and entropy s

$$e \equiv \frac{E}{M}, \qquad v \equiv \frac{V}{M}, \qquad s \equiv \frac{S}{M} \qquad (3.12)$$

The use of specific extensive properties simplifies the analysis of thermodynamic systems utilizing fluids and other materials to produce work or transform energy. By following the changes experienced by a unit mass of material as it undergoes a change within the system, the work and heat quantities per unit mass may be determined. The total work and heat amounts for the system may then be calculated by multiplying the unit quantities by the total mass utilized in the process.

The first and second law properties, energy and entropy, are sometimes not convenient to use in analyzing the behavior of thermodynamic systems. Rather, particular combinations of the properties p, T, v, e, and s turn out to be more helpful. One of these useful properties is called the

---

[3]The third law of thermodynamics is an additional principle that is closely related to the second law. It states that the entropy of all thermodynamic systems is zero at the absolute zero of temperature. Among other things, it is important to the determination of the free energy change in combustion reactions.

[4]In the following discussion, we disregard the kinetic and potential energy components of the total energy as expressed in equation (3.1) as we are considering the properties of a system that is stationary in the earth's gravity field. When it is necessary to take this motion into account, as in a flow through a turbine, we will explicitly add these additional energy components at the appropriate point.

enthalpy (h) and is defined as

$$h \equiv e + pv \tag{3.13}$$

The enthalpy has a simple physical interpretation. Suppose a unit mass of material is surrounded by an environment in which the pressure is fixed and equal to the pressure p of the system. If a small amount of heat, dq, is added to the system, its temperature will rise and it will expand, undergoing an increment of volume dv and performing an amount of work $dw = p\,dv$ on the environment. According to the first law, equation (3.7), the heat and work amounts change the energy e:

$$de = dq - p\,dv$$

$$dq = de + p\,dv = de + d(pv) = d(e + pv) = dh$$

where the equality $p\,dv = d(pv)$ follows from the constancy of p in this process. Thus the amount of heat added in a constant pressure process is equal to the increase in enthalpy of the material. The ratio of the increase in enthalpy, at fixed pressure, to the increment of temperature experienced in this process is called the *constant-pressure specific heat* and is given the symbol c_p,[5]

$$c_p \equiv \left(\frac{\partial h\{p,\ T\}}{\partial T} \right)_p \tag{3.14}$$

When we consider a similar heating at fixed volume, no work is done and the increase in energy de is equal to the heat increment dq. The ratio of the energy increase to the concomitant temperature increase is called the *constant-volume specific heat, c_v,*

$$c_v \equiv \left(\frac{\partial e\{v,\ T\}}{\partial T} \right)_v \tag{3.15}$$

A second property that will be found useful is the *Gibbs' free energy,* given the symbol f and defined by

$$f \equiv h - Ts = e + pv - Ts \tag{3.16}$$

For a process that proceeds at constant temperature and pressure, the second law of thermodynamics requires that the amount of work done by a system cannot exceed the reduction of free energy f. The free energy is a useful thermodynamic function in cases of chemical or phase change. For example, a sample of liquid water and water vapor can be held in equilibrium at the boiling temperature corresponding to the sample pressure. If heat is added while the pressure remains fixed, some liquid is converted to vapor but the temperature remains unchanged. For this heat transfer process at fixed temperature and pressure, the free energy f is unchanged. In Section 3.12 we shall use the free energy to determine the limiting performance of electrochemical cells.

---

[5]In this expression, $h\{p,\ T\}$ is considered a property depending upon the pressure p and temperature T. The partial derivative $\partial h/\partial T$ is taken with respect to T holding p fixed. A similar constraint is implied in equation (3.15) below, where $e\{v,\ T\}$ is a function of volume v and temperature T.

It is possible to express the relationship between these properties in differential form by noting that, for a reversible process, $dq = T\,ds$ by equation (3.10), so that

$$T\,ds = de + p\,dv \tag{3.17}$$

$$= dh - v\,dp \tag{3.18}$$

$$= dh + s\,dT - df \tag{3.19}$$

3.7 STEADY FLOW

Many thermodynamic systems incorporate components through which a fluid flows at a mass flow rate $\dot{m}$ that is invariant in time, which we call *steady flow*. That is true for the compressor, combustor, and turbine of a gas turbine power plant; for the boiler, steam turbine, condenser, and feed pump of a steam power plant; but not for the cylinder of an automobile engine, where the flow is intermittent.[6] If the flow is steady, the first law may be expressed in a form that relates the thermodynamic properties of the inflowing and outflowing fluid streams with the rates $\dot{Q}$ and $\dot{W}$ at which heat is added to and work is done by the fluid within the component in question,

$$\dot{m}\,h_{out} = \dot{m}\,h_{in} + \dot{Q} - \dot{W} \tag{3.20}$$

where the subscripts *out* and *in* identify the thermodynamic state of the fluid at the outlet and inlet of the component.[7] For a boiler, $\dot{Q}$ would be the rate at which heat is added to change the water flow to steam; for a steam turbine, $\dot{W}$ is the mechanical power delivered by the turbine as its shaft rotates to drive an electric generator or other mechanical load. If we divide equation (3.20) by $\dot{m}$, then $q \equiv \dot{Q}/\dot{m}$ and $w \equiv \dot{W}/\dot{m}$ are the heat and work quantities per unit mass of fluid flowing through the device; the change in fluid enthalpy $h_{out} - h_{in}$ is then equal to the sum of these terms,

$$h_{out} - h_{in} = q - w \tag{3.21}$$

Many power plant components belong to one of two categories: adiabatic ($\dot{Q} = 0$) devices that deliver or absorb mechanical power (e.g., turbines, pumps, compressors) or workless ($\dot{W} = 0$) heat exchangers in which a fluid is heated or cooled. The combustion chamber of a gas turbine power plant is an exception to this rule, because it is both adiabatic and workless.

3.8 HEAT TRANSFER AND HEAT EXCHANGE

While the laws of thermodynamics tell us how much work can be generated by adding and subtracting heat from a working fluid, they don't tell us how quickly we can accomplish this task, a

---

[6]When operated at a steady speed, the inflows of fuel and air and outflow of exhaust gas from an automobile engine may be regarded as steady, so that steady flow equation (3.20) below may be applied.

[7]We have omitted in this expression the contributions of the kinetic and potential energies of the fluid, which in many cases are negligible compared to the other quantities.

matter of great practical consequence because the time rate of exchange of work and heat quantities determine the mechanical or thermal power that can be produced. The more power that can be generated from a given mass of material, at a given cost, the more desirable the power system becomes.

Depending upon the circumstances, we may want to augment or diminish the rate at which heat flows from hot to cold environments. For example, a steam power plant boiler is designed to facilitate the rapid heating of the circulating water by the hot combustion gases, the heat being transferred through the wall of the metal tubes within which the water flows and outside of which the hot gases circulate. On the other hand, when heating a building's interior space in winter months, the loss of heat to the cold exterior environment is minimized by installing thermal insulation in the walls.

In most cases of steady heat transfer from a hot to a cold environment, the time rate of heat transfer $\dot{Q}$ may be represented by[8]

$$\dot{Q} = \mathcal{U}A(T_h - T_c) \tag{3.22}$$

where $T_h - T_c$ is the temperature difference between the hot and cold environments, A is the surface area of material that separates the two environments and across which the heat flows, and $\mathcal{U}$ is the *heat transfer coefficient,* a property of the material separating the two environments.[9] To attain high values of $\mathcal{U}$, one should use a thin layer of a material, such as copper, that is a good heat conductor and provide vigorous motion of the hot and cold fluids with which it is in contact. To obtain low thermal conductances, one needs thick layers of thermally insulating material, like foamed plastics.

Heat exchangers are passive devices that accomplish a transfer of heat, usually between two streams of fluids, one hot and the other cold. Typically, one fluid flows inside parallel cylindrical tubes while the other fluid flows outside of them. In a steam boiler, for example, the cold water (or steam) flows inside the tubes while the hot combustion gases flow around them. Similarly, in a steam power plant condenser the cold cooling water flows through the tubes while the hot exhaust steam from the turbine passes outside the tubes, condensing on the cold surface. The use of heat exchangers is often necessary to the functioning of a power plant, as in these examples, but they necessarily exact penalties in the form of loss of mechanical power, increased economic cost, and reduced thermodynamic efficiency. As an example of the latter, consider the design of a condenser that must transfer a fixed amount of heat per unit time, $\dot{Q}$. According to equation (3.22), we could reduce its size (A), and thereby its cost, by increasing the temperature difference $(T_h - T_c)$ and thereby reduce the steam cycle efficiency. Alternatively, we could increase the heat transfer coefficient $\mathcal{U}$ by pumping the cooling water through the tubes at a higher speed, but that would incur an extra pumping power loss. As a consequence, the transfer of heat at finite rates in thermodynamic systems inevitably incurs performance penalties that cannot be reduced to zero except by the expenditure of infinite amounts of capital. Fortunately, these performance

---

[8]This expression is not a thermodynamic law, although it is in agreement with the requirement of the second law that heat can be transferred only from a hot to a cold body, not the reverse.

[9]The product $\mathcal{U}A$ is called the thermal conductance, in analogy with the electrical conductance of an electric circuit, which is the ratio of the electric current (analogous to the heat flux $\dot{Q}$) to the voltage difference (analogous to the temperature difference $T_h - T_c$).

penalties can be limited to acceptable levels at a cost commensurate with that of other components of the system.

3.9 COMBUSTION OF FOSSIL FUEL

The source of energy that is utilized in fossil-fueled power systems is the chemical energy that is released when a fuel is oxidized by burning in air. The most common fossil fuels are hydrocarbons—that is, mixtures of molecules composed of carbon and hydrogen.[10] Upon their complete combustion, the carbon in the fuel is oxidized to carbon dioxide and the hydrogen to water vapor. The energy made available in this oxidation is the net amount released when the carbon and hydrogen atoms are separated from each other and subsequently combined with oxygen to form carbon dioxide and water.

Denoting a hydrocarbon fuel molecule as C_nH_m, where n and m denote the number of carbon and hydrogen atoms in a fuel molecule, the molecular rearrangement accompanying complete oxidation of the carbon and hydrogen may be represented by the reaction

$$C_nH_m + \left(n + \frac{m}{4}\right)O_2 \rightarrow nCO_2 + \left(\frac{m}{2}\right)H_2O \qquad (3.23)$$

For each hydrocarbon molecule, $n + m/4$ diatomic oxygen molecules are required to convert the carbon and hydrogen to n molecules of CO_2 and $m/2$ molecules of H_2O. The ratio of the number of oxygen molecules to the number of fuel molecules, $n + m/4$, is called the *stoichiometric ratio*. It may be expressed alternatively as a mass ratio by multiplying the number of molecules by their molecular masses, yielding

$$\frac{\text{oxygen mass}}{\text{fuel mass}} = \frac{32n + 8m}{12n + 1.008m} \qquad (3.24)$$

This mass ratio lies in the range between $8/3 = 2.667$ (for pure carbon) and 7.937 (for pure hydrogen), being a function of the molar ratio m/n only.

Because fossil fuels invariably are burned in air, the stoichiometric proportions are more usefully expressed in terms of the ratio of air mass to fuel mass by multiplying equation (3.24) by the ratio of the mass of air to the mass of oxygen in air, which is 4.319:

$$\frac{\text{air mass}}{\text{fuel mass}} \equiv (A/F)_{st} = 4.319\left(\frac{32n + 8m}{12n + 1.008m}\right) \qquad (3.25)$$

If less air is available than is required for a stoichiometric proportion, then not all of the carbon or hydrogen will be fully oxidized and some amount of CO, solid C, or H_2 may be present in the products of combustion. In such "rich" mixtures not all of the available chemical energy is released in the (incomplete) combustion process. On the other hand, if extra or excess air is available, then not all of the oxygen available is needed and some will remain unconsumed in the combustion

---

[10]Synthetic fuels made from hydrocarbons may include oxygen-containing components such as alcohols and carbon monoxide.

products, but all of the fuel's chemical energy will have been released in the combustion of this "lean" mixture.

3.9.1 Fuel Heating Value

When a mixture of fuel and air is burned, the temperature of the combustion products formed is much higher than that of the fuel–air mixture. In some instances, heat may be transferred from the hot combustion products to a colder fluid; for example, in a steam boiler, this heat causes the water to warm and then boil to steam. The amount of heat available for this purpose is called the fuel heating value and is usually expressed in energy units per unit mass of fuel.

Consider a combustion chamber that is supplied with a steady flow of a fuel–air mixture (the reactants) at a pressure p_r and temperature T_r. If the fuel is burned at constant pressure p_r and if no heat is lost from the combustion chamber ($\dot{Q} = 0$), then the product gas temperature T_p will be higher than T_r, but the product gas enthalpy $h_p\{T_p, p_r\}$ will exactly equal the reactant stream enthalpy $h_r\{T_r, p_r\}$, by equation (3.20). This process may be illustrated by identifying the reactant and product states as points in the enthalpy–temperature diagram of Figure 3.1, in which the enthalpies of the reactants (h_r) and products (h_p) are shown as functions of temperature, at the pressure p_r, as the upper and lower curve, respectively. The reactant enthalpy can be identified as the point R at the intersection of the upper (reactant) curve and the vertical line at the reactant temperature T_r. The horizontal line through this point then intersects the lower (product) curve at the point P, where the product temperature is T_p, assuring that $h_p\{T_p, p_r\} = h_r\{T_r, p_r\}$. T_p is called the *adiabatic combustion temperature*.

We are now in a position to determine the fuel heating value. If the hot product gases are subsequently cooled at constant pressure to the reactant temperature T_r at the point P', then the heat removed per unit mass of product gas will be equal in magnitude to the reduction in enthalpy of the product gas between T_p and T_r, or $h_p\{T_p, p_r\} - h_p\{T_r, p_r\} = h_r\{T_r, p_r\} - h_p\{T_r, p_r\}$.

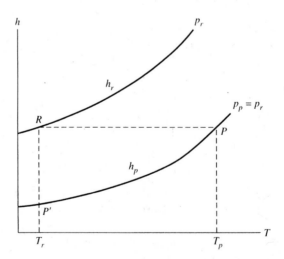

Figure 3.1 The enthalpy h of the reactants (upper curve) and the products (lower curve) of a combustion process, as functions of the temperature T, are related by the fuel heating value. For adiabatic, constant pressure combustion, the products temperature T_p is greater than the reactant temperature T_r.

Multiplying this by the mass flow rate of products, $\dot{m}_p$, divided by the mass flow rate of fuel, $\dot{m}_f$, we obtain the fuel heating value $FHV\{T_r, p_r\}$,

$$FHV\{T_r, p_r\} = \left(\frac{\dot{m}_p}{\dot{m}_f}\right)(h_r\{T_r, p_r\} - h_p\{T_r, p_r\}) \tag{3.26}$$

Common hydrocarbon fuels, such as gasoline or diesel fuel, are mixtures of many hydrocarbons of varying molecular structure. The fuel heating value for such fuels is measured using a calorimeter, to which equation (3.26) is directly applicable. But for pure compounds, such as methane (CH_4), the fuel heating value may be calculated from values of the enthalpy of formation of the fuel and the products of combustion.[11]

Table 3.1 lists the fuel heating value (*FHV*) of some common fuels at 25 °C and one atmosphere of pressure, assuming that the H_2O formed in the product is in the vapor phase. (This is called the *lower heating value, LHV*.) Also listed is the stoichiometric air–fuel ratio $(A/F)_{st}$, the enthalpy difference $h_r - h_p$ at the reference temperature and pressure, and the free energy difference Δf per unit mass of fuel,

$$\Delta f \equiv (\dot{m}_p/\dot{m}_f)(f_r - f_p) \tag{3.27}$$

The fuel heating values of Table 3.1 cover a wide range, from about 10 to 120 MJ/kg. For the saturated hydrocarbons listed, CH_4 to $C_{18}H_{38}$, the range is much smaller, about 44 to 50 MJ/kg. The partially oxygenated fuels, CO, CH_4O, and C_2H_6O, have lower heating values than their parents, C, CH_4, and C_2H_6, because they have less oxidation potential and greater molecular weight. The low value for solid carbon reflects the considerable energy required to convert the carbon atoms from solid to gaseous form.

A different aspect of the fuels is made evident in the fifth column of Table 3.1, which compares the enthalpy difference $h_r - h_p$ at the reference conditions. This enthalpy difference is the chemical energy that is made available in a constant-pressure, adiabatic, stoichiometric combustion process to increase the product temperature to the adiabatic value. If the product gases of the different fuels possessed the same specific heat, then the temperature rise would be proportional to this value. It can be seen that most of these fuels would have approximately the same adiabatic combustion temperature, which turns out to be about 1900 °C, with the exception of hydrogen, which has somewhat higher temperature.

The last column of Table 3.1 lists the fuel heating value per unit mass of fuel carbon. Its reciprocal is the carbon emissions to the atmosphere (in the form of CO_2) per unit of fuel heating value realized when the fuel is burned and the products of combustion are released into the atmosphere. Of the fuels listed, methane provides the most fuel heating value per unit mass of carbon, although all other hydrocarbon fuels possess only about 20% less than methane.[12]

The principal commercially available fuels are coal (anthracite, bituminous, and subbituminous), liquid petroleum fuels (gasoline, diesel fuel, kerosene, home heating fuel, commercial heating fuel), petroleum gases (natural gas, ethane, propane, butane), and wood (hardwood, softwood). The

---

[11] The enthalpy of formation of a compound is the difference between the enthalpy of the compound and that of its elemental constituents (in their stable form), all evaluated at the same reference temperature and pressure.

[12] For a discussion of the significance of carbon emissions, see Chapter 10.

TABLE 3.1 Thermodynamic Properties of Fuel Combustion at 25 °C and One Atmosphere Pressure[a]

Fuel	Symbol	Molecular Weight (g/mol)	FHV^b (MJ/kg fuel)[c]	$(A/F)_{st}$	$(h_r - h_p)^b$ (MJ/kg product)	Δf (MJ/kg fuel)	FHV^b (MJ/kg C)
Pure Compounds[d]							
Hydrogen	H_2	2.016	119.96	34.28	3.400	117.63	NA^e
Carbon (graphite)	$C_{(solid)}$	12.01	32.764	11.51	2.619	32.834	32.764
Methane	CH_4	16.04	50.040	17.23	2.745	51.016	66.844
Carbon monoxide	CO	28.01	10.104	2.467	2.914	9.1835	23.564
Ethane	C_2H_6	30.07	47.513	16.09	2.780	48.822	59.480
Methanol	CH_4O	32.04	20.142	6.470	2.696	22.034	53.739
Propane	C_3H_8	44.10	46.334	15.67	2.779	47.795	56.708
Ethanol	C_2H_6O	46.07	27.728	9.000	2.773	28.903	53.181
Isobutane	C_4H_{10}	58.12	45.576	15.46	2.769		53.142
Hexane	C_6H_{14}	86.18	46.093	15.24	2.838		54.013
Octane	C_8H_{18}	114.2	44.785	15.12	2.778		53.246
Decane	$C_{10}H_{22}$	142.3	44.599	15.06	2.778		52.838
Dodecane	$C_{12}H_{26}$	170.3	44.479	15.01	2.778		52.567
Hexadecane	$C_{16}H_{34}$	226.4	44.303	14.95	2.778		52.208
Octadecane	$C_{18}H_{38}$	254.5	44.257	14.93	2.778		52.102
Commercial Fuels							
Natural gas			36–42				
Gasoline			47.4				
Kerosene			46.4				
No. 2 oil			45.5				
No. 6 oil			42.5				
Anthracite coal			32–34				
Bituminous coal			28–36				
Subbituminous coal			20–25				
Lignite			14–18				
Biomass Fuels							
Wood (fir)			21				
Grain			14				
Manure			13				

[a] Data from Lide, David R., and H. P. R. Frederikse, Eds., 1994. *CRC Handbook of Chemistry and Physics.* 75th ed. Boca Raton: CRC Press; Probstein, Ronald F., and R. Edwin Hicks, 1982. *Synthetic Fuels.* New York: McGraw-Hill; Flagan, Richard C., and John H. Seinfeld, 1988. *Fundamentals of Air Pollution.* Englewood Cliffs, NJ: Prentice-Hall.

[b] H_2O product in vapor phase; heating value is the lower heating value (*LHV*).

[c] 1 MJ/kg = 429.9 Btu/lb mass.

[d] Gas phase, except carbon.

[e] NA, not applicable.

heating values of these fuels vary according to the fuel composition, none of them having a pure molecular composition and some of them including inert components. The unit selling price of these fuels may be based upon the volume (liquids, gases, and wood) or the mass (coal), but the heating value may be a factor in the price. Their heating values are listed in Table 3.1.

In virtually all combustion systems, the water molecules in the products of combustion leaving the device are in the form of vapor, not liquid, because the effluent temperature is high enough and the concentration of water molecules is low enough to prevent the formation of liquid droplets. As a practical matter, the heat of condensation of the water vapor is not available for partial conversion to work, and the effective fuel heating value should be based upon the water product as a vapor, as assumed in Table 3.1. Nevertheless, sometimes a *higher heating value* (*HHV*) is used in the sale of fuel, based upon the assumption that the water product is in the liquid form. To determine this *HHV* for the fuel, we should add to the lower heating value (*FHV* of Table 3.1) the heat of vaporization of water at the reference temperature, expressed as enthalpy per unit mass of hydrogen in water,[13] times the mass fraction of hydrogen in the fuel.[14]

The distinction between higher and lower heating value is primarily a matter of convention. Sellers of fuel like to quote their price in terms of dollars per million Btu of higher heating value, a lower price than that per million Btu of lower heating value. On the other hand, users of fuel who generate electricity prefer to rate their plant efficiency in terms of electrical energy produced per unit of fuel lower heating value consumed, leading to higher efficiencies than when using the higher fuel heating value. As long as the basis of the price or plant efficiency is stated, no confusion should result.

3.10   IDEAL HEAT ENGINE CYCLES

Generating mechanical power from the combustion of fossil fuel is not a straightforward matter. One must utilize the combustion process to change the temperature and/or pressure of a fluid and then find a way to use the fluid to make mechanical work by moving a piston or turning a turbine. The first and second laws of thermodynamics limit the amount of work that can be generated for each unit mass of fuel used, and those limits depend upon the details of how the fuel is used to create power.

To understand the implications of the thermodynamic laws for the conversion of fuel energy to mechanical power, it is convenient to analyze ideal devices in which a fluid is heated and cooled, and produces or absorbs work, as the fluid moves through a cycle. Such a device can be called a *heat engine* in that it exchanges heat with its environment while producing work in a cyclic process. The combustion of fuel is represented in this idealized cycle by the addition of heat from a high-temperature source. Some practical engines, like the gas turbine and the automobile engine, are not heated from an external source. These are termed *internal combustion engines* (ICE). Nevertheless, most of their features can be modeled as ideal heat engine cycles to help us understand their chief attributes.

---

[13]At 25 °C, this value is 21.823 MJ/kg H.

[14]The difference in heating values is a maximum for hydrogen (21.823 MJ/kg fuel), but approaches 3.136 MJ/kg fuel for the heaviest hydrocarbons of Table 3.1.

In this section we consider simple models of heat engines that illustrate the principal features of several practical devices. Of particular importance is the amount of work produced (w) in proportion to the amount of heat that is added (q) to represent the combustion of fuel, whose ratio is called the *thermodynamic efficiency* $\eta_{th}(\equiv w/q)$. But other features are of practical consequence as well, and these are displayed in the analysis.

In all these analyses, we assume that the fluid that produces the mechanical work undergoes reversible processes, so that the entropy change is related to the heat addition by equation (3.10).

3.10.1 The Carnot Cycle

The Carnot cycle is a prototype cycle that has little practical importance but is beautifully illustrative of the second law limits on the simplest of heat engine cycles. It is sustained by two heat reservoirs, a hot one of temperature T_h and a cold one of temperature T_c. (We may think of the hot reservoir as one that is kept warm by heat transfer from a burning fuel source and the cold one as the atmosphere.) Consider the heat engine to be a cylinder equipped with a movable piston and enclosing a fluid of unit mass. The cycle consists of four parts, as illustrated in Figure 3.2: an isothermal expansion during which an amount of heat q_h is added to the engine ($1 \rightarrow 2$ in Figure 3.2); an adiabatic isentropic additional expansion during which the fluid decreases in temperature from T_h to T_c ($2 \rightarrow 3$); an isothermal compression while the system adds a quantity of heat q_c to the cold reservoir ($3 \rightarrow 4$); and finally an isentropic compression to the initial state ($4 \rightarrow 1$). For this cycle the net work w of the piston per cycle is $\oint p\,dv$, and by the first law equation (3.9) we obtain

$$q_h - q_c = w = \oint p\,dv \qquad (3.28)$$

In Figure 3.2 a temperature–entropy plot of the Carnot cycle shows that the entropy increase ($s_2 - s_1$) during heating by the hot reservoir is equal in magnitude to the decrease ($s_3 - s_4$) during cooling, and by the second law (3.10) it follows that

$$\frac{q_h}{T_h} = \frac{q_c}{T_c} \qquad (3.29)$$

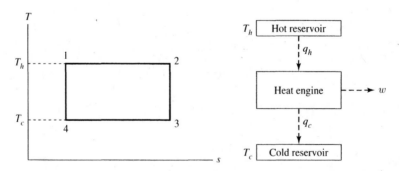

Figure 3.2 The Carnot cycle consists of isothermal and isentropic expansions ($1 \rightarrow 2$, $2 \rightarrow 3$) and compressions ($3 \rightarrow 4$, $4 \rightarrow 1$) of a fluid in a cylinder while absorbing heat q_h from a hot reservoir ($1 \rightarrow 2$), rejecting heat q_c to a cold reservoir ($3 \rightarrow 4$) and producing work w.

By combining these two relations, we may find the thermodynamic efficiency η_{th} of the Carnot cycle to be

$$\eta_{th} \equiv \frac{w}{q_h} = 1 - \frac{T_c}{T_h} \qquad (3.30)$$

The most remarkable aspect of this result is that the thermodynamic efficiency depends only upon the temperatures of the two reservoirs and not at all upon the properties of the fluid used in the heat engine. On the other hand, the amount of net work w that the heat engine delivers does depend upon the fluid properties and the amount of expansion,

$$w = \oint p\, dv = \oint T\, ds \qquad (3.31)$$

where we have used equation (3.17) to show that the net work is equal to the area enclosed by the cycle path in the T–s plane of Figure 3.2.

The conclusion to be drawn from the example of the Carnot cycle is that the thermodynamic efficiency is improved by supplying the heat q_h to the engine at the highest possible temperature T_h. But for a fuel burning in ambient air and supplying this heat to the hot reservoir so as to maintain its temperature, T_h could not exceed the adiabatic combustion temperature T_{ad}. Furthermore, for any $T_h < T_{ad}$ only a fraction of the fuel heating value could be added to the hot reservoir, that fraction being approximately $(T_{ad} - T_h)/(T_{ad} - T_c)$. The resulting thermodynamic efficiency based upon the fuel heating value would then be

$$\eta = \left(1 - \frac{T_c}{T_h}\right)\left(\frac{T_{ad} - T_h}{T_{ad} - T_c}\right) \qquad (3.32)$$

which has a maximum value, when $T_h = \sqrt{T_{ad}T_c}$, of

$$\eta = \frac{\sqrt{T_{ad}/T_c} - 1}{\sqrt{T_{ad}/T_c} + 1} \qquad (3.33)$$

For example, if $T_{ad} = 1900\,°C = 2173$ K and $T_c = 25\,°C = 298$ K, then the maximum thermal efficiency would be 46% when $T_h = 805$ K $= 532\,°C$. This is considerably less than the Carnot efficiency of 86.3% when $T_h = T_{ad}$ for this case.

A possible plan for increasing the efficiency above the value of equation (3.33) would be to employ a large number of Carnot engines, each operating at a different hot reservoir temperature T_h but the same cold reservoir temperature T_c. The combustion products of a constant-pressure burning of the fuel would then be brought into contact with successively cooler reservoirs, transferring heat amounts dh from the combustion gases to produce work amounts dw, where

$$dw = \left(1 - \frac{T_c}{T_h}\right) dh = d(h - T_c s) \qquad (3.34)$$

and where we have made use of the fact that $dh = T_h\, ds$ for a constant-pressure process. The thermodynamic function $h - T_c s$ that appears in equation (3.34) is called the *availability*. In this case the change in availability of the combustion products is the amount of the fuel heat that is available to convert to work by use of an array of Carnot engines. It is the maximum possible work that can be generated if the fuel is burned at constant ambient pressure when the ambient environment temperature is T_c.

If we assume that the constant-pressure specific heat of product gas is constant, then equation (3.34) may be integrated to find the thermodynamic efficiency η,

$$\eta = 1 - \frac{\ln(T_{ad}/T_c)}{(T_{ad}/T_c) - 1} \tag{3.35}$$

For the values of $T_{ad} = 2173$ K and $T_c = 298$ K used above, $\eta = 68\%$. While this is an improvement on the previous case [equation (3.33), $\eta = 46\%$], it is still lower than the Carnot efficiency of 86.3% for $T_h = 2173$ K. Evidently, considerable complication is required to boost ideal thermodynamic efficiencies for fossil-fueled cycles above 50%, and there is little hope of approaching the Carnot efficiency at the adiabatic flame temperature. The Carnot cycle is an important guide to understanding how a simple heat engine might work, but it is not a very practical cycle.

3.10.2 The Rankine Cycle

From the beginning of the industrial revolution until the eve of the twentieth century, most mechanical power generated by the burning of fossil fuel utilized the steam cycle, called the Rankine cycle. In a steam power plant, fuel mixed with air is burned to heat water in a boiler to convert it to steam, which then powers a turbine. This is an external combustion system where the working fluid, water/steam, is heated in pipes that are contacted by hot flue gas formed in the combustion chamber of the furnace. In an efficient steam plant, nearly all the fuel's heating value is transferred to the boiler fluid, but of course only part of that amount is converted to turbine work. The steam cycle is mechanically robust, providing mechanical power even when the boiler and turbine are not perfectly efficient, accounting for its nearly universal use through the end of the nineteenth century.[15]

The thermodynamic processes of the Rankine cycle are illustrated in Figure 3.3, showing (on the left) the changes in temperature and entropy that the working fluid undergoes. In a steam power plant, ambient temperature water is pumped to a high pressure and injected into a boiler ($1 \rightarrow 2$ in Figure 3.3), whereupon it is heated to its boiling point (3), completely turned into steam (4), and then usually heated further to a higher temperature (5). This heating within the boiler occurs at a constant high pressure p_b. The stream of steam flows through a turbine ($5 \rightarrow 6$) undergoing a pressure reduction to a much lower value, p_c, while the turbine produces mechanical power. The low-pressure steam leaving the turbine is cooled to an ambient temperature liquid in the condenser ($6 \rightarrow 1$) and then pumped into the boiler to complete the cycle.

In the idealized Rankine cycle of Figure 3.3, the adiabatic steady flow work per unit mass of steam w_t produced by the turbine is equal to the enthalpy change $h_5 - h_6$ across the turbine, by virtue of the first law equation (3.20). As this is ideally an isentropic process, the enthalpy change

---

[15]Among the earlier American automobiles was the Stanley steamer, powered by a Rankine cycle engine. Although it established a speed record in its day, it was soon surpassed by the more practical gasoline engine.

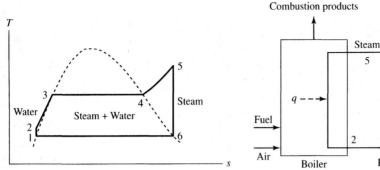

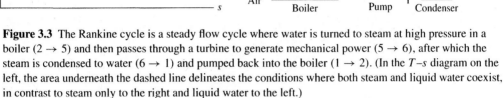

Figure 3.3 The Rankine cycle is a steady flow cycle where water is turned to steam at high pressure in a boiler (2 → 5) and then passes through a turbine to generate mechanical power (5 → 6), after which the steam is condensed to water (6 → 1) and pumped back into the boiler (1 → 2). (In the T–s diagram on the left, the area underneath the dashed line delineates the conditions where both steam and liquid water coexist, in contrast to steam only to the right and liquid water to the left.)

may be expressed, through equation (3.18), as

$$w_t = h_5 - h_6 = \int_6^5 dh = \int_6^5 v\, dp$$

There is a similar expression for the work required to operate the pump. The net work w produced in the cycle may then be expressed as

$$w = \int_{p_c}^{p_b} (v_s - v_w)\, dp \tag{3.36}$$

where v_s and v_w are the specific volumes of the steam in the turbine and water in the pump and p_b and p_c are the boiler and condenser pressures. Because the specific volume of liquid water is so much smaller than that of steam, the power to run the pump is only a tiny fraction of the power produced by the turbine, a mechanically robust attribute of the Rankine cycle.

Because the heating and cooling processes of the ideal Rankine cycle (2 → 5, 6 → 1) occur at constant pressure while the work processes are isentropic, the thermodynamic efficiency may be expressed as

$$\eta_{th} = \frac{(h_5 - h_6) - (h_2 - h_1)}{h_5 - h_2} = 1 - \frac{h_6 - h_1}{h_5 - h_2} = \frac{\oint T\, ds}{\int_2^5 T\, ds} \tag{3.37}$$

There are several aspects of the Rankine cycle that deserve notice. First of all, unlike the Carnot cycle, its thermodynamic efficiency depends explicitly upon the properties of the working fluid, water, as may be seen in Figure 3.3 and equation (3.37). Secondly, the cycle efficiency is increased if the boiler pressure (and steam temperature) is increased. At the same time, high boiler pressures increase the amount of work produced per unit mass of water flowing through the system, reducing the cost of the turbine per unit of power output. The basic cycle is capable of improvements in efficiency by use of internal heat exchange at intermediate pressure levels.

Figure 3.4 A 1500-MW steam turbine with the upper casing removed, showing the rotor stages. From right to left: the high-pressure, intermediate-pressure, and low-pressure stages. (By permission of R. Hoche.)

For Rankine cycles using water as the working fluid, the temperature of the steam from the boiler seldom exceeds 550 °C, and then only for boilers that operate at a high pressure. A high-pressure and high-temperature steam cycle is one for which the steam pressure and temperature exceed the values at the *critical point* of water.[16] The thermodynamic efficiency of a steam cycle is improved if a high pressure and temperature are used. But there are additional steps that may be taken to improve the efficiency of the basic Rankine cycle. One, called *superheating,* is illustrated in Figure 3.3. Here the steam evolved by boiling the water is further heated in the superheater section of the boiler to a temperature T_5 that is higher than the boiling point T_4 corresponding to the boiler pressure $p_4 = p_3$. Another is to reroute steam leaving the high-pressure turbine (see Figure 3.4) back to the boiler for *reheating* to T_5, whereupon it returns to the lower-pressure turbine stages, producing more turbine power than if it had not been reheated. A third improvement involves extracting a fraction of the steam from the high-pressure turbine exhaust and using it to heat the water leaving the feedwater pump, at temperature T_2. This latter is called *regenerative feed water heating*. The net effect of any or all of these alternatives is to increase the average temperature of the working fluid during which heat is added in the boiler. Using the Carnot cycle as the paradigm, this increases the Rankine cycle thermodynamic efficiency. Depending upon the circumstances, employing all these measures can add up to 10 points of efficiency to the basic cycle.

The thermodynamic efficiency of the ideal Rankine cycle is in the range of 30–45%, depending upon the details of the cycle complexity. But actual steam plants have lower-than-ideal efficiencies, for several reasons. The steam turbine and feed water pumps are not 100% efficient, resulting in

---

[16]The critical point is the condition for which the liquid and vapor phases are indistinguishable. For water the critical pressure and temperature are 221.3 bar (= 22.13 MPa = 3210 psi) and 374.2 °C (= 705.6 °F).

less net work than in the ideal cycle. Other mechanical power is required to operate the boiler fans and condenser cooling water pumps, reducing the net power output. The boiler does not transfer to the working fluid (water, steam) all of the fuel higher heating value, because the flue gases exit from the boiler at higher than fuel input temperature and excess air, above that required for stoichiometric combustion, is used. Even the best steam electric power plants seldom exceed 40% thermal efficiency, based upon the ratio of the net mechanical power output divided by the heating value of the fuel supply.

The steam turbine for an electric power plant experiences a large change in pressure between entrance and exit, during which the steam density decreases greatly, requiring ever longer turbine blades to extract power from the flow. Figure 3.4 shows the rotor of a 1500-MW steam turbine, which is divided into three stages, from right to left in the photograph. In the high-, intermediate-, and low-pressure stages the steam pressure is reduced from 71 to 10, 10 to 3, and 3 to 0.1 bar, respectively.[17]

3.10.3 The Otto Cycle

The most ubiquitous fossil-fueled engine is that in the automobile. Unlike the steam plant, the automobile engine does not depend upon heat transfer to the working fluid from an external combustion source. Instead, the fuel is burned adiabatically inside the engine, and the products of combustion produce more work during the expansion stroke than must be invested in the compression stroke, giving a net power output. The combustion products, which are exhausted to the atmosphere, are replaced by a fresh air–fuel charge to begin the next combustion cycle. The working fluid flows through the engine and is not recycled. This is termed an *open cycle*—in contrast to the steam cycle, which is closed.

The thermodynamic heat engine closed cycle that replicates the pressure–volume characteristics of the reciprocating internal combustion engine (ICE) is called the Otto cycle. As sketched in the T–s plane of Figure 3.5, it consists of an isentropic compression from a volume v_e at the

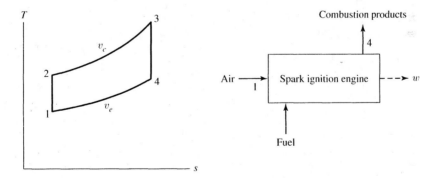

Figure 3.5 The Otto cycle comprises two isentropic compression and expansion strokes of a piston in a cylinder (1 → 2, 3 → 4) interspersed with two constant-volume heating and cooling processes (2 → 3, 4 → 1). It is a model for the spark ignition engine.

[17]For a more detailed description of Rankine cycle power plants using fossil or nuclear fuels, see Chapters 5 and 6.

beginning of the compression stroke to v_c at the end ($1 \rightarrow 2$), followed by a constant-volume heating ($2 \rightarrow 3$) to replicate the adiabatic combustion of the fuel–air mixture, then an isentropic expansion to the maximum volume v_e, after which a constant-volume cooling ($4 \rightarrow 1$) completes the cycle (a process which is absent in the open cycle ICE).

The net cyclic work w, equivalent heat added q, and thermodynamic efficiency η_{th} for the Otto cycle are

$$w = \oint p \, dv = \oint T \, ds$$

$$q = \int_2^3 T \, ds$$

$$\eta_{th} = \frac{\oint T \, ds}{\int_2^3 T \, ds} \tag{3.38}$$

The Otto cycle efficiency is a monotonically increasing function of the volumetric compression ratio, v_e/v_c, and, of course, the thermodynamic properties of the working fluid. For simplifying assumptions about the working fluid, it may be expressed as

$$\eta_{th} = 1 - \frac{1}{(v_e/v_c)^{(c_p/c_v)-1}} \tag{3.39}$$

For a typical gasoline engine compression ratio of 9 and $c_p/c_v = 1.26$, $\eta_{th} = 43.5\%$.

In gasoline engines, the compression ratio is limited by the tendency of the fuel–air mixture to combust spontaneously (called *knock*). A higher compression ratio is used in the diesel engine, where the fuel is injected after the air is compressed, providing a greater thermodynamic efficiency for the diesel engine compared with the gasoline engine.

Because the Otto cycle is a model for the reciprocating ICE, the amount of equivalent heat q is limited by the amount of fuel that can be burned in the fuel–air charge in the cylinder at the end of the compression stroke. The combustible fuel is a maximum when the air/fuel ratio is stoichiometric, which is the normal condition for gasoline engines.[18] The maximum temperature T_3 at the end of combustion is thus the adiabatic combustion temperature of the compressed mixture in the cylinder.[19] Because the reciprocating ICE engine experiences this peak temperature only momentarily and is otherwise exposed to a much lower average temperature, it is able to operate successfully with peak temperatures that exceed the melting point of most materials and secure the favorable thermodynamic efficiencies that accompany such high temperatures.

The thermodynamic efficiencies of automotive engines are noticeably less than the ideal efficiency of equation (3.39). Friction of pistons and bearings, power required to operate valves, cooling pump, and the fuel supply system, pressure losses in intake and exhaust systems, and

---

[18]In gasoline engines the load is varied by changing the pressure p_1 and thus the amount of stoichiometric fuel–air mixture in the cylinder. In diesel engines, the air pressure is fixed but the amount of fuel injected is varied.

[19]In the Otto cycle model of the ICE, combustion occurs at constant volume. The constant-volume fuel heating value and adiabatic combustion temperature are not exactly the same as those for constant-pressure combustion, but the differences are small.

heat loss to the cylinder during the power strokes all combine to reduce the power output compared to the ideal cycle. For four stroke cycle gasoline engines, the low intake pressure experienced at part load is an additional loss that has no counterpart in diesel engines. The best thermal efficiencies for automotive engines are about 28% and 39% for the gasoline and diesel engine, respectively.[20]

The thermodynamic analysis of the Otto cycle does not explain the most salient feature of the reciprocating ICE—that is, that it can be constructed in useful sizes between about 1 kilowatt and 10 megawatts. This is in marked contrast to the steam power plant which, for the generation of electric power, is usually built in units of 100 to 1000 MW. These differences are a consequence of mechanical factors related to the limiting speed of pistons versus turbine blades and other factors unrelated to the thermodynamics of the cycles.[21]

3.10.4 The Brayton Cycle

Since the middle of the twentieth century the gas turbine has become the dominant propulsive engine for large aircraft because of its suitability to high subsonic speed propulsion, light weight, fuel economy, and reliability. But it has made inroads into other uses such as naval vessel propulsion, high-speed locomotives, and, more recently, electric power production. In the latter case, the gas turbine is often used with a Rankine cycle steam plant that is heated by the gas turbine exhaust, the coupled plants being termed a *combined cycle.*

In its simplest form the gas turbine plant consists of a compressor and turbine in tandem, both attached to the same shaft that delivers mechanical power. Situated between the compressor and turbine is a combustion chamber within which injected fuel burns at constant pressure, raising the temperature of the compressed air leaving the compressor to a higher level prior to its entering the turbine. In passing through the turbine, the hot combustion gas is reduced in pressure and temperature, generating more turbine power than is consumed in compressing the air entering the compressor and making available a net mechanical power output from the shaft. The compression, combustion, and expansion processes are adiabatic; and like the reciprocating engine cycles, the gas turbine is an open cycle.

The ideal thermodynamic cycle that models the thermal history of the air and combustion gas flow through a gas turbine power plant is called the *Brayton cycle.* Illustrated in Figure 3.6, it consists of an isentropic compression of air in the compressor from the intake pressure p_i to the compressor outlet pressure p_c ($1 \rightarrow 2$ in Figure 3.6), followed by a constant-pressure heating ($2 \rightarrow 3$) that raises the gas temperature to the value T_3 at the turbine inlet. The combustion gas expands isentropically while flowing through the turbine, its pressure being reduced from p_c to p_i ($3 \rightarrow 4$).

In the Brayton cycle, the steady flow work produced by the turbine and that absorbed by the compressor are each equal to the change in enthalpy of the fluid flowing through them. Per unit mass of fluid, the net work w of the gas turbine plant is the difference between the turbine work

---

[20]Under average operating conditions, the thermal efficiency is less than these maximum values because the engine operating conditions are selected to optimize vehicle performance rather than efficiency.

[21]For a detailed description of the Otto cycle, see Chapter 8.

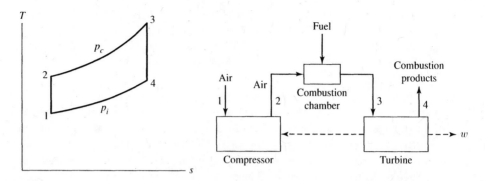

Figure 3.6 The Brayton cycle models the gas turbine cycle, where air is compressed from the inlet pressure p_i in a compressor to the outlet pressure p_c ($1 \to 2$), then burned with fuel to a higher temperature at the constant pressure p_c ($2 \to 3$), and subsequently expanded in the turbine ($3 \to 4$), producing net work w.

and the compressor work,

$$w = (h_3 - h_4) - (h_2 - h_1) \tag{3.40}$$

The heat q added to the fluid leaving the compressor, which replicates the temperature rise caused by adiabatic combustion, is just the increase in enthalpy in the constant pressure process,

$$q = h_3 - h_2 \tag{3.41}$$

As a consequence, the thermodynamic efficiency η_{th} of the Brayton cycle is

$$\eta_{th} = \frac{w}{q} = \frac{(h_3 - h_4) - (h_2 - h_1)}{h_3 - h_2} = 1 - \frac{h_4 - h_1}{h_3 - h_2} = \frac{\oint T\, ds}{\int_2^3 T\, ds} \tag{3.42}$$

where the final expression on the right of equation (3.42) follows from the equality of constant-pressure heat addition and $\int T\, ds$.

The thermodynamic efficiency of the ideal Brayton cycle depends upon the pressure ratio $p_2/p_1 = p_3/p_4$ and the thermodynamic properties of air and combustion products. For simplifying assumptions about these properties, this efficiency may be expressed as

$$\eta_{th} = 1 - \frac{1}{(p_2/p_1)^{(1 - c_v/c_p)}} \tag{3.43}$$

showing that the efficiency increases with increasing pressure ratio. For example, if the pressure ratio $p_2/p_1 = 10$ and $c_p/c_v = 1.3$, then $\eta_{th} = 41.2\%$.

There are several practical problems in building a successful gas turbine power plant. The power produced by the turbine, and usually that absorbed by the compressor, is each greater than the net power output, so that the total power of this machinery is considerably greater than the net output power. The aerodynamic efficiencies of the compressor and turbine need both be high so that as much net power is produced as possible. The thermodynamic efficiency of the cycle can be improved by increasing the turbine inlet temperature, but the latter is limited by the

Figure 3.7 A model of a gas turbine for use in a combined cycle power plant, the compressor on the right and the turbine on the left. The combustion occurs in two stages, the first between the compressor exit and the first turbine stage and the second immediately after the first turbine stage.

high-temperature strength of the turbine blades. It was not until the development of efficient aerodynamic blade designs and high-temperature turbine materials that the gas turbine plant became practical and economically feasible in the twentieth century.

For the simple Brayton cycle, the best thermodynamic efficiencies are about 33%. By use of heat exchange between the hot exhaust gas and the compressed gas entering the combustion chamber, this efficiency may be increased by about four percentage points. Economic factors may require that the plant operate at maximum power output, for which the efficiency would be somewhat lower than these values.

The compressor and turbine of a gas turbine power plant are usually built into a single rotor, as shown in Figure 3.7, with the combustion chamber sandwiched between the compressor and turbine. This is the arrangement for aircraft gas turbines, which must be as light as possible. The rotor shaft delivers the net power difference between the turbine and compressor powers to the electric generator.

3.10.5 Combined Brayton and Rankine Cycles

The combustion products gas stream leaving the gas turbine carries with it that portion of the fuel heating value that was not converted to work. This hot stream of gas may be used to generate steam in a boiler and produce additional work without requiring the burning of more fuel. The use of a gas turbine and steam plant to produce more work from a given amount of fuel than either alone could produce is called a combined cycle.

The thermodynamic efficiency η_{cc} of a combined cycle power plant may be determined as a function of the component efficiencies, η_g and η_s, of the gas turbine and steam cycles. For the gas turbine, the work w_g is equal to $\eta_g q_f$, where q_f is the fuel heat added per unit mass of combustion products. The amount of heat that can be utilized in the steam cycle is just $q_f - w_g = q_f(1 - \eta_g)$. The steam cycle work output w_s is therefore η_s times this heat, or $\eta_s q_f(1 - \eta_g)$. Thus we find the combined cycle efficiency,

$$\eta_{cc} \equiv \frac{w_g + w_s}{q_f} = \eta_g + \eta_s(1 - \eta_g) = \eta_g + \eta_s - \eta_g \eta_s \tag{3.44}$$

The efficiency of the combined cycle is always less than the sum of the efficiencies of the component cycles. Nevertheless, the combination is always more efficient than either of its components. For example, if $\eta_g = 30\%$ and $\eta_s = 25\%$, then $\eta_{cc} = 47.8\%$.

In the combined cycle gas plus steam power plant, the thermal efficiency of the steam cycle component is considerably lower than that for the most efficient steam-only power plant, because the gas turbine exhaust gas is not as hot as the combustion gas in a normal boiler and because the gas turbine requires much more excess air than does a steam boiler. Both restrictions limit the steam cycle efficiency, but nevertheless the combined cycle plant provides an overall fuel efficiency that is higher than that for any single cycle plant.

The combined cycle power plant burning natural gas or jet fuel is often the preferred choice for new electric generating plants, rather than coal-fueled steam plants, for a variety of reasons that overcome the fuel price differential in favor of coal. These reasons are mostly financial and environmental, the latter including the reduced air pollutant emissions, especially carbon dioxide.[22]

3.11 THE VAPOR COMPRESSION CYCLE: REFRIGERATION AND HEAT PUMPS

The use of mechanical power to move heat from a lower temperature source to a higher temperature sink is the thermodynamic process that underlies the functioning of refrigerators, air conditioners, and heat pumps. The process is the reverse of a heat engine in that power is absorbed, rather than being produced, but it still observes the restrictions of the first and second laws of thermodynamics that the heat and work quantities balance [equation (3.9)] and that the ratio of the heat quantities is related to the temperatures of the heat source and sink.

The most common form of refrigeration system employs the *vapor compression cycle*. The refrigeration equipment consists of an evaporator, a vapor compressor, a condenser, and a capillary tube connected in series in a piping loop filled with the refrigerant fluid. The refrigerant is chosen so as to undergo a change of phase between liquid and vapor at the temperatures and pressures within the system. The fluid flows through these components in the order listed, being propelled by the pump. The purpose of the capillary tube is to reduce the pressure of the refrigerant fluid flowing from the condenser to the evaporator, resulting in a lowering of its temperature. The fluid states in the ideal vapor compression cycle are illustrated in the temperature–entropy diagram on the left of Figure 3.8. Vapor leaving the evaporator (1) is compressed isentropically ($1 \to 2$) by the compressor to a higher pressure, where it enters the condenser at a temperature T_2 that is higher than the environment to which heat will be transferred from the condenser. The condenser, a heat

---

[22]For further discussion, see Section 5.3.1.

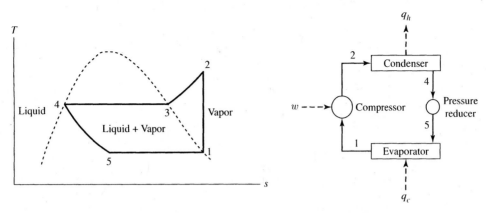

Figure 3.8 The vapor compression cycle for refrigeration begins with an isentropic compression (1 → 2) of the vaporized fluid leaving the evaporator, followed by a constant-pressure cooling in the condenser (2 → 4) to form a saturated liquid (4). The liquid leaving the condenser undergoes an adiabatic pressure decline (4 → 5), entering the evaporator as a cold liquid–vapor mixture (5), whereupon it absorbs heat from the refrigerated space (5 → 1). (In the T–s diagram on the left, the area beneath the dashed line delineates the conditions where both vapor and liquid refrigerant coexist, in contrast to vapor only to the right and liquid to the left.)

exchanger, condenses the vapor to liquid form (2 → 4) by transferring heat to the atmosphere or other environmental sink. The liquid refrigerant leaving the condenser (4) passes through a small-diameter capillary tube, undergoing a viscous pressure drop to enter the evaporator at a lower pressure (5). In this adiabatic, constant-enthalpy process, the fluid temperature decreases and some of the liquid changes to the vapor form. The liquid–vapor mixture then passes through the evaporator, a heat exchanger that absorbs heat from the refrigerated space while changing the liquid portion of the refrigerant to a vapor, completing the cycle.

In heat engine cycles that produce work from the combustion of fuel, the thermodynamic efficiency measures the ratio of the output (work) to the fuel input (heat). The second law of thermodynamics assures that the output is always less than the input, so that the thermodynamic efficiency is less than 100%. For refrigerators and air conditioners, however, the desired output (heat removed from the refrigerated space) is not necessarily less than the input (compressor work). Nevertheless, we may form the ratio of output to input, which is called the *coefficient of performance* (*COP*), using this as a figure of merit for the performance of these devices.

The coefficient of performance of the vapor compression cycle may be determined in terms of the changes in the thermodynamic states of the refrigerant fluid, such as those illustrated in Figure 3.8. The work w required to compress a unit mass of refrigerant equals its change in enthalpy $h_2 - h_1$. The heat q_c absorbed by the refrigerant from the refrigerated space is equal to its change in enthalpy $h_1 - h_5$, which also equals $h_1 - h_4$ because the process 4 → 5 is one of unchanging enthalpy. As a consequence, the coefficient of performance is

$$COP \equiv \frac{q_c}{w} = \frac{h_1 - h_4}{h_2 - h_1} = \frac{h_1 - h_4}{(h_2 - h_4) - (h_1 - h_4)} \tag{3.45}$$

The coefficient of performance is greatest when the temperature difference between the refrigerated space and the environment is least. It decreases monotonically as this temperature difference becomes larger, and becomes zero at a sufficiently large value of the temperature difference, depending upon the characteristics of the refrigerant fluid.

A heat pump is a refrigerator operating in reverse; that is, it delivers heat to an enclosed space by transferring it from an environment having a lower temperature. Heat pumps are commonly used to provide wintertime space heating in climates where there is need for summertime air conditioning. The same refrigeration unit is used for both purposes by redirecting the flow of air (or other heat transfer fluid) between the condenser and evaporator.

For a heat pump, the coefficient of performance $(COP)_{hp}$ is defined as the ratio of the heat q_h transferred to the higher-temperature sink, divided by the compressor work w,

$$(COP)_{hp} \equiv \frac{q_h}{w} = \frac{q_c + w}{w} = 1 + \frac{q_c}{w} = \frac{h_2 - h_4}{h_2 - h_1} = \frac{h_2 - h_4}{(h_2 - h_4) - (h_1 - h_4)} \tag{3.46}$$

where we have used the first law relation that $q_h = q_c + w$ and equation (3.45) to simplify the right-hand side of (3.46).

The heat pump's coefficient of performance is always greater than unity, for the heat output q_h always includes the energy equivalent of the pump work w. But in very cold winter climates, $(COP)_{hp}$ may not be very much greater than unity and the heat delivered would not be much greater than that from dissipating the compressor's electrical power in electrical resistance heating of the space, a much less capital intensive system. It is the year-round use of refrigeration equipment for summertime air conditioning and wintertime space heating that justifies the use of this expensive system.

3.12 FUEL CELLS

In Section 3.10 we considered several different systems for converting the energy of fuel to mechanical energy by utilizing direct combustion of the fuel with air, each based upon an equivalent thermodynamic cycle. In these systems, a steady flow of fuel and air is supplied to the "heat engine," within which the fuel is burned, giving rise to a stream of combustion products that are vented to the atmosphere. The thermal efficiency of these cycles, which is the ratio of the mechanical work produced to the heating value of the fuel, is usually in the range of 25% to 50%. This efficiency is limited by the combustion properties of the fuel and mechanical limitations of the various engines. Thermodynamically speaking, the combustion process itself is an irreversible one, and it accounts for a large part of the failure to convert more of the fuel energy to work.

Is there a more efficient way to convert fuel energy to work? The second law of thermodynamics places an upper limit on the amount of work that can be generated in an exothermic chemical reaction, such as that involved in oxidizing a fuel in air. In a chemical change that proceeds at a *fixed temperature and pressure*, the maximum work that can be extracted is equal to the decrease in Gibbs free energy of the reactants as they form products in the reaction. For most fuels the change of Gibbs free energy f, defined in equation (3.16), is only slightly different from the fuel heating value (see Table 3.1), but this limit is certainly much greater than the work produced by practical heat or combustion engines. But thermodynamics alone does not tell us how it might be possible to capture this much greater amount of available energy in fuels.

An electrochemical cell is a device that can convert chemical to electrical energy. It consists of an electrolyte filling the space between two electrodes. In a *battery,* the electrodes are

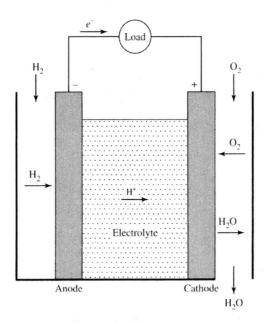

Figure 3.9 A sketch of a hydrogen–oxygen fuel cell. Hydrogen and oxygen are supplied to porous electrodes separated by an electrolyte in which the electric current is carried by hydrogen ions. In the external electric circuit, current is carried by a matching electron flow. The product of oxidation, water, evolves from the cathode.

chemically dissimilar, which causes an electric potential difference between them. In a *fuel cell,* the electrodes are chemically similar but one is supplied with a fuel and the other with an oxidant, generating an electric potential difference between them. By closing the electric circuit external to an electrochemical cell, a current may be drawn from the cell, generating electrical power. The electrical energy consumed in the external circuit is generated by chemical changes within the cell. The electrical energy delivered to the external electric circuit is never greater than the reduction of free energy of the accompanying chemical changes within the cell.

The structure of a fuel cell, sketched in Figure 3.9, is amazingly simple. Two porous metal electrodes are separated by a space filled with an electrolyte, a fluid or solid in which the fuel or oxidant can dissociate into ionic components. Fuel and oxidant are supplied to separate electrodes, diffusing through the porous material to the electrolyte. At the anode (the negative electrode), electrons are transferred to the electrode from the electrolyte as positive ions are formed; at the cathode (the positive electrode), electrons are emitted to the electrolyte to form negative ions or to neutralize positive ions. If the electrodes are connected by electrically conducting wires to an external load, as shown in Figure 3.9, an electric current will flow (in the counterclockwise direction in Figure 3.9) and electrical work will be expended on the load because the cathode electric potential is greater than that of the anode. Inside the fuel cell, the electric current completing the circuit is carried by ions moving through the electrolyte.

The chemical reaction that generates the electrical energy expended in the external load occurs partially at each electrode. Taking as an example the hydrogen–oxygen fuel cell shown in Figure 3.9, the surface reaction at the anode when a hydrogen fuel molecule is ionized upon entering the electrolyte is

$$H_2 \rightarrow 2H^+\{\Phi_{el}\} + 2e^-\{\Phi_a\} \qquad (3.47)$$

where Φ_{el} is the electric potential of the electrolyte and Φ_a is that of the anode. In this reaction the electron moves into the anode at its potential Φ_a while the ion moves into the electrolyte at the potential Φ_{el}. At the cathode, the oxidizing reaction is

$$2H^+\{\Phi_{el}\} + 2e^-\{\Phi_c\} + \frac{1}{2}O_2 \rightarrow H_2O \qquad (3.48)$$

where Φ_c is the cathode electric potential. The net effect of both of the electrode reactions in the production of water and the movement of a charge through the external circuit, found by adding (3.47) and (3.48), is

$$H_2 + \frac{1}{2}O_2 \rightarrow H_2O + 2e^-\{\Phi_a\} - 2e^-\{\Phi_c\} \qquad (3.49)$$

In this overall reaction the hydrogen and the oxygen molecules produce water molecules. In the process, for each hydrogen molecule two electrons flow from the low anode potential to the high cathode potential in the external circuit, producing electrical work. If q_e is the magnitude of the charge of an electron and m_{H_2} is the mass of a hydrogen molecule, then the electrical work work per unit mass of fuel in the reaction is $w = (2q_e/m_{H_2})(\Phi_c - \Phi_a)$. Multiplying the numerator and denominator of the first factor of this expression by Avogadro's number (see Table A.3), we find

$$w = \left(\frac{2\mathcal{F}}{\mathcal{M}_{H_2}}\right)(\Phi_c - \Phi_a) \qquad (3.50)$$

where $\mathcal{F} = 9.6487E(4)$ coulomb/mole is the Faraday constant and $\mathcal{M}_{H_2}$ is the molecular mass (g) of a mole of diatomic hydrogen (H_2).

The second law limits the electrode potential difference $\Phi_c - \Phi_a$ that can be achieved, because the work w cannot exceed the free energy change Δf available in the oxidation reaction

$$w \le (\Delta f)_{H_2}$$
$$(\Phi_c - \Phi_a) \le \frac{(\Delta f)_{H_2} \mathcal{M}_{H_2}}{2\mathcal{F}} \qquad (3.51)$$

where (3.50) has been used to eliminate w in the second line of (3.51). The right-hand side of (3.51) is thus the maximum possible electrode potential difference. For a hydrogen–oxygen fuel cell at 20 °C and one atmosphere of pressure, using the values of Table 3.1, this is calculated to be 1.225 V.

The maximum potential difference of a fuel cell, determined by the free energy change of the fuel oxidation reaction as in equation (3.51), is only reached when the cell is operated in a reversible manner by limiting the current to extremely low values, in effect zero current or open external circuit. For finite current draw, there will be a voltage drop at the electrode surfaces and within the electrolyte that is needed to move the fuel ions to the cathode at a finite rate, along with a corresponding decline in the electrode potential difference. The thermodynamic efficiency η_{fc} of a fuel cell may then be defined as the ratio of the actual electric work w delivered by the cell to

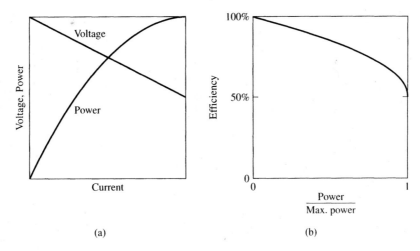

Figure 3.10 (a) Fuel cell voltage and power as a function of current and (b) efficiency as a function of power.

the maximum work Δf,

$$\eta_{fc} \equiv \frac{w}{\Delta f} \qquad (3.52)$$

The high efficiencies of fuel cells, compared to heat engines utilizing the direct combustion of fuel with air, stems from the electrode processes where the electrostatic energy binding molecules can be converted directly to electrostatic energy of the ions and electrons that move in the cell circuit. In contrast, in an adiabatic combustion process the fuel energy is converted to random kinetic and potential energy of product molecules, which cannot be fully recovered in subsequent flow processes.

Figure 3.10(a) illustrates how the fuel cell voltage and power vary with the current when the electrolyte provides the only resistance to current flow within the cell. As the current increases, the cell voltage drops linearly from its maximum value given in equation (3.51). The power output, which is the product of the current times the voltage, reaches a maximum when the voltage has fallen to 50% of its maximum value. The fuel cell efficiency, given in equation (3.52), declines with increasing power, as shown in Figure 3.10(b), to 50% at maximum cell power. When operating at part load, fuel cells can have significantly higher efficiencies than do combustion engines.

To maintain the fuel cell temperature at a fixed value, heat must be removed. The magnitude of the heat removed per unit mass of fuel, $|q_{fc}|$, is determined from the first law of thermodynamics applied to this steady flow process to be

$$|q_{fc}| = FHV - w \geq FHV - \Delta f \qquad (3.53)$$

where FHV is the fuel heating value and the inequality follows from the second law constraint. In fuel cells used to generate electric utility power, the heat removed from the fuel cell may be used to generate additional electricity in a Rankine cycle plant, provided that the fuel cell operating temperature is sufficiently high. This combined cycle plant can achieve very high thermal efficiencies.

3.13   FUEL (THERMAL) EFFICIENCY

The energy embodied in fuels can serve many purposes: generating mechanical or electrical energy, propelling vehicles, heating working or living spaces, creating new materials, refining ores, cooking food, and so on. In this chapter we have focused on the production of mechanical work as an example of the constraints imposed on fuel energy use by the laws of thermodynamics. Nevertheless, these laws apply universally to all energy transactions.

The use of fuel to produce mechanical or electrical power, for whatever purpose, accounts for more than half of all fossil fuel consumption. It is for this reason that the considerations of this chapter provide important information needed to evaluate the environmental consequences of current use of this type and especially of options for reducing fossil fuel use in the future. Whatever the beneficial use of the mechanical energy produced, be it vehicle propulsion, material processing, fluid pumping, and so on, the initiating step of converting fuel energy to mechanical form is an essential ingredient that has important economic and environmental consequences. A measure of the influence upon these consequences is the efficiency with which the fuel energy is converted to mechanical form.

A practical measure of the efficiency of converting fuel energy to work is the ratio of the work produced to the heating value of the fuel consumed, which we may call the *fuel efficiency* η_f (or alternatively, *thermal efficiency*). Usually we use the lower heating value *LHV* of the fuel for this purpose, as this measures the practical amount of fuel energy available. Fuel efficiency is useful because we can readily calculate the fuel mass consumption rate $\dot{m}_f$ of an engine of given power output $\mathcal{P}$ if we know its fuel efficiency,

$$\dot{m}_f = \frac{\mathcal{P}}{\eta_f(LHV)} \tag{3.54}$$

Sometimes the fuel efficiency is expressed differently. The ratio of fuel consumption rate $\dot{m}_f$ to engine power $\mathcal{P}$ is called the *specific fuel consumption*. From equation (3.54), we can conclude that the specific fuel consumption is the inverse of the product $\eta_f(LHV)$.

Table 3.2 summarizes the range of values of fuel (thermal) efficiencies of current technologies for producing mechanical or electrical power. It can be seen that none of these exceeds 50%. These

TABLE 3.2   Fuel (Thermal) Efficiencies of Current Power Technologies

Type	Efficiency
Steam electric power plant	
Steam at 62 bar, 480 °C	30%
Steam at 310 bar, 560 °C	42%
Nuclear electric power plant	
Steam at 70 bar, 286 °C	33%
Automotive gasoline engine	25%
Automotive diesel engine	35%
Gas turbine electric power plant	30%
Combined cycle electric power plant	45%
Fuel cell electric power	45%

efficiencies reflect the constraints of the laws of thermodynamics, the limitations of materials, and the compromises inherent in achieving economical as well as efficient systems. While there is room for improvement, only modest increases above the values in Table 3.2 can be expected from extensive development efforts.

3.14 SYNTHETIC FUELS

A synthetic fuel is one that is manufactured from another fuel so as to enhance its usefulness while retaining as much of the original heating value as possible. Typical examples are oil produced from coal, oil shale, or tar sands; gas from coal, oil, or biomass; alcohols from natural gas or biomass; and hydrogen from coal, oil, or natural gas. Some liquid fuels, such as gasoline, are partially synthetic in that the refining process produces components that are synthesized from petroleum constituents and added to the natural fractions of petroleum that ordinarily comprise the liquid fuel. Major advantages of a synthetic fuel, other than its form as liquid, gas, or solid that might enhance its transportability and convenience of storage, are (a) the removal of base fuel constituents such as sulfur, nitrogen, and ash that lead to harmful air pollutants and (b) the ability to burn the fuel in special devices such as gas turbines and fuel cells. A major disadvantage is the cost of synthesizing the fuel and the loss of its heating value, both of which raise the cost of synthetic fuel heat. This cost factor has been the major obstacle to widespread production and use of synthetic fuels.

Synthetic fuels are formed in steady flow reactors supplied with fuel and other reactants, usually at elevated pressures and temperatures. The chemical reactions that generate the synthetic fuels are aided by the use of catalysts that enhance the reaction rates to economically practical levels. Where the synthesizing reactions are endothermic, heat must be added to maintain the reactor temperature. Thus the synthesis proceeds at fixed temperature and pressure.

As an example of a synthetic fuel, consider the production of synthesis gas from coal by the reaction

$$C_{solid} + (H_2O)_g \rightarrow (CO)_g + (H_2)_g \tag{3.55}$$

where 12 kilograms of solid carbon and 18 kilograms of steam react to form 28 kilograms of carbon monoxide and 2 kilograms of hydrogen. But the fuel heating value of the carbon is (see Table 3.1) 12 kg × 32.76 MJ/kg = 393 MJ, while that of the synthesis gas is 18 kg × 10.10 MJ/kg + 2 kg × 141.8 MJ/kg = 465 MJ, so that this process would be endothermic in the amount of 2.4 MJ/kg of products at 25 °C. Thus additional coal would have to be burned to provide this amount of heat to maintain the synthesis. In addition, the second law requires that the free energy of the product gas not exceed that of the reactant mixture because no mechanical work is invested in the process. As a consequence, practical synthesis results in a reduction of heating value in the synthesized fuel.[23]

Fossil fuels are the final synthesis step in the photosynthetic production of carbohydrates from carbon dioxide and water in plants, deriving their chemical energy from sunlight. In the photosynthesis reaction, energetic solar photons provide the large increase in free energy of the carbohydrate molecules, in contrast to the thermal reforming synthesis of 3.55 where the product molecules have lower free energy. Any reforming of natural fossil fuels to synthetic form will

---

[23]Synthetic fuel may be produced in advanced power plant cycles. See Section 5.3.2.

TABLE 3.3   Thermal Efficiencies of Synthetic Fuel Production

Fuel	Product	Efficiency[a] (%)
Coal	Synthesis gas	72–87
Coal	Methane	61–78
Coal	Methanol	51–59
Coal	Hydrogen	62
Oil	Hydrogen	77
Methane	Hydrogen	70–79
Coal, oil, or gas	Hydrogen (electrolytic)	20–30
Oil shale	Oil and gas	56–72
Methanol	Oil and gas	86
Wood	Gas	90
Corn	Ethanol	46
Manure	Gas	90

[a]Thermal efficiency is the ratio of the heating value of the synthetic product divided by the heating value of the parent fuel.

result in a loss of fuel heating value and an increase of carbon emissions per unit of synthetic fuel heating value.

Table 3.3 summarizes the thermal efficiencies (the ratio of synthetic fuel heating value to that of the parent fuel) for several synthetic fuel production processes. With but few exceptions, these efficiencies lie within the range of 60% to 90%. Most conversion processes require high process temperatures and pressures, need catalytic support to improve the production rate, and consume mechanical power to provide for the requisite pressurization and heat transfer processing. The economic and energy costs of synthetic fuel production can only be justified when there are compensating gains attending the use of synthetic fuels, such as the suitability for use in fuel cells or convenience of storage and transport.

Synthetic nuclear fuels can be produced in nuclear reactors. Uranium-238, which is not a fissionable nuclear fuel, can be converted to plutonium-239, which can be used to fuel a nuclear fission reactor. See Section 6.2 for a description of this process.

3.14.1   The Hydrogen Economy

Hydrogen has been promoted as an environmentally friendly synthetic fuel that can be used in a fuel cell to generate electrical power at high efficiency while emitting no air pollutants. Two possible sources of hydrogen fuel are the reforming of methane and the electrolysis of water:

$$CH_4 + 2H_2O \rightarrow CO_2 + 4H_2 \tag{3.56}$$

$$2H_2O \rightarrow 2H_2 + O_2 \tag{3.57}$$

Both of these reactions require additional energy to bring them to completion. The first requires combustion of additional methane to supply the heat needed for the reforming of methane to hydrogen. The second cannot be effectuated by heating alone, because there is a great increase in free energy. Instead, the free energy increase is provided by electric power in an electrolytic cell and only a small amount of heat is involved. Producing electrolytic hydrogen is very energy inefficient when the electricity is generated by burning a fossil fuel, because the heating value of the hydrogen will be less than one-third of the heating value of the fuel burned in the electric power plant.

Proponents of hydrogen as a substitute for fossil fuel in vehicles or power plants point to its lack of carbon dioxide emissions. But if hydrogen is produced by reforming a fossil fuel, there is no net reduction in carbon dioxide emissions; in most circumstances there will be an increase in emissions and costs. If electrolytic hydrogen is produced by electricity, there is no reduction in carbon dioxide emissions as long as some electricity is produced in fossil fuel plants. In addition, storing and transporting hydrogen is expensive, so that its production by electricity is most economically accomplished at the point of use.

On the other hand, hydrogen production from a fossil fuel provides a path for CO_2 recovery and sequestration (see Section 10.4.4). In this scheme a fossil fuel is converted to a noncarbon fuel, H_2, while the CO_2 that is formed in the conversion process, such as equation (3.56) above, can be recovered and sequestered under ground or in the ocean, preventing its emission into the atmosphere that would have followed from direct combustion of the fossil fuel. In this manner, 60% to 80% of the heating value of a fossil fuel may be utilized while reducing or eliminating the emissions of CO_2.

3.15 CONCLUSION

Nearly 86% of the world's energy is supplied by the combustion of fossil fuel. While the processes by which the energy of this fuel is made available for human use in the form of heat or mechanical power are circumscribed by the principles of thermodynamics, the technologies employed are the consequences of human invention.

In this chapter we have described the implications of the first and second law of thermodynamics for the functioning of selected technologies for producing mechanical power, with particular attention to the efficiency of conversion of fuel energy to useful work. We found that the fuel energy that can be made available by burning fuel in air, called the fuel heating value, appears as the sum of work produced by a heat or combustion engine and heat rejected to the surrounding environment, as required by the first law of thermodynamics. But only a fraction of the fuel energy can be converted to work, according to the second law of thermodynamics, with the magnitude of that fraction depending upon the detailed operation of the technology being used. It is extremely difficult to convert more than half of the fuel heating value to work, but very easy to convert all of it to heat alone.

Most mechanical power is produced in steam power plants, where it is converted to electrical form for distribution to end consumers. Water/steam is circulated within a closed loop, being heated in a boiler by the combustion of fuel and then powering a steam turbine. The most efficient steam plants convert about 40% of the fuel energy to mechanical power.

The gas turbine engine, developed initially for aircraft propulsion and utilizing combustion of fuel in the air steam that flows through the engine, is one prominent form of internal combustion engine. Unlike the steam engine, it does not require the exchange of heat with an external combustion system. Its performance is limited by the strength of the turbine blades that must endure high combustion temperatures. When used to generate electricity, it's efficiency is only about 30%. By combining with a steam plant, called combined cycle, the efficiency of the combination is about 45%.

The other prominent form of internal combustion engine is the common automobile engine, either gasoline (spark ignition) or diesel (compression ignition). Here the cyclic nature of the engine allows much higher combustion temperatures than in the gas turbine, along with higher efficiencies of 25% to 35%.

There is an alternative to producing mechanical power by the above methods. The electro-chemical cell, or fuel cell, supplied with fuel and oxidant, can convert part of the fuel energy directly to electrical power. Operating much like a battery, electrochemical reactions at positive and negative electrodes generate an electric current within the cell that flows from the negative to positive terminal, as in an electric generator. These reactions are sustained by the oxidation of the fuel, which occurs at moderate temperatures well below those encountered in mechanical engines. Fuel cell efficiencies are about 45%.

The science of thermodynamics provides a guide for improvements in the performance of combustion systems, but cannot substitute for the invention and ingenuity needed to bring them about.

PROBLEMS

Problem 3.1

It takes 2.2 million tons of coal per year to fuel a 1000-MW steam electric plant that operates at a capacity factor of 70%. If the heating value of coal is 12,000 Btu/lb, calculate the plant's thermal efficiency and heat rate. (The latter is defined as the ratio of fuel heat in Btu to electric output in kWh.)

Problem 3.2

Given a pressure ratio $p_2/p_1 = 12$ across a gas turbine and a specific heat ratio $c_p/c_v = 1.35$ of the working gas fluid, calculate the ideal thermal efficiency of a Brayton cycle gas turbine plant. Explain why real gas turbine plants achieve only 25% to 35% thermal efficiency.

Problem 3.3

A Rankine cycle ocean thermal power plant is supplied with heat from a flow of warm ocean surface water, which is cooled by 15 °C as it passes through the boiler. If the plant electrical output is 10 MW and the thermal efficiency is 5%, calculate the volume flow rate of warm water, in m^3/s, supplied to the boiler.

Problem 3.4

A synthetic fuel consists of 50% by weight CO and 50% by weight H_2. Calculate its fuel heating value, MJ/kg fuel and MJ/kg product, and stoichiometric air–fuel ratio using the data of Table 3.1.

Problem 3.5

A combined cycle power plant has a gas turbine cycle thermodynamic efficiency of 30% and a steam cycle efficiency of 30%. Calculate the combined cycle thermodynamic efficiency, the ratio of gas turbine power to steam turbine power, and the fraction of the fuel heat that is removed in the condenser of the steam plant.

Problem 3.6

In synthesizing hydrogen for use in a fuel cell, methane may be used in the overall reaction

$$CH_4 + 2H_2O \rightarrow CO_2 + 4H_2$$

which is endothermic. Using the data of Table 3.1, calculate the amount of heat that must be supplied for this synthesis, per unit mass of methane consumed, if it proceeds at 25 °C and one atmosphere of pressure. If this heat is supplied by burning additional methane in air, calculate the total kilograms of methane consumed per kilogram of hydrogen produced.

Problem 3.7

The potential difference $\Delta\Phi$ of a hydrogen fuel cell, as a function of the cell current density i amp/cm^2, is found to be

$$\Delta\Phi = (1.1 \text{ volt})(1 - 0.5\,i)$$

Calculate the maximum power output of the cell in W/m^2, and calculate the values of i, $\Delta\Phi$, and the cell efficiency at maximum power.

Problem 3.8

A PEM fuel cell using methanol as fuel is supplied with a methanol–water mixture at the anode. The anode reaction produces hydrogen ions in the electrolyte and anode electrons according to the reaction

$$CH_4O + H_2O \rightarrow 6H^+\{\Phi_{el}\} + 6e^-\{\Phi_a\} + CO_2$$

At the cathode, the cathodic electrons and electrolytic hydrogen ions combine with oxygen to form water molecules,

$$6H^+\{\Phi_{el}\} + 6e^-\{\Phi_c\} + 3O_2 \rightarrow 3H_2O$$

The combination of these reactions is the oxidation of methanol while producing a current flow in the external circuit across the electrode potential difference $\Phi_a - \Phi_c$,

$$CH_4O + 3O_2 \rightarrow 2H_2O + CO_2 + 6e^-\{\Phi_a\} - 6e^-\{\Phi_c\}$$

(a) Using the data of Table 3.1, calculate the maximum potential difference $\Phi_a - \Phi_c$ that the fuel cell can generate. (b) Calculate the ratio of the maximum electrical work per unit of fuel mass (J/kg) to the fuel heating value.

BIBLIOGRAPHY

Aschner, F. S., 1978. *Planning Fundamentals of Thermal Power Plants.* New York: John Wiley & Sons.

Black, William Z., and James G. Hartly, 1985. *Thermodynamics.* New York: Harper and Row.

Flagan, Richard C., and John H. Seinfeld, 1988. *Fundamentals of Air Pollution.* Englewood Cliffs: Prentice-Hall.

Haywood, R. W., 1991. *Analysis of Engineering Cycles,* 4th edition. Oxford: Pergamon Press.

Holman, J. P., 1980. *Thermodynamics,* 3rd edition. New York: McGraw-Hill.

Horlock, J. H., 1992. *Combined Power Plants Including Combined Cycle Gas Turbine (CCGT) Plants.* Oxford: Pergamon Press.

Kordesch, Karl, and Gunter Simader, 1996. *Fuel Cells and Their Applications.* New York: VCH Publishers.

Lide, David R., and H. P. R. Frederikse, Eds., 1994. *CRC Handbook of Chemistry and Physics,* 75th edition. Boca Raton: CRC Press.

Probstein, Ronald F., and R. Edwin Hicks, 1982. *Synthetic Fuels.* New York: McGraw-Hill.

Saad, Michael A., 1997. *Thermodynamics: Principles and Practice.* Upper Saddle River: Prentice-Hall.

Spalding, D. B., and E. H. Cole, 1973. *Engineering Thermodynamics.* London: Edward Arnold.

Wark, Kenneth, Jr., 1988. *Thermodynamics,* 5th edition. New York: McGraw-Hill.

Electrical Energy Generation, Transmission, and Storage

4.1 INTRODUCTION

Prior to the industrial revolution, human and animal power had provided the bulk of the mechanical work needed in an agricultural society to provide food, clothing, and shelter for human settlements. Through invention and technological development, wind, tidal, and river flows provided mechanical power for milling of grain, sawing of timber, and ocean transportation of goods. But the invention of the steam engine in the early years of the industrial revolution greatly expanded the amount of mechanical power available in industrializing countries for the manufacture of goods and the transportation of people and freight, giving rise to economic growth. By the late nineteenth century the forms of mechanical power generation had evolved to include the steam turbine and gasoline and diesel engines and their uses in ocean and land vehicles. By that time the dominant fuels that produced mechanical power were coal and oil rather than wood. Although human and animal power, as well as the renewable power of wind and stream, were still significant at the dawn of the twentieth century, fossil-fueled mechanical engines were clearly the major and rapidly growing source of industrial energy.

A technological development that greatly augmented the usefulness of mechanical power in manufacturing and, eventually, commercial and residential settings was the nineteenth-century invention of the electric generator and motor that converts mechanical and electrical power from one form to the other with little loss of energy. Unlike mechanical power, which is generated from fossil fuel at the site of power use, as in a manufacturing plant, railroad locomotive, or steamship, electric power generation made possible the transmission of electrical power from a central location to distant consumers via electric transmission lines, which greatly increased the usefulness of the electrical form of work. Together with the end-use inventions that make electrical power so useful, such as the electric light and electric communication devices, the production of electric power has grown so that it constitutes nearly one-third of energy use in current industrialized societies.

Today in the United States, most electric power is generated in large power plants where fossil or nuclear fuel provides the heat needed to generate mechanical work in a steam cycle, with the mechanical power being converted to electrical power in the electric generator, or alternator, as 60-cycle synchronous alternating current (AC) electricity.[1] In transmitting the power to distant

---

[1]Electric utility plants are interconnected with each other by transmission lines so that electric power may be reliably supplied to all customers. This requires synchronization of the generators and standardization of voltages among participating plants. For a description of these plants, see Chapters 5 and 6.

consumers, the voltage is increased by electrical transformers and the power is merged into a network of high-voltage transmission lines that joins the outputs of many power plants to supply the myriads of end-users who are connected to the transmissions lines by a network of distribution lines at lower voltage. More recently, electric power is being generated in smaller plants, often employing gas turbine engines, either alone or in combination with steam turbines (called combined cycle power plant), as the mechanical power source. In addition, some electricity is generated in industrial or commercial plants that utilize the waste heat from the power production process for process or space heating (called cogeneration). The transmission and distribution systems that connect all these sources to each other and to the users of electric power is usually owned and managed by public utility companies.[2]

Some electric power is produced from nonthermal sources of energy. The most prominent of these is hydropower, where mechanical power is produced by hydroturbines supplied with high-pressure water from a reservoir impounded by the damming of a river. The energy source is the difference in gravitational energy of the higher-level water behind the dam compared with the lower-level water downstream of the dam, this difference in level being called the head.[3] Hydropower may be generated as well from the rise and fall of ocean tides. Some electric power is now generated by the mechanical power of wind turbines extracting energy from the wind.

Solar insolation is being used to generate electric power either directly, utilizing photovoltaic cells that convert the energy of solar photons to electric power, or indirectly by supplying heat to thermal steam or vapor engines that drive electric generators. Solar insolation provides the energy needed to grow plants and may thereby be utilized indirectly to generate electric power when biomass crops are used as fuel in thermal power plants.

Nonthermal and solar sources of electric power are termed renewable, in contrast to mineral fuels, such as fossil and nuclear, that are extracted from the earth or ocean. Renewable energy systems are discussed in a Chapter 7.

Thermochemical systems, such as batteries and fuel cells, convert the chemical energy of reactant molecules directly to electricity without an intermediate step where mechanical power is generated. While they currently contribute very little to the amount of electric power generation, they are obviously important in such applications as portable communication devices and to the development of electric drive road vehicles.[4]

The physical principles of electricity and magnetism, which explain how stationary and moving electric charges generate electric and magnetic fields, how mechanical forces are exerted on electric currents flowing in electrical conductors in the presence of magnetic fields, and how electric current is induced to flow by electric fields in conducting materials, provide the basis for the mechanism by which electrical power is generated, transmitted, and utilized.[5] These interactions are those of thermodynamic work, using the parlance of the science of thermodynamics explained in Chapter 2.

---

[2]In many U.S. states, the ownership of transmission and distribution lines is being separated from the ownership of generation facilities as a part of the government's deregulation of electric utility industry.

[3]The ultimate source of this power is solar insolation that evaporates ocean water which is subsequently precipitated to the land drainage basin of the reservoir.

[4]See Section 8.6 for a discussion of electric drive vehicles.

[5]Electric communication systems consume only a small proportion of electrical power in modern economies. Despite their great importance, we will not explore here their interesting technologies.

Insofar as electromagnetic physics deals with forces on charges and currents, it may be regarded as explaining a form of work or power, reasonably called electrical work or power, to distinguish it from mechanical work or power, in which the forces are caused by mechanical contact or a gravitational field. Correspondingly, we may regard that the work required to move a charge or current-carrying conductor in an electric or magnetic field results in a change of electric or magnetic energy of the field, so that energy may be said to be stored in the field, in analogy with the change in potential energy of a mass in the earth's gravity field.

There are circumstances where it is desirable to store mechanical or electrical work generated from mineral fuel consumption for use at later times. A common example is the storage of electrical energy in an automobile's electric storage battery to supply the power needed to start the engine in a subsequent use. In the automobile engine itself a flywheel stores rotational energy produced by the power stroke of the pistons, returning it during the compression stroke.

One of the most prominent uses of electric energy storage is that employed by electric utility systems to even out diurnal variations in the demand for electric power by storing energy during nighttime hours when demand is low and excess power is available and then restoring this energy to the system during daytime hours of peak demand. This permits the central electric generating plants to operate at constant power and best efficiency all day long, lowering the cost of electricity generation.

The principle involved is illustrated in Figure 4.1 depicting the diurnal variation of electric power demand in a typical electric utility system. Starting at midnight, the demand declines to a minimum in the early morning hours and then increases to a daily maximum in the late afternoon or early evening hours, after which it declines again to its midnight level. The utility power suppliers usually match this demand by turning on individual plants during the rising portion of the demand and then taking them off line as the demand declines. However, a sufficient number of plants must be run continuously to supply the minimum demand, called the base load. Supplying the base load can be accomplished more efficiently than supplying the variable load between the minimum and peak load where plants operate at less-than-optimal conditions, both economically

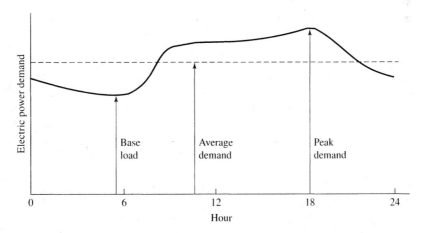

Figure 4.1 A typical diurnal demand for electric power has an early morning minimum and a late afternoon maximum, with the former defining the base load demand that is met by continuously operating plants.

and thermodynamically. If electric storage systems are used to store energy during times of less-than-average demand and return this energy during times of above-average demand, then fewer power plants will be needed to satisfy the diurnal demand and those fewer plants may be operated at optimal conditions.

The most common form of storing energy for use in central electric power systems is the pumped storage system where water is pumped from a lower- to a higher-level reservoir, with the energy being stored in the earth's gravitational field. This energy is later recovered when the stored water flows through a turbogenerator to the lower-level reservoir. Other schemes have been proposed as well, including the use of electric storage batteries, electric capacitors and inductors, and flywheels. These latter are also possible sources of emergency electric power for essential purposes, like hospital operating room power and computer power.

Electric power produced from some sources of renewable energy, such as photovoltaic cells, wind turbines, and tidal power systems, may not be available at times that synchronize with electric power demand, as illustrated in Figure 4.1. Energy storage may be necessary for the successful use of these systems, especially where they cannot be tied into an electric power grid that will provide power at times when the renewable source is absent.

In recent years a renewed interest in electrically propelled automobiles and other vehicles has arisen as a consequence of a desire to improve vehicular energy efficiency and reduce air pollutant emissions from ground transportation vehicles. Most of the proposed systems incorporate, in part, some energy storage in the form of electric storage batteries, flywheels, or electric capacitors. For such applications, the weight, volume, and cost of sufficient energy storage to provide acceptable vehicle performance is generally much greater than that for conventional hydrocarbon-fueled vehicles, posing obstacles to their widespread use. Compared to stationary energy storage systems, mobile ones must meet much more stringent requirements.

This chapter treats the physical principles that determine how electrical power is generated, transmitted, and stored. We first consider how mechanical work is converted to electrical work in an electrical generator, along with the inverse of that process in an electric motor. We next show how electric power flows in a transmission line that connects the electric generator with the end-user of the electric power, and we explain how some of that power is lost in resistance to the flow of current in electric wires. Finally, we describe the kinds of energy storage systems that are used to provide electrical or mechanical power for use at times when it is not otherwise available. Emphasis is placed on the energy and power per unit mass and volume of these systems, the cost per unit energy stored, and the efficiency of recovery of the stored energy, because these are the parameters that determine their use to replace the conventional systems where mineral fuel is consumed as needed to supply electrical or mechanical power and heat.

4.2   ELECTROMECHANICAL POWER TRANSFORMATION

The electric generator and electric motor are the principal devices by which mechanical and electrical power are converted from one form to the other. In the United States, almost all utility electric power is generated by steam, gas, hydro, or wind turbines driving an electric generator. About 60% of this electric power is converted by electric motors to mechanical power for residential, commercial and industrial use. Fifty percent of all mechanical power produced by fossil and nuclear fuel consumption is used to generate electric power.

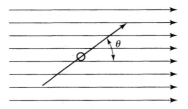

Figure 4.2 A magnetized needle in a magnetic field requires a counterclockwise torque to hold it in place at an angle θ.

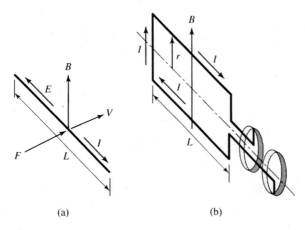

(a)   (b)

Figure 4.3 (a) A wire carrying a current I and moving perpendicular to a magnetic field B at a speed V is subject to a restraining force F and experiences an electric field E. (b) A sketch of a simple armature circuit showing how the current loop is connected to slip rings that deliver the current to an external circuit.

The magnetic interaction that underlies the operation of an electric motor or generator is illustrated in Figure 4.2, showing a magnetized compass needle placed in a magnetic field. The needle seeks to align itself with the magnetic field lines, as does a compass needle in the earth's magnetic field. If we hold the needle stationary at an angle θ from the direction of the magnetic field, we must apply a torque T that is equal to the product of the needle's magnetic dipole moment M, the strength of the magnetic field B, and $\sin\theta$. Alternatively, work can be done by the needle if it rotates to align itself with the magnetic field, in the amount $MB(1 - \cos\theta)$, but work can be generated only for a half revolution of the needle, at most. To make an electric motor of this device, we must reverse the direction of the needle's magnetization every half revolution. This can be accomplished by surrounding the needle with a coil of wire through which a current flows from an external circuit, with the current being reversed each half revolution. The basic elements of both motor and generator are a magnetizable rotor and a stationary magnetic field, either or both of which are connected to external electric circuits that adjust the amount and direction of the magnetic fields.

The physical principles underlying both the electric motor and electric generator are illustrated in Figure 4.3(a). A wire of length L carrying a current I in the presence of a magnetic field B is subject to a restraining force F,

$$F = IBL \qquad (4.1)$$

In addition, if the wire moves at a velocity V in the direction of the force, it experiences an electric field E in the direction opposite to that of the current in the amount[6]

$$E = VB \qquad (4.2)$$

In a generator or motor, wires attached to a rotating armature move through a magnetic field established by a field coil or permanent magnet. The force F applied to a generator armature wire delivers mechanical power $\mathcal{P}$ at a rate FV while the electrical power produced when the current I flows in the direction of the potential increase EL is IEL. Utilizing equations (4.1) and (4.2), these powers are found to be equal:[7]

$$\mathcal{P} = FV = IBLV = IEL \qquad (4.3)$$

Neglecting any electrical and mechanical losses, the mechanical power input (FV) to an ideal generator then equals the electrical power output (IEL).

If we reverse the direction of the velocity V, the direction of the mechanical power is also reversed; that is, there is a mechanical power output from the device. Simultaneously, the electric field is reversed and electric power is now an input. In this mode the device is an electric motor. Neglecting losses, equation (4.3) defines the equality of input electrical power to output mechanical power for an electric motor.

Figure 4.3(b) shows how a rectangular loop of wire attached to a rotating armature and connected to slip rings can deliver the current to an external stationary circuit via brushes that contact the rotating slip rings. For this geometry, the peripheral velocity V of the armature wires is $2\pi rf$, where f is the armature's rotational frequency and r is the distance of the wire from the armature axis, so that the ideal output and input power $\mathcal{P}$ is, using equation (4.3),

$$\mathcal{P} = 2IBL(2\pi rf) = I(4\pi frLB) \qquad (4.4)$$

where the factor of 2 arises from the return leg of the circuit. Because the electrical power equals the product of the current I times the potential difference $\Delta\phi$ across the electrical output terminals, the factor $(4\pi frLB)$ in equation (4.4) is equal to $\Delta\phi$.[8]

In the case of the armature circuit of Figure 4.3(b), the electric potential changes algebraic sign as the armature loop rotates through 180 degrees, thus producing alternating current (AC) power

---

[6]The general forms of equations (4.1) and (4.2), using boldface characters to represent vector quantities, are, respectively, $\mathbf{F} = -(\mathbf{I} \times \mathbf{B})L$ and $\mathbf{E} = -\mathbf{V} \times \mathbf{B}$.

[7]Here we are ignoring any resistive loss in the wire.

[8]The potential increase $\Delta\phi$ of an electric generator is called an *electromotive force*. The generator's internal current I flows in the direction of increasing electric potential ϕ, whereas the current in an external circuit connected to the generator flows in the direction of decreasing potential. A source of electromotive force, such as a generator or battery, is a source of electric power for the attached external circuit.

(in the case of a generator), provided that the magnetic field B maintains its direction. Various forms of armature and magnetic field circuits give rise to the several types of alternating current and direct current motors and generators.

The AC synchronous generator in an electric utility power plant rotates at a precise speed so as to produce 60-cycle AC power. All power plants that feed into a common transmission line must adhere to the same frequency standard. Because generators have an integral number of magnetic poles, their rotational speeds are an integral submultiple of 60 Hz. The generators maintain a fixed voltage in the transmission and distribution systems, while the current varies to suit the power needs of electricity customers.

Because of electrical and mechanical losses in electric motors and generators, the output power is less than the input power. The ratio of the output to the input powers is the efficiency η of the device,

$$\eta \equiv \frac{\text{output power}}{\text{input power}} \tag{4.5}$$

According to the first law of thermodynamics, the difference between the input and output powers appears as a heat flow from the device to the environment, in the amount of $(1 - \eta)$ times the input power. Both electric motors and generators must be cooled when operating in order to prevent the overheating of internal parts.

The electrical resistance R of the armature wire is a source of inefficiency in both generators and motors. The electrical power lost in overcoming this resistance, I^2R, increases as I^2 whereas the power increases as I, as in equation (4.3). This electrical loss thereby becomes a larger fraction of the power output as the latter is increased. As a consequence, the electrical efficiency of motors and generators is least at full power. For economic efficiency we would like to obtain the maximum power for a given investment, which means operating at maximum power and, therefore, least electrical efficiency.

The efficiency of electric motors and generators is greater for large than for small machines. Figure 4.4 exhibits this increase of efficiency with size for energy-efficient induction motors. The

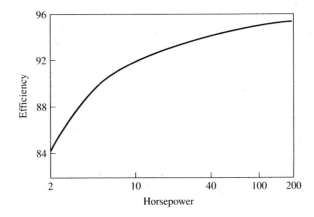

Figure 4.4 The efficiency of energy-efficient induction motors as a function of motor power. (Data from Andreas, John C., 1992. *Energy-Efficient Electric Motors,* 2nd edition. New York: Marcel Dekker.)

Figure 4.5 The stator of a large induction generator, with the armature removed, showing the complex wiring and magnetic circuits needed for high efficiency.

size effect is related to the lower rotational speed of larger machines and the necessarily more complex structure of the magnetic and electric circuits. Figure 4.5 shows the stator of a large induction generator and the complex electric and magnetic circuit features that are needed to achieve high efficiency.

To achieve good efficiency and meet the practical limits on magnetic field strength, the power per unit volume or mass of a motor or generator is limited, being of the order of 1E(5) W/m^3 and 30 W/kg. Because of the similarity of the structure of these devices, the material amounts per unit volume are approximately independent of size and hence power output.

4.3 ELECTRIC POWER TRANSMISSION

The vast growth of electric power production and consumption in the twentieth century was a consequence of the ability to transmit electric power from the source of production, the central power plant, to the many consumers in residential, commercial, and industrial locations far removed from the site of the power plant. Even though there is little energy storage in the transmission and distribution system and the power plant must produce sufficient power to meet the instantaneous demands of the many consumers, this system works remarkably well, especially when many power plants are connected in a network that permits sharing the power demand in an economical and reliable manner.

The components of an electric power system are sketched in Figure 4.6. A gas or steam turbine powers an electric generator supplying alternating current to a transformer that steps up the line voltage to a high value in the transmission line, usually 60–500 kilovolts. After transmission from the power plant to the vicinity of load centers, a step-down transformer reduces the voltage to a lower value, 12–35 kilovolts, for the distribution system that delivers power to the end-users. For residential use, a further reduction to 120–240 volts is required.

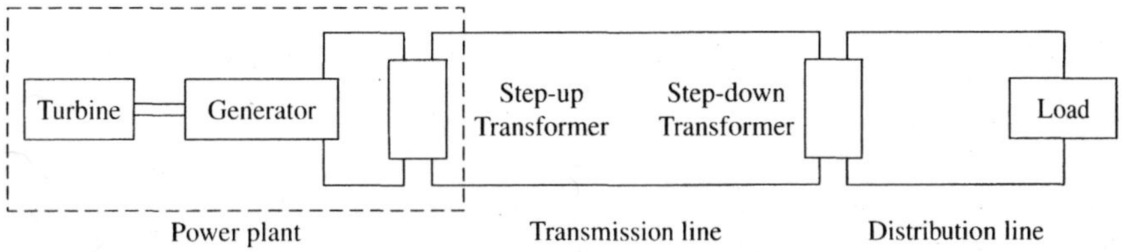

Figure 4.6 A sketch of the elements of an electric power system for generating, transmitting, and distributing power to end-users.

Electric power is transmitted via electric cable consisting of two wires that conduct electrical current at different electric potentials.[9] In a *direct current* (DC) system the electric current in each wire flows in one direction only. On the other hand, in an *alternating current* (AC) system, where the potential difference and current are sinusoidal functions of time, the currents in each wire reverse direction every half-cycle. In either case, the instantaneous electrical power $\mathcal{P}_{el}$ transmitted in the cable is the product of the potential difference $\Delta\Phi$ between the wires and the current I flowing in them,

$$\mathcal{P}_{el} = (\Delta\Phi)I \tag{4.6}$$

The time-averaged power $\overline{\mathcal{P}_{el}}$ is

$$\overline{\mathcal{P}_{el}} = \sqrt{\overline{(\Delta\Phi)^2}\ \overline{(I)^2}}\cos\phi \tag{4.7}$$

where the overline indicates a time-averaged value and ϕ is the phase angle between the current I and the potential difference $\Delta\Phi$ for alternating current.

In both DC and AC transmission systems there is a loss of electric power in the transmission line in the amount I^2R, where R is the electrical resistance of the line. To minimize this loss, the line resistance can be reduced by using large size copper wire and the current minimized, for a given power, by increasing the potential difference. Long-distance transmission lines operate at hundreds of thousands of volts to reduce the transmission loss. But high voltages are impractical and unsafe for distribution to residences and commercial users so the voltage in AC distribution systems is reduced to much lower levels by transformers. Because DC power cannot be transformed easily to a different voltage, its use is restricted to special applications, such as electric rail trains. The use of AC power predominates in modern electrical power systems. Nevertheless, high-voltage DC ($\sim$ 500 kilovolt) transmission lines may be used for very long distance transmission ($\sim$ 1500 kilometers) to reduce energy loss.

The energy efficiency of electric power transmission and distribution is almost entirely determined by economic choices. To increase the efficiency, more money must be invested in copper wire and transformer cores, which is only justified if the value of the saved electric energy exceeds the amortization costs of the increased investment. Transmission and distribution losses in electric power systems are usually held to 5–10%.

---

[9]Electric utility transmission and distribution lines usually include more than two wires, one of which is a common ground return.

4.3.1 AC/DC Conversion

Although nearly all electric power is generated, transmitted, and utilized in AC form, there are important uses for DC power that usually require the conversion of an AC power input to DC form. The most prominent of these are communication systems, such as the telephone and computer, where digital circuits use DC power. Another common DC system is that of the automobile, where alternating current generated in the engine driven alternator is converted to 12-volt DC power that charges the battery and supplies power for lights, fans, radio, and so on.

A *rectifier* is an electrical circuit device that converts AC to DC power. It consists of diodes that permit current to flow in one direction only, thereby transforming the AC current to DC form. As in most electrical devices, some power is lost in this transformation. Every computer has an internal or external power supply that converts household AC to the DC power needed for digital circuits.

There are some sources of electric power that are DC in nature, most notably photovoltaic cells and fuel cells. Where this power is fed to electric utility AC transmission or distribution lines, the DC power first must be transformed to AC form. The electric device that accomplishes this conversion is called an *inverter*. While it is possible to achieve this conversion mechanically by employing a DC motor to power an AC generator, inverters are generally electrical circuit components that accomplish the same purpose. There is some power loss that accompanies this conversion.

4.4 ENERGY STORAGE

There is very little energy stored in the electric utility system that supplies electric power to consumers. The electric power must be supplied at the same rate that it is being utilized by the utility's customers. Although there is some energy temporarily stored (and removed) each half-cycle in the transmission and distribution lines and transformers, it is not available for supplying power when demand exceeds supply during a sustained period. However, there are systems that will fulfill this need and that of other applications, such as storing energy that could be used for electric drive vehicles. In this section we will describe the principles that form the basis for the construction of such devices.

4.4.1 Electrostatic Energy Storage

A capacitor is a device for storing electric charge at an elevated potential. As sketched in Figure 4.7(a), it consists of two electrically conducting plates, of area A, separated by an electrically insulating dielectric medium of equal area and thickness h. Positive and negative charges, in equal amounts Q, stored on the plates induce an electric potential difference $\Delta\phi$ between them. The charge and potential difference are related by Coulomb's law for this configuration,

$$Q = \left(\frac{\epsilon A}{h}\right) \Delta\phi \equiv C \, \Delta\phi \qquad (4.8)$$

where ϵ is the *electric permittivity* of the dielectric medium and $C \equiv \epsilon A/h$ is the *capacitance* of

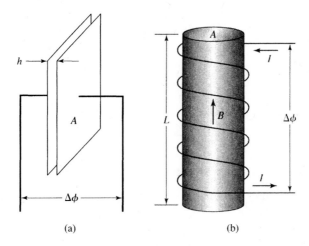

Figure 4.7 The electric capacitor (a) and inductor (b) are devices for storing electrical energy.

the device.[10] If an increment dQ of charge is moved from the negative plate to the positive one through the potential increase $\Delta\phi$, via the circuit external to the capacitor, an amount of electrical work $\Delta\phi\,dQ$ is done in this process, increasing the capacitor free energy by the amount dF,

$$dF = \Delta\phi\,dQ = \frac{Q\,dQ}{C} = \frac{1}{2C}d(Q^2)$$

$$F = \frac{Q^2}{2C} = \frac{C\,(\Delta\phi)^2}{2} \tag{4.9}$$

We may determine the free energies per unit volume and mass in terms of the electric field $E = \Delta\phi/h$ in the capacitor as

$$\frac{F}{V} = \frac{(\epsilon A/h)\Delta\phi^2}{2Ah} = \frac{\epsilon E^2}{2}$$

$$f = \frac{\epsilon E^2}{2\rho} \tag{4.10}$$

where f is the free energy per unit mass, ρ the mass density (kg/m^3) of the dielectric medium, and we neglect the mass of the electrodes.

To obtain high-energy storage densities, we should choose a material with a high electric permittivity ϵ and an ability to withstand a high electric field E without breakdown—that is, without conducting a current that would short-circuit the capacitor internally. Such materials are composed of molecules that have a permanent electric dipole moment and that are not easily ionized in the presence of strong electric fields. For example, polypropylene has a permittivity

---

[10]The dimension of the electrical charge is the coulomb, and that of the capacitance is the farad = coulomb/volt. The electric permittivity has the dimensions of farad/meter = coulomb/volt meter. (See Table A.1.)

$\epsilon = 2E(-11)$ farad/meter, a dielectric strength of $E = 6.5E(9)$ V/m, and density $\rho = 1E(3)$ kg/m^3, so that the free energy per unit volume is $F/V = 4.3E(8)$ J/m^3 and the free energy per unit mass is $f = 4.3E(5)$ J/kg.

High-energy density capacitors are being developed for potential electric vehicle use, with energy densities of about 10 Wh/kg $= 0.036$ MJ/kg. The electric power input and output requires high and variable electric potentials requiring power conditioning equipment to deliver the lower voltage and higher current needed for traction motors. Electric failure of the capacitor dielectric can present safety problems.

4.4.2 Magnetic Energy Storage

It is possible to store energy in the magnetic field produced by a current flowing in a conducting wire. In the sketch of Figure 4.7(b), a magnetic inductor consists of a long cylinder of material, of cross-sectional area A and length L, around which a coil of electric wire having N turns carries a current I. Ampere's law relates the *magnetic induction* B[11] in the material to the current flowing in the wire,

$$B = \frac{\mu N I}{L} \qquad (4.11)$$

where μ is the magnetic permeability of the material.[12] To determine the energy stored in the inductor, we first note that an increment of magnetic induction dB, caused by an increment of current dI in a time interval dt, is related to the potential difference $\Delta\phi$ between the ends of the coil by Faraday's law of magnetic induction,

$$A\frac{dB}{dt} = \frac{\Delta\phi}{N} \qquad (4.12)$$

During this time interval, electric power of amount $\Delta\phi I$ is expended in increasing the magnetic induction, so that the free energy increase is

$$dF = \Delta\phi I \, dt = N A I \, dB = \frac{AL}{2\mu} \, dB^2 = \frac{\mu AL}{2} \, d\left(\frac{NI}{L}\right)^2$$
$$\frac{F}{V} = \frac{B^2}{2\mu} = \frac{\mu(NI)^2}{2L^2} \qquad (4.13)$$

This relation may be expressed alternatively in terms of the *inductance* $\mathcal{L}$ of the coil,[13] which is the ratio $\Delta\phi/(dI/dt) = \mu N^2 A/L$ by equation (4.13), so that

$$F = \frac{\mathcal{L}I^2}{2} \qquad (4.14)$$

---

[11]The units of magnetic induction are weber/meter^2 = volt second/meter^3. (See Table A.1.)

[12]The units of magnetic permeability are henry/meter = volt second/ampere meter. (See Table A.1.)

[13]The unit of inductance is the henry = volt second/ampere. (See Table A.1.)

The storage of magnetic energy is complicated by two effects. Maintaining the coil current I requires the expenditure of electric power to overcome its resistive losses, unless a superconducting coil is used. There is a practical upper limit to the magnetic induction B related to the fact that the coil windings are subject to an outward force per unit length proportional to IB that must be supported by a containing structure having a volume comparable to AL. This limits magnetic induction to values less than about 6 weber/m^2 and corresponding energy densities of 1.43E(7) J/m^3 and 1.8E(3) J/kg.

Superconducting magnetic energy storage systems have been proposed for use with electric utility systems. Designs project energy storage densities of 1 MJ/m^3, prospective costs of \$180/kWh = \$50/MJ, and an energy efficiency of 95%.

4.4.3 Electrochemical Energy Storage

We are familiar with the employment of batteries to supply electric power for a myriad of uses: flashlights, portable radios, watches, hearing aids, heart pacemakers, toys, and so on. Batteries that convert the energy of reactant chemicals to electrical work and are then discarded when discharged are called *primary batteries.* In contrast, *secondary batteries,* whose chemical constituents can be reconstituted by recharging with electrical power, are commonly termed storage batteries. The most common storage battery is that used in the automobile, primarily for supplying engine starting power.

A storage battery consists of two electrodes, a positive one and a negative one, each of different chemical composition, immersed in an electrolyte. The electrodes provide the store of chemical energy that is converted to electrical work as the battery is discharged. The electrolyte provides an internal electric current of negatively charged anions and/or positively charged cations closing the external electric circuit that consumes or provides electric work as the battery is discharged or charged. In contrast with the fuel cell, which also converts chemical energy to electrical work but which requires an external source of fuel and oxidant flow, the storage battery has only a limited amount of chemical energy stored within it that, once consumed, brings to an end its supply of electrical work. In this manner it resembles the other energy storage systems that we have been discussing in this chapter.

To illustrate the principle of operation of a storage battery, we will consider the example of the lead-acid storage battery. In its completely charged configuration, it consists of a positive electrode composed of lead dioxide (PbO_2), a negative one of pure lead (Pb), and an electrolyte that is a concentrated solution of sulfuric acid (H_2SO_4) in water. The sulfuric acid is dissociated into hydrogen cations (H^+) and hydrosulfate anions (HSO_4^-). The conversion of chemical energy to electrical work occurs partially at each electrode. At the negative electrode, a hydrosulfate ion at the electrolyte potential Φ_{el} reacts with the lead to form lead sulfate ($PbSO_4$), a hydrogen ion (H^+) at the electrolyte potential Φ_{el}, and two conducting electrons in the negative electrode at its electric potential Φ_n:

$$Pb + HSO_4^- \{\Phi_{el}\} \rightarrow PbSO_4 + H^+ \{\Phi_{el}\} + 2e^- \{\Phi_n\} \qquad (4.15)$$

At the positive electrode, two electrons at its potential Φ_p and a lead dioxide molecule combine with three hydrogen ions and a hydrosulfate ion at the electrolyte potential Φ_{el} to form a lead sulfate molecule and two water molecules:

$$PbO_2 + 2e^- \{\Phi_p\} + 3H^+ \{\Phi_{el}\} + HSO_4^- \{\Phi_{el}\} \rightarrow PbSO_4 + 2H_2O \qquad (4.16)$$

The net reaction in the storage battery is the sum of the cathode and anode reactions, equations (4.15) and (4.16):

$$Pb + PbO_2 + 2(H^+ \{\Phi_{el}\} + HSO_4^- \{\Phi_{el}\}) \rightarrow 2PbSO_4 + 2H_2O + 2(e^- \{\Phi_n\} - e^- \{\Phi_p\}) \quad (4.17)$$

In the discharge reaction of the lead-acid battery, equation (4.17), both electrode surfaces are each converted to lead sulfate while sulfuric acid is removed and water is added to the electrolyte, diluting it.[14] When both electrode surfaces are completely covered with lead sulfate, there is no difference between the electrodes and no further electric work may be drawn from it.

In the discharge reaction, equation (4.17), the electrical work of moving the two electrons through the external circuit equals $2q_e(\Phi_p - \Phi_n)$, where q_e is the magnitude of the electron charge. When this discharge occurs very slowly at fixed temperature and pressure, the electrical work is then equal to the reduction of free energy per unit mass $(\Delta f)_{la}$ when the reactants $Pb + PbO_2 + 2(H^+ + HSO_4^-)$ are converted to the products $2PbSO_4 + 2H_2O$. As a consequence, the battery potential difference $\Phi_p - \Phi_n$ can be written as

$$\Phi_a - \Phi_c = \frac{(\Delta f)_{la} \mathcal{M}_{la}}{2\mathcal{F}} \qquad (4.18)$$

where $\mathcal{M}_{la}$ is the molecular weight of the reactants and $\mathcal{F}$ is the Faraday constant.[15] At usual conditions, the lead-acid battery potential difference is 2.05 V. In the reaction of equation (4.17), the free energy change per unit mass of reactants, $(\Delta f)_{la}$, is 171 Wh/kg = 0.6 MJ/kg.

The advantage of the storage battery is that the discharge reactions, such as those of equations (4.15) and (4.16) for the lead-acid battery, can be reversed in direction by imposing an external potential difference between the positive and negative electrodes that exceeds the equilibrium value of equation (4.18). In so doing, the products of the discharge reaction are converted back to reactants, renewing the energy stored in the battery, with the energy increment being supplied by electrical work from the external charging circuit. In the discharging and recharging of a storage battery there is usually a net loss of energy because more electrical work is required to charge the battery than is recovered in its discharge. This energy loss of the charge–discharge cycle is dependent upon the charging and discharging rates: When these rates are higher, the loss is greater. Some of the loss is caused by resistive heating resulting from the internal ion current of the battery, while the rest is a consequence of electrochemical irreversibilities at the electrode surfaces.

The electrode reactions that release or absorb electrical work occur at the electrode surface. The electrodes of storage batteries have a porous, sponge-like structure to maximize the ratio of surface area to electrode volume. In this way the amount of energy stored per unit mass of electrode material may be maximized and the cost of energy storage minimized.

---

[14]The degree of discharge of the lead-acid battery can be determined by measuring the specific gravity of the electrolyte, which decreases as the acid content is diluted.

[15]See Table A.3.

TABLE 4.1   Properties of Battery Systems

Battery Type	Energy/Mass (Wh/kg)	Peak Power/Mass (W/kg)	Cost ($/kWh)	Efficiency (%)
Lead acid	40	250	130	80
Nickel/cadmium	50	110	300	75
Nickel/metal hydride	80	250	260	70
Sodium/sulfur	190	230	330	85
Lithium-ion	100	250	200	95

The lead-acid storage battery is a well-developed technology. There are about 200 million lead-acid batteries installed in U.S. road vehicles, each storing about 1 kWh of energy, for a total of about 7E(4) GJ of electric energy stored. This is about 1% of the daily electric energy produced in the United States. If storage batteries were used to level the daily electric utility supply, a very large increase in battery production, above that required for the automotive market, would be needed to satisfy this requirement.

The energy efficiency of lead-acid batteries is about 75% at low (multi-hour) charge and discharge rates. Manufacturing cost is about $50/kWh = $14/GJ. Battery life is usually about 1000 full charge–discharge cycles, which makes storage batteries more expensive than pumped storage for electric utility storage systems.

If lead-acid batteries are overcharged, they emit hydrogen gas, which can be an explosion hazard.

Other well-developed storage batteries employ an alkaline electrolyte, a metal oxide positive, and a metal negative electrode. The nickel–cadmium battery, commonly used in portable electronic equipment, consists of a NiOOH positive and Cd negative electrode, with a KOH electrolyte. Such batteries can provide more energy storage per unit weight than lead-acid batteries, but at greater economic cost. Table 4.1 lists several electric storage battery systems and their properties.

4.4.4   Mechanical Energy Storage

The common form of storing energy for use in electric utility systems is that of pumped hydropower. It consists of a normal hydroelectric plant that is supplied with water impounded behind a dam at high elevation and discharging to a body of water at a lower elevation through a turbine that drives an electric generator. But unlike the usual hydropower plant, the water flow may be reversed and pumped from the lower to the higher reservoir using electric power available from the utility system during times of low demand. Operating on a diurnal cycle, the pumped storage system undergoes no net flow of water but does not deliver as much electrical energy as it uses during the pumping part of the cycle because its components (turbogenerator and pump–motor) are less than 100% efficient.

Energy in a pumped hydropower system is stored by lifting a mass of water through a vertical distance h in the earth's gravitational field. The gravitational energy stored in the reservoir water, per unit mass of water, is gh, while the energy per unit volume is gh/ρ, where g is the gravitational acceleration. For $h = 100$ m, $gh = 9.8E(2)$ J/m^3 and $gh/\rho = 0.98$ J/kg. These are extremely low energy storage densities, requiring very large volumes of water to store desirable amounts of energy.

Because water is essentially a free commodity, the cost of storage is related to the civil engineering costs of the reservoir and power house.

This system is by far the most commonly used energy storage in electric power utility systems. The largest pumped storage system in the United States is the Luddington, Michigan plant that stores 15 GWh (= 5.4E(4) GJ) of energy, providing 2000 MW of electrical power under an hydraulic head of 85 m. But the total U.S. pumped storage power is about 2% of the total U.S. electric power, implying an energy storage of 170 GWh (= 6E(5) GJ). Because normal hydropower provides about 11% of U.S. electrical power, pumped storage hydropower is not a negligible application of hydropower machinery.

The overall energy efficiency of hydropower storage is about 70%; that is, the energy delivered to the electric power system is 70% of that withdrawn during the storage process. This creditably high value stems from the fixed speed and hydraulic head of the turbomachinery used to fill and withdraw water stored in the reservoir.

Hydroelectric machinery is safe, reliable, and cheap to operate. The capital cost of pumped storage installations is significantly related to the capital cost of the civil works (dam, reservoir, etc.). If the capital cost is $500/kW of power, then the capital cost of pumped storage energy is $23/GJ.

Flywheel energy storage systems are being developed for use in road vehicles and for emergency electric power supplies. In this device, an axially symmetric solid material is rotated about its axis of symmetry at a high angular speed Ω. The material at the outer edge of the flywheel rim, whose radius is R, moves at the speed ΩR and has the kinetic energy per unit mass of $(\Omega R)^2/2$ and the kinetic energy per unit volume of $\rho(\Omega R)^2/2$, where ρ is the flywheel mass density. If the flywheel rim has a radial thickness small compared with R, then it would experience a tangential stress $\sigma = \rho(\Omega R)^2$, which is twice the kinetic energy per unit volume. In other words, the maximum kinetic energy per unit volume equals $\sigma/2$ and the kinetic energy per unit mass is $\sigma/2\rho$, where σ is the maximum allowable tangential stress in the flywheel. For high-strength steel, where $\sigma = 9E(8)$ Pa and $\rho = 8E(3)$ kg/m^3, the kinetic energy per unit volume is $4.5E(8)$ J/m^3 and the kinetic energy per unit mass is $5.63E(4)$ J/kg.

The energy input and output are usually in the form of electric power. These systems have very high rotational speeds that decline as energy is withdrawn from them. Their overall energy efficiency is comparable to other forms of storage. Energy storage densities are about 50 Wh/kg = 0.18 MJ/kg. To reach these energy storage densities, high-strength-to-weight materials, such as carbon fiber, are used. In the event of a stress failure, the flywheel components become dangerous projectiles moving at the flywheel peripheral speed, so safety can be a problem.

4.4.5 Properties of Energy Storage Systems

For some energy storage systems the storage capacity per unit volume or mass (J/m^3, J/kg) are important characteristics. For example, in electric-drive highway vehicles the mass or volume of a battery or flywheel system needed to supply enough traction energy for a desirable trip length may be too great for a practical design. Also, the capital cost of the energy storage system is relatable to its mass and affects the dollar cost per unit of stored energy ($/MJ). Values of these properties for the energy storage systems considered above are listed in Table 4.2.

Among the systems listed in Table 4.2, there is a large range in stored energy per unit volume and per unit mass. Yet even the best of them does not approach the high energy density of a hydrocarbon

TABLE 4.2 Properties of Energy Storage Systems

Type	Energy/Volume (J/m^3)	Energy/Mass (J/kg)	Cost ($/MJ)	Efficiency (%)
Capacitor	4E(7)	4E(4)		95
Inductor	1E(7)	2E(3)	50	95
Pumped hydro	1E(3)	1	25	70
Lead-acid storage battery	3E(7)	2E(5)	15	75
Flywheel	2E(8)	2E(5)		80

fuel, based upon its heating value, of 5E(7) J/kg. It is this great energy density of fossil fuel that makes highway vehicles, ships, and aircraft such productive transportation devices. On the other hand, the energy needs of a portable digital computer can be satisfied easily by high-grade storage batteries having a mass that is small compared to the computer mass.

The capital cost of energy storage is not greatly different among the systems listed in Table 4.2. If we value electric energy at three cents per kilowatt hour ($= 2.8E(-2)$ $/MJ), then the capital cost of a typical storage system is about 3000 times the value of the stored energy. If the stored energy is discharged each day, as it would be in a supply leveling system for an electric utility, the capital charge for storage would about equal the cost of electric energy, thus doubling the cost of the stored energy in this instance. The economic cost of storing energy is an important factor limiting its use in electric utility systems.

The efficiency of energy storage systems—the ratio of the output energy to the input energy—is exhibited in the last column of Table 4.2. These are relatively high values, reflecting the best that can be done with optimum energy management.

4.5 CONCLUSION

The generation and transmission of electric power is a necessary component of the energy supply of modern nations. Most such power is generated in thermal plants burning fossil or nuclear fuel, or in hydropower plants. In either case, steam, gas, or hydro turbines supply mechanical power to electric generators that feed electric power via transmission and distribution lines to the end consumer. The electricity generation and distribution process is very efficient, with overall percentage losses being in the single-digit numbers.

There is very little electric energy stored in the generation and transmission system, so electric power must be generated at the same rate as it is consumed if line voltages and frequencies are to be maintained. The diurnal pattern of electricity demand requires that the electric power network be capable of supplying the peak demand, which may be 25% or more above the average demand. Pumped hydroelectric plants are used to store energy that may be used to supply daily peak electric power demand.

Electric energy may be stored for various purposes in storage batteries, capacitors, and inductors, but the only significant amount of energy storage of these types is that of lead-acid batteries in motor vehicles. Cost and weight of storage batteries has limited their use for primary power in vehicles.

PROBLEMS

Problem 4.1

From Figure 4.1, estimate the ratio of peak demand/average demand and that of minimum demand/average demand. If the electricity supply system must have a capacity 20% above the peak demand, estimate the system capacity factor—that is, the ratio of average demand to system capacity.

Problem 4.2

A generator armature coil of length $L = 10$ cm and half-width $r = 2$ cm, as shown in Figure 4.3(b), is placed in a magnetic field of strength $B = 1$ weber/m^2. If the coil rotates at a frequency of 60 revolutions per second and a current I is 10 amperes, calculate the peripheral velocity V, the force F on each length L of the coil and the corresponding electric field E, the electric potential difference $\Delta \Phi$ across the coil slip rings, and the external torque T applied to the armature.

Problem 4.3

A capacitor for storing 100 MJ of energy in an electric drive vehicle utilizes a dielectric of thickness $h = 0.1$ mm and electric permittivity $\epsilon = 2\mathrm{E}(-11)$ farad/meter. The maximum electric field E is $3\mathrm{E}(9)$ volts/meter. For this capacitor, calculate the electric potential $\Delta \Phi$, the required capacitance C, the capacitor area A, and the volume of dielectric material.

Problem 4.4

A typical automotive lead-acid battery stores 100 ampere-hours of charge. Calculate the charge in coulombs. Assuming that the battery can discharge its full charge at an electrode potential of 12 volts, calculate the energy stored in the fully charged battery.

Problem 4.5

A pumped storage plant is being designed to produce 100 MW of electrical power over a 10-hour period during drawdown of the stored water. The mean head difference during this period is 30 m. Calculate the amount of electrical energy to be delivered and the required volume of water to be stored if the hydropower electric generator system is 85% efficient.

Problem 4.6

The characteristics of a battery powered electric vehicle, listed in column 1 of Table 8.3, give an electric energy storage of 18.7 kWh and a battery mass of 595 kg. If the batteries were replaced by an electric, magnetic, or flywheel energy storage system of characteristics given in Table 4.2, calculate the mass and volume of these alternative systems. Would any of these alternatives be practical?

BIBLIOGRAPHY

Andreas, John C., 1992. *Energy-Efficient Electric Motors.* 2nd edition. New York: Marcel Dekker.

Dunn, P. D., 1986. *Renewable Energies: Sources, Conversion and Applications.* London: Peter Peregrinus.

Howes, Ruth, and Anthony Fainberg, Eds., 1991. *The Energy Sourcebook. Guide to Technology, Resources, and Policy.* New York: American Institute of Physics.

Panofsky, Wolfgang K. H., and Melba Phillips, 1962. *Classical Electricity and Magnetism.* Reading: Addison-Wesley.

Rand, D. A. J., R. Woods, and R. M. Dell, 1998. *Batteries for Electric Vehicles.* Somerset: Research Studies Press.

Ter-Gazarian, A., 1994. *Energy Storage for Power Systems.* Stevenage: Peter Peregrinus.

Fossil-Fueled Power Plants

5.1 INTRODUCTION

In Chapter 2 we saw that fossil-fueled electric power plants worldwide consume 55.5% of the annual supply of fossil fuel, of which more than 80% is in the form of coal. Thus, fossil-fueled power plants are major contributors to the anthropogenic emissions of CO_2 and other pollutants, such as SO_2, NO_x, products of incomplete combustion, and particulate matter (PM).

Because power plants, for reasons of economy of scale, are usually built in large, centralized units, typically delivering in the range of 500–1000 MW of electric power, much effort is being spent on their efficiency improvement and environmental control, since these efforts could result in significant global reduction of pollutant emissions and conservation of fossil fuel reserves.

Almost all fossil-fueled power plants work on the principle of a heat or combustion engine, converting fossil fuel chemical energy first into mechanical energy (via steam or gas turbines), then into electrical energy, as described in Chapters 3 and 4.[1] Most large-scale power plants use the Rankine steam cycle, in which steam is produced in a boiler heated by a combustion gas; the steam drives a steam turbine that drives a generator of electricity.[2] Usually, steam turbine plants provide the base load for a regional grid in combination with nuclear plants. On the other hand, peak loads are sometimes supplied by gas turbine plants that work on the Brayton cycle.[3] In those plants, natural gas is burned, and the combustion products directly drive a gas turbine, which in turn drives a generator.

The best steam cycle power plants can achieve a thermal efficiency above 40%; the U.S. average is 36% and the worldwide average is about 33%.[4] Gas turbine power plants achieve a thermal efficiency in the 25–30% range. More advanced power plants exist today that use a combination of Brayton and Rankine cycles. Combined cycle power plants can achieve an efficiency of about 45%. The relatively low thermal efficiency of power plants is due to two factors. The first is a consequence of the Second Law of Thermodynamics, whereby in a heat engine cycle, after performing useful

---

[1]The exception would be a power plant that uses a fuel cell for the conversion of fossil fuel chemical energy directly into electrical energy. However, very few power plants exist today working on the principle of a fuel cell. The workings of a fuel cell electricity generator are described in Chapter 3.

[2]For a detailed description of the Rankine cycle, see Section 3.10.2.

[3]For explanation of base and peak load, see Chapter 4. The Brayton cycle is described in Section 3.10.4.

[4]Thermal (or fuel) efficiency is defined as the ratio of electrical energy output to fossil fuel energy input (see Section 3.13). An equivalent definition is the *heat rate*. This is the amount of British thermal units or gigajoules of fuel energy spent per kilowatt hour of electricity produced.

work, the residual of the fuel heat needs to be rejected to a cold reservoir, usually a surface water (river, lake or ocean) or the atmosphere via a cooling tower. The second factor is due to parasitic heat losses through walls and pipes, frictional losses, and residual heat escaping with the flue gas into the atmosphere. It is an unfortunate fact of electricity generation that power plants use only about 25–50% of the input chemical energy of fossil fuels to generate electricity; the rest is wasted; that is, it goes down the river and up in the air, so to speak.

In this chapter we describe the workings of fossil fueled power plants and their components: fuel storage and preparation, burner, boiler, turbine, condenser, and generator. We focus in particular on emission control technologies that power plants are required by law in most countries to install in order to safeguard public health and the environment. We also review briefly advanced cycle power plants that have the promise of increasing thermal efficiency and reducing pollution.

5.2 FOSSIL-FUELED POWER PLANT COMPONENTS

In a fossil-fueled power plant the chemical energy inherent in the fossil fuel is converted first to raise the enthalpy of the combustion gases; that enthalpy is transferred by convection and radiation to a working fluid (usually, water/steam); the enthalpy of the working fluid is converted to mechanical energy in a turbine; and finally the mechanical energy of the turbine shaft is converted to electrical energy in a generator.

The major components of a fossil fueled power plant are as follows:

- Fuel storage and preparation
- Burner
- Boiler
- Steam turbine
- Gas turbine
- Condenser
- Cooling tower
- Generator
- Emission control

5.2.1 Fuel Storage and Preparation

Coal is delivered to a power plant by rail or, in the case of coastal or riverine plants, by ship or barge. Usually, power plant operators like to have several weeks of coal supply on site, in case of delivery problems or coal mine strikes. Because a 1000-MW power plant, having a thermal efficiency of 35%, consumes on the order of 1E(4) metric tons of coal per day, the coal mounds near the plant may contain up to 3E(5) metric tons of coal. Some coal-fired power plants are situated right near coal mines (so-called mine-mouth plants). Even these plants store at least a month's supply of coal near the plant.

When coal arrives by rail, it is usually carried by a unit train, consisting of a hundred wagons filled with coal, at 100 tons per wagon. The wagons are emptied by a rotary dump, and the coal is carried by conveyors to a stockpile, or directly to the power plant.

Coal is delivered to a plant already sized to meet the feed size of the pulverizing mill (see below), in the order of a few to ten centimeters per coal lump. In the United States and many other countries, coal is washed at the mine. Washing of coal removes much of the mineral content of the coal (including pyritic sulfur), thus reducing its ash and sulfur content and improving its heating value per unit mass. In preparation for washing, the coal is crushed at the mine mouth to less than a centimeter per *nut* or *slack*.

Most modern steam power plants fire pulverized coal. The raw coal from the stockpile is delivered on a conveyor belt directly to a pulverizing mill. Such mills are either of the rotating ring, rotating hammer, or rotating ball type. The mill reduces the raw coal lumps to particles smaller than 1 millimeter. The pulverized coal is stored in large vertical silos from whence it is blown pneumatically into the burners at a rate demanded by the load of the plant.

In oil-fired power plants, oil is stored in large tanks (the "tank farm"), to which oil is delivered either by pipeline, by railroad tankers, or by tanker ship or barge if the plant is located near navigable waters. Power plants like to have at least a 30-day supply of oil in their tanks. For a 1000-MW plant with 35% efficiency, this can amount to over 1E(5) metric tons of oil. The oil is purchased from refineries in the form it is combusted in the burners, with specified sulfur, nitrogen, and ash content as well as other properties, such as viscosity and vapor pressure.

In natural gas-fired power plants, gas is delivered to the power plant by pipeline at high pressure (compressed natural gas, CNG). Some gas-fired power plants use liquefied natural gas (LNG). Liquefied gas is transported in huge (up to 1.25E(5) m^3) refrigerated tankers at $-164\,°C$. The LNG is stored in refrigerated tanks until used.

5.2.2 Burner

The role of the burner is to provide a thorough mixing of the fuel and air so that the fuel is completely burned. Ignition is accomplished by a spark-ignited light oil jet until the flame is self-sustaining. In the combustion chamber a pulverized coal particle or atomized oil droplet burns in a fraction of a second. The coal particle or oil droplet burns from the outside to the core, leaving behind incombustible mineral matter. The mineral matter is called ash. In modern pulverized coal and atomized oil fired power plants, more than 90% of the mineral matter forms the so-called *fly ash*, which is blown out of the boiler by forced or natural draft and is later captured in particle collectors. About 10% of the mineral matter falls to the bottom of the boiler as *bottom ash*. When the bottom of the boiler is filled with water, the bottom ash forms a wet sludge, which is sluiced away into an impoundment. Some of the fly ash, however, is deposited on the water pipes lining the boiler. This forms a slag which hinders heat transfer. The slag needs to be removed from time to time by blowing steam jets against it or by mechanical scraping.

Coal burns relatively slowly, oil burns faster, and gas burns the fastest. For complete combustion (*carbon burn-out*), excess air is delivered—that is, more air than is required by a stoichiometric balance of fuel and the oxygen content of air. Pulverized coal requires 15–20% excess air; oil and gas 5–10%.

A typical burner for pulverized coal is depicted in Figure 5.1. The central coal impeller carries the pulverized coal from the silo in a stream of primary air. Tangential doors (*registers*) built into the wind box allow secondary air to be admixed, generating a fast burning turbulent flame. The impeller is prone to corrosion and degradation and has to be replaced once a year or so.

The burners are usually arranged to point nearly tangentially along the boiler walls. In such a fashion a single turbulent flame ensues from all four burners in a row, facilitating the rapid and

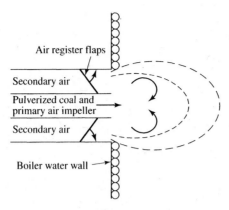

Figure 5.1 Pulverized coal burner.

complete burn-out of the fuel. Depending on the power output of the boiler, as many as six rows of burners are employed, totaling 24 burners.

Some power plants employ cyclone furnaces, especially for poorer grades of coal with a high ash content. In a cyclone furnace the combustion of the pulverized coal is accomplished in a water-cooled horizontal cylinder located outside the main boiler wall. The hot combustion gases are conveyed from the cyclone furnace to the main boiler. The advantage of a cyclone burner is that the majority of the mineral matter forms a molten ash, called *slag,* which is drained into the bottom of the boiler, and only a smaller portion exits the boiler as fly ash. Thus, smaller and less expensive particle collectors are required. The disadvantage is that at the high temperatures experienced inside the cyclone furnace, copious quantities of NO_x are formed. Nowadays, plants equipped with cyclone furnaces require the installation of flue gas denitrification devices for reducing NO_x concentrations in the flue gas, largely vitiating the cost savings of cyclone furnaces in terms of coal quality and ash content.

Some older power plants and smaller industrial boilers employ stoker firing. In stoker-fired boilers, the crushed coal is introduced into the boiler on an inclined, traveling grate. Primary air is blown from beneath the grate, and secondary overfire air is blown above the grate. By the time the grate traverses the boiler, the coal particles are burnt out, and the ash left behind falls into a hopper. The carbon burn-out efficiency is lower in stoker than in pulverized coal burners because of poorer mixing of coal and air that is achievable in stoker-fired boilers. Therefore, stoker-type boilers have a lower thermal efficiency compared to pulverized coal boilers.

5.2.3 Boiler

The boiler is the central component of a fossil-fueled steam power plant. Most modern boilers are of the water wall type, in which the boiler walls are almost entirely constructed of vertical tubes that either carry feed water into the boiler or carry steam out of the boiler. The first water wall boiler was developed by George Babcock and Stephen Wilcox in 1867. The early water wall boilers were used in conjunction with reciprocating piston steam engines, such as used in old locomotives. Only in the twentieth century, with the advent of the steam turbine, and its requirement for large steam pressures and flows, has the water wall boiler been fully developed. In a modern water wall boiler the furnace and the various compartments of the boiler are fully integrated.

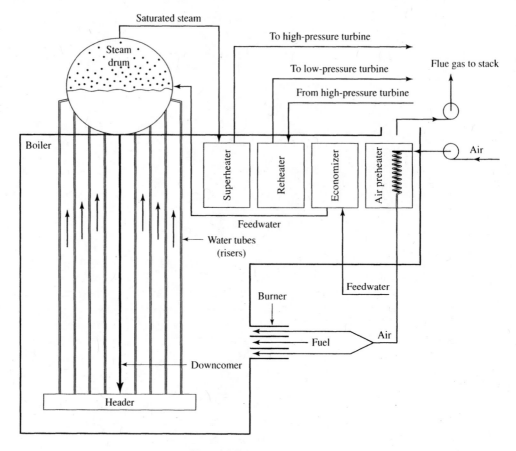

Figure 5.2 Boiler, schematic.

Figure 5.2 shows a schematic flow diagram of a common water wall boiler. Water from the high pressure *feed water heater* at a temperature of 230–260 °C is further heated in the *economizer* section of the boiler to 315 °C, then flows into the *steam drum,* which is mounted on top of the boiler.[5] The steam drum measures typically 30 m in length and 5 m in diameter. In the steam drum liquid water is separated from the steam, usually by gravity. From the steam drum, liquid water flows down the *downcomer* tubes into the *header*. From there, the hot pressurized water flows upward (because of a negative density gradient) through the *riser* tubes, where the actual boiling of water into steam occurs. The separated steam passes another section of the boiler, called the *superheater,* where its temperature is raised to 565 °C at a pressure of 24 MPa. At this point the temperature and pressure are higher than the critical temperature ($T_c = 374\,°C$) and pressure ($p_c = 22$ MPa) of water. The supercritical steam drives the high-pressure turbine. The exhaust steam from the high-pressure turbine flows through the *reheater* section of the boiler, where the temperature is raised again to about 500 °C at a pressure of 3.7 MPa. This steam drives the

---

[5]The feed water is first heated by steam bled from the low-pressure turbine (not shown in Figure 5.2).

low-pressure turbine. The superheater and reheater sections of the boiler are usually situated past a bend in the boiler, called the *neck*. In order to optimize thermal efficiency, the combustion air is preheated to a temperature of 250–350 °C in the *air preheater* section of the boiler. Near the burners, heat is transferred from the combustion gases to the boiler tubes by radiation. Away from the burners, heat is transferred mainly by convection. Coal and oil flames are highly luminous in the visible portion of the spectrum because of the radiation from unburnt carbon and ash particles. Natural gas flames are less visible because of the absence of particles in the flame. However, most of the radiative transfer of heat from all flames occurs in the nonvisible infrared portion of the spectrum. The theoretical (Carnot) thermodynamic efficiency of a heat engine that works between a temperature differential of 838 K (565 °C) and 298 K (25 °C)—that is, between the temperature of the superheater and the condensation temperature of water in the condenser—is $\eta = (T_H - T_L)/T_H = 64\%$. However, as mentioned in the introduction, typical efficiencies of steam power plants are in the 33–40% range. Because heat is added to the water and steam at all temperatures between these limits, the Rankine cycle efficiency is necessarily lower than the Carnot value. Furthermore, parasitic efficiency degradation occurs because of heat losses through the walls of the boiler, ducts, turbine blades and housing, and frictional heat losses.

5.2.4 Steam Turbine

Steam turbines were first employed in power plants early in the twentieth century. They can handle a much larger steam flow, much larger pressure and temperature ratios, and a much larger rotational speed than can reciprocating piston engines. Today virtually all steam power plants in the world employ steam turbines. The steam turbine arguably is the most complex piece of machinery in the power plant, and perhaps in all of industry. There are only a score of manufacturers in the world that can produce steam (and gas) turbines.

The antecedent of the steam turbine is the water wheel. Just as water pushes the blades of a water wheel, steam pushes the blades of a steam turbine. Considering the high pressure and temperature of the steam, that the turbine must be leak proof, the enormous centrifugal stresses on the shaft, and the fact that steam condenses to water while expanding in the turbine, thereby creating a two phase fluid flow, one gets an idea of the technological problems facing the designer and builder of steam turbines.

The development of the modern steam turbine can be attributed to Gustav deLaval (1845–1913) of Sweden and Charles Parsons (1854–1931) of England. deLaval concentrated on the development of an impulse turbine, which uses a converging–diverging nozzle to accelerate the flow speed of steam to supersonic velocities. That nozzle still bears his name. Parsons developed a multistage reaction turbine. The first commercial units were used for ship propulsion in the last decade of the 1800s. The first steam turbine for electricity generation was a 12 MW unit installed at the Fisk power plant in Chicago in 1909. A 208-MW unit was installed in a New York power plant in 1929.

5.2.4.1 Impulse Turbine

In an impulse turbine a jet of steam impinges on the blades of a turbine. The blades are symmetrical and have equal entrance and exit angles, usually 20 °. Steam coming from the superheater, initially at 565 °C and over 20 MPa, when expanded through a deLaval nozzle will have a linear velocity of about 1650 m s⁻¹. To utilize the full kinetic energy of the steam, the blade velocity should be about 820 m s⁻¹. Such a speed would generate unsustainable centrifugal stresses in the rotor. To

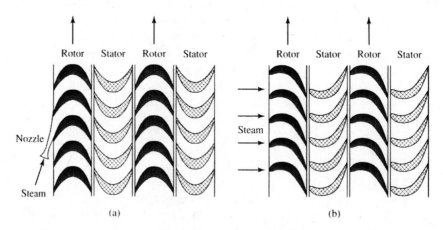

Figure 5.3 (a) Impulse turbine; (b) reaction turbine, schematic.

reduce the rotor speed, turbines usually employ compounding or staging. In a staged turbine, two or more rows of moving blades (*rotors*) are separated by rows of stationary blades (*stators*), as in Figure 5.3(a). Each pair of stator and rotor blades is called a stage. When the steam kinetic energy is divided among n stages, the linear blade velocity of the rotors will be $1/2n$ that of a single rotor.

The force exerted on a rotor blade is $F = \dot{m}(v_s - v_b)$ newtons, where $\dot{m}$ is the mass flow rate of steam through the blade (kg s^{-1}), v_s is the tangential velocity of the steam jet (m s^{-1}), and v_b is the blade speed. The power generated by the blade is $P = F v_b$ watts. It can be shown that maximum power, obtained for $v_b = v_s/2$, is $P_{max} = \dot{m} v_s^2/4$.

5.2.4.2 Reaction Turbine

A reaction turbine consists of rows of fixed (stator) and moving (rotor) blades. The blades are shaped to form a converging nozzle [see Figure 5.3(b)]. Within the converging blades the steam pressure, density and temperature decline while converting its enthalpy to kinetic energy. The steam pressure drops steadily through all rows of blades, stationary and moving, but the steam velocity oscillates, depending on location within the blade formation. In a reaction turbine the optimum blade velocity is $v_{b,opt} = v_s \cos\theta$, where θ is the leaving angle of the blades, and the maximum power obtained is $P_{max} = \dot{m} v_{b,opt}^2$.

In reaction turbines the pressure drops across the moving blades. This makes them less suitable for high-pressure steam because of leakage around the blade tips and consequent loss of efficiency. Therefore, impulse turbines are usually used for high-pressure steam, and reaction turbines are usually used for intermediate and low-pressure steam. In addition, the high-pressure impulse stage can reduce the steam flow to the turbine by closing off some of the first-stage nozzles; this flow control is absent in reaction stages. In Figure 3.4 we have seen a photo of a 1500-MW steam turbine complex for high-, intermediate-, and low-pressure stages.

In both the impulse and reaction turbines, efficiency losses are due to supersaturation, fluid friction, leakage, and heat transfer losses. Supersaturation occurs primarily in reaction turbines when according to thermodynamic equilibrium in the expansion process steam ought to condense, releasing latent heat of condensation. Instead, steam remains for a while in a supercooled state before reverting to thermodynamic equilibrium. This results in a sudden release of energy, called the condensation shock. It is an irreversible process, causing loss in efficiency and energy availability.

Fluid friction occurs throughout the turbine, in the steam nozzles, along the blades, and along the rotor disks that carry the blades. In addition, the rotor and blade rotation impart a centrifugal action on the steam, causing it to be dragged along the blades. When the blades are not properly designed, flow separation may ensue, further increasing turbine losses.

Heat transfer losses are caused by conduction, convection, and radiation. Conduction is the result of heat transfer between metal parts of the turbine. Convection is the heat transfer between steam and the metal parts. Radiation is the heat given off by the turbine casings to the surroundings. Heat transfer losses are highest in the high-temperature, high-pressure sections of the turbines.

In addition, there are frictional losses in bearings, governor mechanisms, and reduction gearing. Also, turbines must supply power for accessories, such as oil pumps. The combined efficiency losses and syphoning of auxiliary power amount to 10–20%; that is, turbines convert only 80–90% of the available steam enthalpy into mechanical energy that drives the generator.

5.2.5 Gas Turbine

In a gas turbine plant where oil, natural gas, or synthesis gas may be used as a fuel, the hot combustion gases are directly used to drive a gas turbine, rather than transferring heat to steam and driving a steam turbine. This requires a different turbine, appropriate for the much higher temperature of the combustion gases and their different thermodynamic properties compared to steam. Gas turbines are easily brought on line and have flexible load match. But their cycle efficiencies are lower than those of steam plants, and the fuel is more expensive. Therefore, gas turbines are mostly used for peak load production and for auxiliary power, such as during major plant outages. However, recently many natural gas-fueled gas turbine plants have been installed in the United States and other countries; but these usually employ the combined cycle mode, which has a higher efficiency than the single cycle mode. Gas turbines operate on the principle of the Brayton cycle, which was described in Chapter 3. Compressed air enters a combustion chamber, where liquid or gaseous fuel is injected. The combustion of the fuel increases the temperature of the combustion gas, producing a net work output of the turbine–compressor system. The temperature of the combustion gases is on the order of 1100–1200 °C, which is the maximum tolerable by present-day steel alloys used for gas turbine blades. Even at these temperatures, thermal stresses and corrosion problems are manifested, so that turbine blade cooling from the inside or outside of the blades by air or water is necessary.

Gas turbines are of the reaction type, where blades form a converging nozzle in which the combustion gases expand, thus converting enthalpy to kinetic energy. As in steam turbines, staged turbines are employed, consisting of several rows of moving and fixed blades.

The working fluid in gas turbines, composed of nitrogen, excess oxygen, water vapor, and carbon dioxide, is not recycled into the compressor and combustion chamber but is, instead, vented into the atmosphere. In some systems, a part of the energy still residing in the exhaust gas is recovered in heat exchangers to heat up the air entering the combustion chamber in order to enhance the overall thermal efficiency of the Brayton cycle, but eventually the exhaust gas is vented. This is in contrast to steam turbines where the working fluid, steam, is recycled into the boiler as condensed water.[6]

---

[6]Gas turbines have many applications other than for electricity generation. They are used for pipeline pumping of natural gas, ship propulsion, and foremost for airplane propulsion in turbojet aircraft. Here, air is compressed in the compressor, and jet fuel (kerosene) is added to the combustion chamber. The combustion gases drive the turbine that supplies power to the compressor and auxiliary systems (e.g., electricity generation), and the turbine exhaust gases pass a deLaval nozzle to provide forward thrust to the airplane.

5.2.6 Condenser

In heat engine cycles, after performing useful work, the working fluid must reject heat to a cold body. Steam turbine electric power plants work on the principle of heat engines. They reject a significant amount of heat into the environment. Between 1.5 and 3 times as much heat is rejected as the plant produces work in the form of electricity. A 1000-MW electric power plant working at 25% efficiency rejects 3000 MW of heat to the environment, whereas one working at 40% efficiency rejects 1500 MW. Some of that heat is added to the environment by the condensing system in a steam cycle, and the rest is added by the discharge to the atmosphere of the hot flue gas vented through the smoke stack.

In the Rankine steam cycle, after expansion in the turbine, the steam is first condensed into water in a condenser, then the water is recycled into the boiler by means of a feed pump. The circulating cooling water of the condenser rejects its heat to the atmosphere by means of a cooling tower or to a nearby surface water. The condenser serves not only the purpose of condensing the high-quality feed water of the boiler, but also to lower the vapor pressure of the condensate water. By lowering the vapor pressure, a vacuum is created which increases the power of the turbine.

There are two types of condenser: *direct contact* and *surface contact* condensers. A direct contact condenser is depicted in Figure 5.4(a). The turbine exhaust passes an array of spray nozzles through which cooling water is sprayed, condensing the steam by direct contact. The warm condensate is pumped into a cooling tower where updrafting air cools the condensate that flows in tubes. The cooled condensate is recycled into spray nozzles. Because the cooling water is in direct contact with the steam exhaust, its purity must be maintained, just like that of the feed water. This makes the process more expensive, and the majority of power plants use a surface contact condenser, depicted in Figure 5.4(b). It is essentially a shell-and-tube type heat exchanger. The turbine exhaust passes an array of tubes in which the cooling water flows. Because for large power plants very large volumes of steam need to be condensed, the contact surface area can reach 100,000 m^2 for a 1000-MW plant. The design of a properly functioning condenser involves complicated calculations of heat transfer. The tubes are surrounded by fins to increase the heat transfer area. The incoming steam needs to be deaerated of noncondensables, mainly air that leaked into the system. The oxygen of

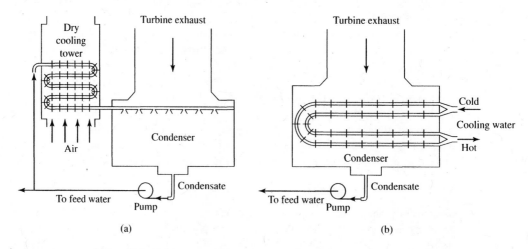

Figure 5.4 (a) Direct contact condenser; (b) surface condenser, schematic.

air is corrosive. Also, noncondensables in the condensate would increase the condenser pressure and thereby reduce the steam turbine power.

5.2.7 Cooling Tower

The bulk of the heat rejection from the steam cycle occurs either to a surface water or to the atmosphere. In the past, most power plants were located near a river, lake, or ocean. In those plants the hot water from the condenser is directly discharged to the surface water by means of a diffuser or indirectly into a canal that leads to the surface water. The discharge of warm water into the surface water can cause thermal pollution and possible harm to aquatic organisms. Also, contaminants that leach into the discharge water from pipes and ducts may pollute the surface waters. As a consequence, environmental protection agencies in many countries mandate that heat rejection occur into the atmosphere via cooling towers.

There are two types of cooling towers: *wet* and *dry*. There are also combinations of wet and dry towers, as well as combinations of cooling towers and surface water cooling.

5.2.7.1 Wet Cooling Tower

In a wet cooling tower the hot condensate or cooling water is cooled by transfer of sensible heat to the atmosphere and evaporation of part of the recirculated water itself, thereby absorbing from the air and water the latent heat of vaporization. The cooling tower is usually visible as a gigantic spool-like structure constructed near a fossil or nuclear power plant. The spool configuration is advantageous from a structural standpoint; it requires less concrete for its size, and it is more resistant to strong winds.

A typical wet tower schematic with natural draft is depicted in Figure 5.5. Hot water from the condenser is sprayed over a latticework of closely spaced slats or bars, called *fill* or *packing*. Outside

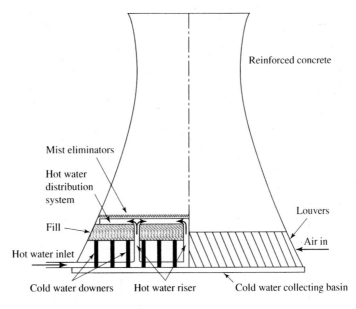

Figure 5.5 Wet cooling tower, schematic.

air is drawn in through louvers in the bottom of the tower by natural draft. Heat is transferred from the cooling water to the air directly, and the water is further cooled by evaporation of part of the water. Cold water falls by gravity into the collecting basin from whence it is recirculated into the condenser. A mist eliminator is placed above the fill. Nevertheless, a mist plume usually forms above the tower, especially in cold weather. This white plume is often mistaken as pollution; in fact it is just composed of water droplets or ice crystals.[7]

A 1000-MW power plant working at 33% efficiency in a hot climate evaporates 0.63 m^3 s^{-1} water, which is 1.3% of the recirculating water. In a cold climate the same plant would evaporate only 1% of the recirculating water. This amounts to 2E(7) or 1.5E(7) m^3 y^{-1} water that the power plant must withdraw from a surface water, well, or municipal water in hot or cold climate, respectively.

5.2.7.2 Dry Cooling Tower

In a dry cooling tower, the recirculating water flows through finned tubes over which cold air is drawn. In a dry tower all heat rejection is to the atmosphere. The advantages are that no water evaporates, and a dry tower is less expensive to maintain. The disadvantages are that it is more expensive to build and the back pressure on the turbines is higher, causing plant efficiency loss. Nevertheless, in arid areas where no make-up water is available, dry cooling towers are employed.

5.2.8 Generator

The generator is the heart of the power plant. That is where electricity is generated. Compared to the boiler, turbines, condenser, cooling tower, and other auxiliary equipment, the generator occupies a small fraction of the total plant outlay. Its noise level is also negligible compared to the hum and drum of the coal mills, burners, pumps, fans, and turbines.

The electromagnetic theory of the generator is described in Chapter 4. Briefly, the shaft of the turbine turns conducting coils within a magnetic field. This induces an electric current to flow in the coils. The electric power output of the generator equals the mechanical power input of the shaft, minus minor resistive losses in the coils and frictional losses. In order to prevent the overheating of the generator induced by these losses, generators are cooled by high conductivity gases, such as hydrogen or helium.

The generator produces an alternating current (AC): 60 Hz in the United States and Canada, 50 Hz in most other countries. The shaft of the turbine must rotate at a precise speed in order to produce the exact frequency of the AC. The voltage produced by the generator is enhanced by step-up transformers and then transmitted into the grid. Because resistance losses are proportional to the square of the current but linearly proportional to the voltage, it is advantageous to transmit power at low current and high voltage. Long-distance transmission usually occurs in the hundreds of kilovolt range. At the user, the voltage is reduced by step-down transformers to 110 or 220 V.

---

[7]However, some solids dissolved in the make-up water may produce a particle plume downwind of the cooling tower.

5.2.9 Emission Control

If left uncontrolled, power plants can emit quantities of air pollutants that cause ambient pollutant levels to exceed standards designed to protect human health and the environment. Suppose a 1000-MW power plant burns coal with a 10% mineral content and 2% sulfur content (not unusual). It is a base-loaded plant that works at 100% capacity and 35% thermal efficiency. The coal has a heating value of 12,000 Btu lb^{-1} ($\approx$ 28 MJ kg^{-1}). All the mineral content exits the smoke stack as particles (fly ash), and the sulfur exits as sulfur dioxide SO_2. This plant would emit 3.2E(5) t y^{-1} of particles and 1.3E(5) t y^{-1} of SO_2 (sulfur dioxide has twice the molecular weight of sulfur). In addition, the plant would emit copious quantities of nitrogen oxides, products of incomplete combustion (PIC), carbon monoxide, and volatile trace metals. Clearly, such an uncontrolled power plant could present a major risk to human health and the environment. Therefore, in most countries, environmental regulations require that the operator of the power plant install emission control devices for these pollutants. These devices contribute significantly to the capital and operating cost of the plant, and reduce to some degree the thermal efficiency, because the devices syphon off some of the power output of the plant. These costs are passed on to the customers as added cost of the electricity. The control devices also produce a stream of waste, because what is not emitted into the atmosphere usually winds up as a solid or liquid waste stream.

5.2.9.1 Control of Products of Incomplete Combustion and Carbon Monoxide

The control of PIC and CO is relatively easy to accomplish. If the fuel and air are well-mixed, as is the case in modern burners, and the fuel is burnt in excess air, the flue gas will contain very little, if any, PIC and CO. It is in the interest of power plants to achieve a well-mixed, fuel-lean (air-rich) flame, not only for reducing the emission of these pollutants, but also for complete burn-out of the fuel, which increases the thermal efficiency of the plant. PIC and CO emissions do occur occasionally, especially during start-ups and component breakdowns, when the flame temperature and fuel–air mixture is not optimal. Under those conditions a visible black smoke emanates from the smoke stack. These occurrences should be rare and should not contribute significantly to ambient concentrations of these pollutants.

5.2.9.2 Particle Control

Particles, also called particulate matter (PM), would be the predominant pollutant emanating from power plants were it not controlled at the source. This stems from the fact that coal, and even oil, contains a significant fraction by weight of incombustible mineral matter. In older, stoker-fed and cyclone burner plants, the mineral matter accumulates in the bottom of the boiler as bottom ash and is discarded as solid waste or taken up in water and sluiced away. In modern pulverized coal-fired plants the majority ($\approx$ 90%) of the mineral matter is blown out from the boiler as fly ash. The fly ash contains (a) a host of toxic metals, such as arsenic, selenium, cadmium, manganese, chromium, lead, and mercury, and (b) nonvolatile organic matter (soot), including polycyclic aromatic hydrocarbons (PAHs); these would pose a public health and environmental risk if emitted into the atmosphere. For that reason, most countries instituted strict regulations on particle emissions from power plants.

In the United States, power plants built between 1970 and 1978 had to meet a standard for PM emissions of a maximum of 0.03 lb per million Btu heat input (0.013 kg/GJ). For power plants built after 1978, there is no numerical standard, but the so-called Best Available Control Technology (BACT) standard applies (see Section 9.2.1). Presently, BACT for power plants is an electrostatic precipitator.

Electrostatic Precipitator. The ESP was invented in the early 1900s by F. G. Cottrell at the University of California, Berkeley, in order to collect acid mist in sulfuric acid manufacturing plants. It was soon applied for collecting dust in cement kilns, lead smelters, tar, paper and pulp mills, and other factories. Beginning in the 1930s and 1940s, ESP was applied to coal-fired power plants. The installation of early ESPs preceded environmental regulations; they were installed to protect the owners from possible liability suits because the particle emissions could cause a health hazard.

The ESP works on the principle of charging particles negatively by a corona discharge and attracting the charged particles to a grounded plate. A schematic of an ESP is given in Figure 5.6. Several charging wires are suspended between two parallel plates. A high negative voltage, on the order of 20 to 100 kV, is applied to the wires. This causes an electric field to be established between the wires and the plates along which electrons travel from the wires toward the plates. This is called a *corona discharge*. The electrons collide with gas molecules, primarily with oxygen, creating negative ions. The molecular ions keep traveling along the field lines, colliding with particles and transferring the negative charge to the particles. Now, the particles migrate to the plates where their charge is neutralized. The neutral particles are shaken off the plates by rapping them periodically. The particles fall into a hopper from whence there are carted away.

The collection efficiency of an ESP depends on many factors, primarily the particle diameter (the smaller the diameter, the less the efficiency), the plate area, the volumetric flow rate of the flue gas passing between the plates, and the particle migration speed toward the plates. The efficiency

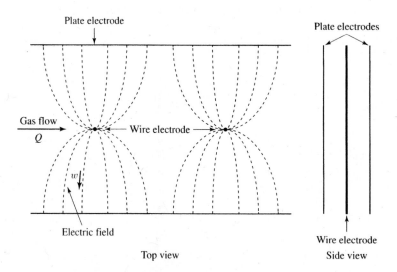

Figure 5.6 Electrostatic precipitator, schematic.

is calculated from the Deutsch equation

$$\eta = 1 - \exp(-wA/Q) \tag{5.1}$$

where w is the migration speed, A is the total area of the plates, and Q is the volumetric flow rate. The migration speed is approximated from $w \approx 0.05d_p$ m s^{-1} when the particle diameter d_p is given in μm. Actually, w is a function of the voltage and the electrical resistivity of the particles. Ironically, the higher the sulfur content of the coal, the lower the resistivity of the fly ash and the higher the migration speed. However, the higher the sulfur content of the coal, the higher acid gas emissions, so there is a trade-off.

For example, consider a 1000-MW coal-fired power plant with a thermal efficiency of 36% using coal with a heating value of 12,000 Btu lb^{-1} (7.75 kWh kg^{-1}). This plant would consume coal at a rate of 99.6 kg s^{-1}. Assuming that the atomic composition of the combustible part of coal is CH and the coal is burned with 20% excess air, the combustion generates flue gas at the rate of $Q = 1260$ m^3 s^{-1}. If particles of 1-μm diameter are to be collected with 95% efficiency, we need a plate area $A = 75,500$ m^2. If each plate measures 20×5 m, and two sides of the plate are collecting particles, we need 377 plates. Figure 5.7 gives the collection efficiency of an ESP with the above specifications as a function of particle diameter. It is seen that for particle diameters greater than 1 μm, the efficiency approaches 100%, but submicron particles are not collected efficiently at all.

A cutaway drawing of a large ESP is shown in Figure 5.8. Typically the overall width of an ESP is about 20–30 m, and the overall length is 18–20 m. Typical plate dimensions are 20×5 m, plate spacing is 0.25 m, and the number of plates are in the hundreds. After exiting from the boiler, the flue gas is transferred to the ESP in large ducts. From the ESP, the cleansed flue gas enters the smoke stack. The ESP, together with its housing, transformers, hoppers, and truck bays, may occupy

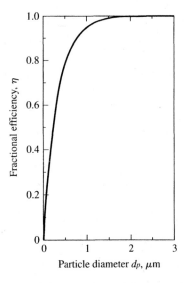

Figure 5.7 Particle collection efficiency of electrostatic precipitator as a function of particle diameter calculated from the Deutsch equation (5.1). Flue gas volumetric flow rate $Q = 1260$ m^3 s^{-1}, plate area $A = 75,500$ m^2, and drift velocity $w = 0.05 \, d_p$ m s^{-1}, where d_p is given in micrometers.

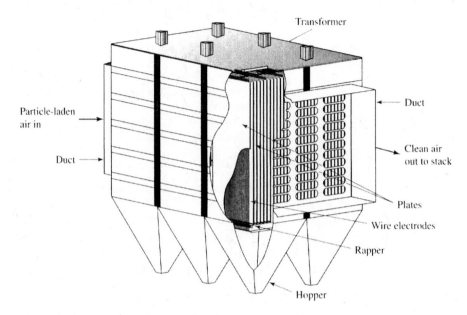

Figure 5.8 Electrostatic precipitator, cutaway.

a structure half the size of the boiler of a power plant. The power drawn by the ESP, including the fans that drive the flue gas through it, and the mechanical energy for rapping the plates consume less than 0.1% of the plant's power output. The levelized cost (amortized capital and operating cost) of running an ESP may add 6–10% to the generating cost of electricity.[8]

The relatively low efficiency of small-particle collection by existing ESPs poses a problem to the operators of existing power plants because environmental protection agencies in several countries, notably the United States, plan to introduce new ambient standards for particles less than 2.5 μm in diameter. The new standard may require retrofitting of power plants with devices that collect more efficiently small particles, such as a fabric filter.

Fabric Filter. A fabric filter, also called *baghouse,* works on the principle of a domestic vacuum cleaner. Particle-laden gas is sucked into a fabric bag, the particles are filtered out, and the clean gas is vented into the atmosphere. The pore size of the fabric can be chosen to filter out any size of particles, even submicron particles, albeit at the expense of power that is required to drive the flue gas through the pressure drop represented by the fabric pores.

A typical fabric filter schematic is shown in Figure 5.9. Long, cylindrical tubes (bags) made of the selected fabric are sealed at one end and open at the other end. The sealed end of the tubes are hung up-side down from a rack that can be shaken mechanically. The particle-laden flue gas enters through the bottom, open end of the tubes. The clean flue gas is sucked through the fabric of the tubes by fan-induced draft or special pumps and is vented through the smoke stack. A baghouse for a large power plant may contain several thousand tubes, each up to 4 meters high, 12.5 to 35 cm in

---

[8]Data on emission control efficiency and cost are obtained from "Air pollution control costs for coal fired power station," International Energy Association Coal Research, IEAPER/17, London, UK (1995).

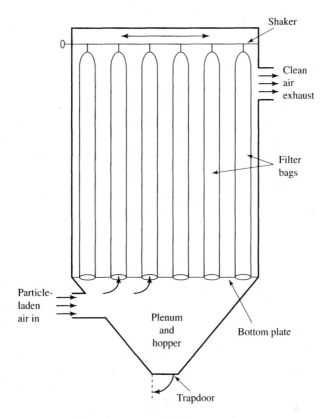

Figure 5.9 Baghouse with mechanical shakers, schematic.

diameter. It is necessary to distribute the incoming flue gas equally to all tubes, which is done in the *plenum*. The tubes provide a large surface area per unit of gas volumetric flow rate. The inverse is called the air-to-cloth or filtering ratio, which is equal to the superficial gas velocity; it typically ranges between 0.5 and 4 cm s^{-1}.

The collected particles can be removed by mechanical shaking of the tubes or by the reverse jet method. Mechanical shaking is induced by a cam-driven moving rack on which the tubes hang. In the reverse jet method, a strong air flow is blown from the outside of the tubes toward the inside, dislocating the particle cake that has built up. The removal of particles is not always complete, because particles cling to the fabric and are lodged in the pores. This causes frequent break downs and requires replacement of the fabric tubes about once a year. Ironically, the more clogged the tubes, the better the removal efficiency, but at the expense of pumping power that has to be supplied in order to maintain the superficial gas velocity. The removed particles fall into the bottom hopper from whence they are carted away by trucks.

The flue gas at the entrance to the baghouse has a relatively high temperature of 300–350 °C. Also, the flue gas may contain corrosive gases and moisture. These conditions require a heat and corrosion resistant fabric. Usually, fiberglass is chosen for coal-fired power plants, whereas other fabrics, natural or synthetic, are applicable for other facilities, like cement kilns, ferrous and nonferrous smelters, and paper and saw mills.

Fabric filters with a high collection efficiency for small particles are expected to have a higher capital and operating cost than ESPs.

5.2.9.3   Sulfur Control

As living organisms contain sulfur in their cellular make-up, this sulfur is mostly retained in the fossilized remnants of these organisms. Coal can contain up to 6% by weight of sulfur, and oil up to 3%. However, most coals and crude oils contain a much lower percentage of sulfur. Generally, bituminous, subbituminous, and lignite coals are used in power plants. They contain 0.7–3% by weight of sulfur. Residual oil used in power plants contains 0.7–2% sulfur. Without sulfur emission control devices, the oxidized sulfur, mainly sulfur dioxide SO_2, minor quantities of SO_3, and sulfuric acid H_2SO_4, would be emitted through the smoke stack into the environment. The oxides of sulfur are precursors to acid deposition and visibility impairing haze (see Sections 9.2.3 and 9.2.7). Because coal-fired power plants emit the majority of all sulfur oxide emissions worldwide (industrial boilers, nonferrous smelters, diesel and home heating oil make up the rest of sulfur emissions), operators of these plants are required to limit the emissions by either switching to low-sulfur containing fuels or installation of sulfur emission control devices.

There are basically three approaches to reducing sulfur emissions: before, during, and after combustion of the fossil fuel.

Before Combustion

Coal Washing. When coal is removed from the coal seams in underground or surface mines, there is always some mineral matter included in the coal. The majority of mineral matter is composed of silicates, oxides, and carbonates of common crustal elements, such as calcium, magnesium, aluminum, and iron, but some of it contains pyrites, which are sulfides of iron, nickel, copper, zinc, lead, and other metals. Because the specific gravity of mineral matter, including the pyrites, is greater than that of the carbonaceous coal, a part of the mineral matter can be removed by "washing" the coal. Coal washing not only reduces the sulfur content of coal, but also reduces its ash content, thereby increasing its heating value (Btu/lb or J/kg) and putting a lesser load on the particle-removing systems.

Coal washing is usually performed at the mine mouth. Typically, the crushed raw coal is floated in a stream of water. The lighter coal particles float on top, and the heavier minerals sink to the bottom. The wet coal particles are transferred to a dewatering device, generally a vacuum filter, centrifuge, or a cyclone. The coal can be further dried in a hot air stream.

One problem with coal washing is that the stream containing mineral matter may contain dissolved toxic metals and it can be acidic. Strict regulations have been introduced in the United States and elsewhere to prevent dumping of these toxic acidic streams into the environment without prior treatment.

In the United States, about 50% of all coals delivered to power plants are washed. These are mainly coals from eastern and midwestern shaft and strip mines. Western coals generally have a low mineral and sulfur content, so washing is not necessary. Coal washing can remove up to 50% of the pyritic sulfur, which is equivalent to 10–25% removal of the total sulfur content of the coal.

Coal Gasification. Coal can be converted by a chemical process into a gas, called synthesis gas, or syngas for short. In the process of coal gasification, most of the sulfur can be eliminated before the final stage of gasification. The clean, desulfurized syngas can be used to fuel a gas turbine or combined cycle power plant. The process of coal gasification will be described in Section 5.3.2.

Oil Desulfurization. Refineries can reduce the sulfur content of crude oil almost to any desired degree. This is usually done in a catalytic reduction–oxidation process called the Claus process. First, sulfur compounds in the oil are reduced to hydrogen sulfide by blowing gaseous hydrogen through the crude oil in presence of a catalyst:

$$RS + H_2 \rightarrow H_2S + R \tag{5.2}$$

where R is an organic radical. Next, H_2S is oxidized to SO_2 by atmospheric oxygen, and simultaneously SO_2 is reduced by H_2S to elemental sulfur, also in presence of a catalyst:

$$H_2S + \frac{3}{2}O_2 \rightarrow H_2O + SO_2 \tag{5.3}$$

$$2H_2S + SO_2 \rightarrow 2H_2O + 3S \tag{5.4}$$

The elemental sulfur is an important byproduct of oil refining and can be often seen as a yellow mound within the refinery complex. It is a major raw material for sulfuric acid production.

Even though the sulfur in crude oil is a salable byproduct of oil refining, refineries charge more for low sulfur oil products. Thus, utilities pay a premium price for low sulfur oil, and we pay higher rates for electricity generated by the oil burning power plant. However, it is usually cheaper for a utility to buy low sulfur oil than to remove SO_2 from the stack gas.

During Combustion

Fluidized Bed Combustion. Fluidized bed combustion (FBC) is the burning of coal (or any other solid fuel) imbedded in a granular material, usually limestone, riding on a stream of air. The primary aim of the development of FBC was not specifically to reduce SO_2 emissions, but rather to enable the combustion of all sorts of fuel, including nonpulverizable coal, municipal solid waste, industrial and medical waste, wood, tar, and asphaltene (residue of oil refining). The admixed limestone acts as a sorbent, extracting sulfur and other impurities from the fuel.

A schematic of a FBC with simultaneous coal and sorbent injection is shown in Figure 5.10. The FBC is a cylindrical retort with a grate in the bottom. Crushed coal, 6–20 mm in size, together with limestone ($CaCO_3$), is blown pneumatically over the grate. Combustion air is blown from beneath the grate. After ignition, the burning coal–limestone mixture floats over the grate, riding on an air cushion. The burning mixture acts dynamically like a fluid, hence the name fluidized bed combustion. Boiler tubes are immersed into the fluidized bed where there is direct heat transfer from the bed to the tubes. Another bank of boiler tubes is mounted above the bed, to which the heat transfer occurs by convection and radiation from the flue gases.

The sulfur is captured by the sorbent to form a mixture of calcium sulfite ($CaSO_3$) and calcium sulfate ($CaSO_4$) particles. These particles, together with unreacted $CaCO_3$ and unburnt coal particles, are carried with the flue gas into a cyclone. The larger particles are separated in the cyclone and are recycled into the fluidized bed for reburning. Smaller particles exit the cyclone and are removed in an electrostatic precipitator or baghouse.

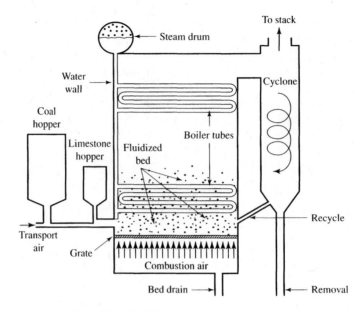

Figure 5.10 Fluidized bed combustor, schematic.

The capture efficiency of SO_2 in FBC is dependent on the intimate mixing of coal and limestone and is also dependent on their quality. Calcined limestone containing CaO is found to be a better sorbent than raw limestone. On the average, FBC with limestone injection captures only about 40–60% of the sulfur in coal.

The advantages of FBC are as follows:

- Low-quality coal and other fuels can be used.
- It has a thermal efficiency comparable to that of pulverized coal combustion with emission controls.
- Sorbent is cheap (limestone).
- Combustion temperature is lower (800–900 °C) than that in pulverized coal combustion (1900–2000 °C), which reduces thermal NO_x formation.

The disadvantages of FBC are as follows:

- Sulfur capture efficiency is only about 40–60%, and it may not meet emission standards for SO_2.
- It has a constant capacity and cannot follow load.
- It has start-up difficulties.
- It causes corrosion and fouling of boiler tubes.
- Because of the stringent requirement for proper mixing of fuel, limestone, and air, it cannot be scaled to large power plant loads; it is suitable only for 10- to 100-MW-capacity boilers.

In summary, FBC is best suited for combustion of low-quality fuels, such as nonpulverizable coal, crude oil residues, wood, and municipal and industrial waste. Its sulfur emission reduction capacity is limited. It is difficult to scale FBC to the large load requirements of centralized power plants.

After Combustion

The removal of sulfur oxides from the flue gas after combustion of the fuel in a furnace or boiler is called flue gas desulfurization (FGD). There are several methods of FGD: sorbent injection and wet and dry scrubbers.

Sorbent Injection. In sorbent injection (SI), a sorbent, usually dry sintered $CaCO_3$ or CaO, or a slurry thereof, is injected into the flue gas in the upper reaches of the boiler, past the neck. The sorption of SO_2 proceeds similarly as in FBC by forming a mixture of calcium sulfite and sulfate. The capture efficiency is dependent on many factors: the temperature, oxygen and moisture content of the flue gas, time of contact between sorbent and SO_2, and the characteristics of the sorbent (e.g., sintered sorbent, porosity, admixture of other sorbing agents). The resulting particles, consisting of hydrated calcium sulfite and sulfate and unreacted sorbent, in addition to the fly ash, need to be captured in an electrostatic precipitator or baghouse.

Sorbent injection can be retrofitted to existing coal-fired power plants, albeit the particle removal system may have to be upgraded to collect the considerable larger load of particles. The sulfur capture efficiency is on the average 50%, which may be adequate for meeting emission reduction quotas for *existing* power plants, but not enough for emission standards of *new* power plants.

Wet Scrubber. In a wet scrubber the flue gas is treated with an aqueous slurry of the sorbent, usually limestone ($CaCO_3$) or calcined lime (CaO), in a separate tower. A schematic of a wet scrubber is shown in Figure 5.11. After exiting the electrostatic precipitator, the flue gas enters an absorption tower where it is sprayed through an array of nozzles with a slurry of the sorbent. The following sequence of reactions takes place between SO_2 and the sorbent slurry:

$$CaCO_3 + SO_2 + \frac{1}{2}H_2O \rightarrow CaSO_3 \cdot \frac{1}{2}H_2O + CO_2 \tag{5.5}$$

$$CaSO_3 \cdot \frac{1}{2}H_2O + \frac{3}{2}H_2O + \frac{1}{2}O_2 \rightarrow CaSO_4 \cdot 2H_2O \tag{5.6}$$

The water molecules that are attached to calcium sulfite and sulfate are called water of crystallization. Hydrated calcium sulfate is similar to natural gypsum.

The formed mixture of hydrated calcium sulfite and sulfate, together with some unreacted limestone, falls to the bottom of the wet scrubber in the form of a wet sludge, from where it is transferred into a funnel-shaped thickener. The top liquor of the thickener is decanted into an overflow tank, from where it is recycled to make up a fresh slurry of limestone. The settled thick sludge is pumped into a vacuum filtering system where it is dewatered as much as possible. Both hydrated $CaSO_3$ and $CaSO_4$ are difficult to dewater; they form a gelatinous sludge. This sludge may be thickened further with fly ash coming from the ESP, then disposed into a landfill.

Above the spray nozzles, a mist eliminator condenses water. The clean flue gas enters a reheater (to add buoyancy), and then it exits through the smoke stack.

While perfected over the past decades, the wet scrubber still poses many operational problems. The spray nozzles tend to clog; sludge often clings to the bottom and side of the absorption tower,

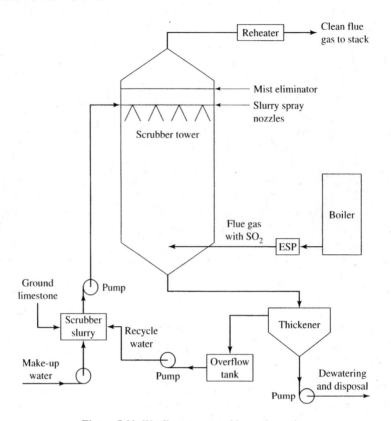

Figure 5.11 Wet limestone scrubber, schematic.

from where it has to be removed mechanically; the slurry of lime and limestone is highly corrosive; the dewatering system is prone to breakdowns; the dewatered sludge is difficult to transport to the disposal site. Frequent outages may still be experienced, and because power plants cannot afford to install dual systems, during scrubber outages the SO_2 containing flue gas is simply bypassed directly into the smoke stack.

The sorption of SO_2 by limestone is far more complete in the aqueous slurry of the wet scrubber than in fluidized bed combustion or sorbent injection. A well-designed wet scrubber can remove as much as 90–99% of the sulfur in the flue gas. Its power requirements (pumps, filters, reheater, etc.) may syphon off 2–3% of the power plant's electrical output, thereby reducing the overall thermal efficiency by the same amount. Its amortized capital and operating cost may add 10–15% to the electricity generating cost.

Dry Scrubber. The chemical reaction mechanism in the dry scrubber is similar to that in the wet scrubber; that is, $CaCO_3$, CaO, or both are used to absorb SO_2 from the flue gas, forming a mixture of calcium sulfite and sulfate. The difference is that in the dry scrubber the sorbent is introduced as a very fine spray of an aqueous slurry. The hot flue gas is blown countercurrent against the slurry spray. The proportions of slurry and flue gas are carefully metered, so that the slurry completely evaporates within the scrubber. In such a fashion, a dry powder of calcium sulfite, sulfate, and unreacted sorbent is created. Here, the particle removal system, usually a fabric filter, is installed *downstream* of the scrubber, in contrast to the wet scrubber, where the particle removal system, usually an ESP, is *upstream* of the scrubber. Thus, the dry scrubber does not create a gelatinous sludge that is difficult to transport and dispose of, but it does impose a larger load on the particle removal system.

The dry scrubber SO_2 removal efficiency is not as high as that of the wet scrubber, amounting to 70–90%. Its capital and operating costs are somewhat lower than those of a wet scrubber.

In the United States, all new coal-fired power plants must install a scrubber, either a wet or a dry scrubber, depending on the sulfur content of the coal. At present, in the United States about 25% of coal-fired power plants have installed a scrubber. In Germany and Japan practically all coal-fired power plants have installed a scrubber.

5.2.9.4   Nitrogen Oxide Control

The other major category of pollutants that emanates from fossil fuel combustion is nitrogen oxides, called NO_x, which includes nitric oxide NO, nitrogen dioxide NO_2 (and its dimer N_2O_4), nitrogen trioxide NO_3, pentoxide N_2O_5, and nitrous oxide N_2O. Other than NO and NO_2, the other oxides are emitted in minuscule quantities, so that NO_x usually implies the sum of NO and NO_2.

NO_x is a pernicious pollutant because it is a respiratory tract irritant and it is a precursor to photo-oxidants, including ozone, and acid deposition (see Section 9.2.6). At this time, the U.S. EPA plans to introduce a stricter ambient standard for ozone, which necessarily implies a more stringent control of NO_x emissions from both stationary and mobile sources, including electric power plants.

Coal and oil contain organic nitrogen in their molecular structure. When burnt, these fuels produce the so-called *fuel* NO_x. In addition, all fossil fuels produce *thermal* NO_x. This results from the recombination of atmospheric nitrogen and oxygen under conditions of the high temperatures prevailing in the flame of fossil fuel combustion:

$$N_2 + O_2 \leftrightarrow 2NO \tag{5.7}$$

The recombination involves intermediate radicals, such as atomic oxygen and nitrogen, and organic radicals, which are formed at the high temperatures. As the combustion gases cool, the formed NO does not revert to N_2 and O_2, as it would if thermodynamic equilibrium prevailed at the gas stack temperatures. As the flue gas traverses the stack, a part of the NO oxidizes into NO_2 and other nitrogen oxides.

Coal and oil combustion produce both fuel and thermal NO_x, whereas natural gas produces only thermal NO_x. As a rule of thumb, coal and oil produce about equal amounts of fuel and thermal NO_x. The flue gas of uncontrolled coal and oil combustion contains thousands of parts per million by volume of NO_x, whereas that of natural gas contains half as much.

Because organic nitrogen cannot be removed prior to combustion of the fuel, NO_x emission control can only be achieved during and after combustion.

During Combustion

Low-NO_x Burner. A low-NO_x burner (LNB) employs a process called *staged combustion*. The fact that NO_x formation is a function of air-to-fuel ratio (by weight) in the flame is exploited in LNB. This ratio affects the flame temperature and the availability of free radicals that participate in the NO_x formation process. A plot of NO_x concentration in the flame versus air-to-fuel ratio

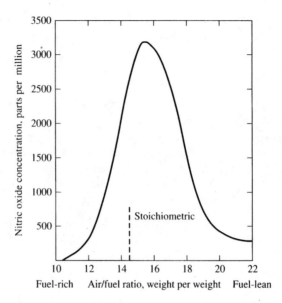

Figure 5.12 Nitric oxide concentrations in flue gas versus air-to-fuel mass ratio.

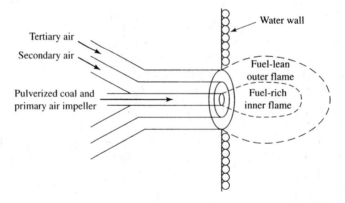

Figure 5.13 Low-NO$_x$ burner, schematic.

is presented in Figure 5.12. It is seen that under stoichiometric conditions (air/fuel ratio ≈ 15)— that is, when exactly as much air oxygen is present as necessary for complete combustion of the fuel—maximum NO$_x$ is formed. Less NO$_x$ is formed both under fuel-rich and fuel-lean combustion conditions. (Fuel-rich conditions are to the left in Figure 5.12; fuel-lean conditions are to the right.)

A schematic of a LNB is presented in Figure 5.13. Fuel (e.g., pulverized coal) and air is injected through the central annulus of the burner. The air/fuel ratio is less than stoichiometric, that is, fuel-rich. This produces a luminous flame, with some of the pulverized coal left unburnt, but also with low NO$_x$ formation, according to the left-hand side of Figure 5.12. Secondary and tertiary air arrives through outer annuli, creating an outer flame envelope that is fuel-lean. Here, all the unburnt carbon burns up and less NO_x is formed, according to the right-hand side of Figure 5.12. The net result is complete burn-out of the fuel and less NO_x formation, circumventing the peak NO_x formation of Figure 5.12.

The retrofitting of boilers with LNB is relatively easy and inexpensive to accomplish. The incremental cost of electricity production using LNB is only 2–3%. In fact, most fossil-fueled power plants and industrial boilers have installed LNBs. The problem is that LNB can reduce NO_x formation only by 30–55% compared to regular burners. Because of the prevailing problems of acid deposition and high ozone concentrations in urban-industrial regions of continents, the pressure is mounting on operators of fossil-fueled power plants and industrial boilers to further reduce NO_x emissions by more effective means than LNB.

After Combustion

Selective Catalytic Reduction. In a selective catalytic reduction (SCR) process either ammonia or urea is injected into a catalytic reactor through which the flue gas flows. The following reaction takes place when ammonia is injected:

$$4NO + 4NH_3 + O_2 \rightarrow 4N_2 + 6H_2O \tag{5.8}$$

Thus, NO is reduced by ammonia and ammonia is oxidized by NO and O_2 to form elemental nitrogen. The catalyst is a mixture of titanium and vanadium oxides dispersed on a honeycomb structure. The SCR reactor is placed between the economizer and air preheater sections of the boiler, where the flue gas temperature is 300–400 °C. The reaction is 80–90% complete; thus, 10–20% of the NO_x escapes through the smoke stack, and 10–20% of the unreacted ammonia also escapes. This is called *ammonia slip*. While ammonia is a toxic gas, by the time the flue gas plume disperses to the ground, its concentration is not considered harmful.

The catalyst is easily "poisoned," especially if the flue gas contains fly ash and sulfur oxides. (Note that the catalytic reactor is upstream from the ESP and scrubbers.) Therefore, the catalyst requires frequent replacement, which is one of the major cost elements in operating a SCR. For SCR a major reconstruction is necessary, because the catalytic reactor must be inserted between the economizer and air preheater sections of the boiler. The incremental cost of electricity production in a coal-fired power plant using SCR is about 5–10%.

Selective Noncatalytic Reduction. A reduction of NO can be accomplished at a higher temperature without a catalyst by using a process of selective noncatalytic reduction (SNCR). In this process, urea is used instead of ammonia. An aqueous solution of urea is injected into the superheater section of the boiler, where the temperature is about 900–1000 °C, which is sufficiently high for the reaction to proceed to near completion:

$$4NO + 4CO(NH_2)_2(aq) + O_2 \rightarrow 4N_2 + 4CO_2 + 2H_2O \tag{5.9}$$

Many operators of electric power plants prefer this option because it needs no catalyst. However, urea is more expensive than ammonia. This option is also more amenable to retrofitting of existing plants because the urea injector can be mounted directly onto the wall of the boiler.

SNCR can be used in conjunction with low-NO$_x$ burners. The final reduction of NO$_x$ is in the 75–90% range, and the incremental cost of electricity production is 3–4.5%.

5.2.9.5   Toxic Emissions

Coal and (to a lesser extent) oil contain mineral matter that during the combustion process may produce toxic vapors and particles. Particles larger than 1–2 microns in diameter are almost all captured by the particle removal system, either by an electrostatic precipitator or a fabric filter. Smaller particles and vapors may escape through the smoke stack and pollute the environment. Mercury, selenium, cadmium, and arsenic are semivolatile toxic metals that in part may escape through the smoke stack as vapors. In particular, mercury emissions are causing concern because excessive mercury concentrations have been found in some lakes and coastal waters. Fish and other aquatic organisms may bioaccumulate mercury and pass it on through the food chain to humans. Currently, in the United States, Europe, and Japan, studies are being conducted on the extent of the mercury problem and on the technologies that could be employed in order to reduce mercury emissions from power plants.[9]

Another problem associated with coal-fired power plants is radon emissions. Radon is a disintegration product of uranium, the minerals of which may cling to the coal. Radon is a radioactive gas that emits alpha particles. A part of the omnipresent radon in the atmosphere may be traced to coal-fired power plants.

5.2.10   Waste Disposal

Coal-fired power plants produce a significant amount of solid waste. Oil-fired plants produce much less waste, and gas-fired plants produce practically none. We calculated before that a 1000-MW plant fired with coal that contains 10% by weight mineral matter produces about 3.2E(5) t y^{-1} of fly ash. If the coal contains 2% by weight of sulfur, and if that sulfur is removed by flue gas desulfurization using a wet limestone scrubber, another 3-4E(5) t y^{-1} of wet sludge is created containing hydrated calcium sulfite, calcium sulfate, and unreacted limestone. While some plants succeed in selling, or at least giving away, the fly ash for possible use as aggregate in concrete and asphalt or as road fill material, the scrubber sludge has practically no use.[10] For some coals the fly ash and the scrubber sludge may contain toxic organic and inorganic compounds. In that case, the waste needs to be disposed of in a secure landfill. The landfill must be lined with impenetrable material, so that leaching into the soil and groundwater is to be prevented. Typical liner materials are natural clays or synthetic fabrics. Because transport costs can be substantial, the landfill area should be in the vicinity of the power plant. The mixed fly ash and sludge from a 1000-MW coal-fired power plant would fill about 10–20 acres (4–8 hectares) about 1 foot (0.305 m) deep every year. Thus, a new coal-fired power plant must be situated where a suitable landfill area is available.

---

[9]Municipal solid waste incinerators are the principal sources of mercury and other toxic metals emissions because electrical switches, batteries, and fluorescent light bulbs are often thrown into the trash that is being incinerated.

[10]It is cheaper to produce gypsum wall boards from mined gypsum than from scrubber sludge.

5.3 ADVANCED CYCLES

We have seen that the best fossil-fueled steam power plants can achieve a thermal efficiency of close to 40%, but the average in the United States is 36%, and the worldwide average is 33%. Power plants that work on the gas turbine principle achieve even less, on the order of 25–30% thermal efficiency. This means that 60–75% of the fossil fuel heating value that powers them goes to waste. Furthermore, the emissions per kilowatt hour of electricity produced is inversely proportional to thermal efficiency: The lower the efficiency, the more the plant pollutes and the higher the CO_2 emissions. Thus, great efforts and money are spent by private and government agencies to improve power plant thermal efficiencies by developing advanced cycle power plants. In the United States, research and development is sponsored by the Electric Power Research Institute and the U.S. Department of Energy.

5.3.1 Combined Cycle

We described the thermodynamic principles of a combined cycle in Section 3.10.5. A power plant schematic using a gas turbine combined cycle (GTCC) is shown in Figure 5.14. In the first cycle, called the *topping* cycle, a suitable fluid fuel, usually natural gas, powers a gas turbine. The still hot exhaust gas of the gas turbine passes through a heat exchanger, called a heat recovery boiler (HRB), and then to the stack. In the HRB, feed water is boiled into steam that powers a steam turbine, called a *bottoming* cycle. Sometimes, more fuel (heat) is added to the gas turbine exhaust gas in a combustion chamber before the hot gases enter the HRB. The combination of the two cycles can achieve a thermal efficiency of 45%, an improvement over either a single-cycle steam turbine or a gas turbine power plant.

A problem with the combined cycle is that the primary fuel, natural gas, is more expensive per unit heating value than coal. Also, gas reserves will not last as long as coal reserves (see Chapter 2). Combined cycle power plants are suitable where gas supplies are plentiful and cheap and where environmental regulations impose a heavy technical and financial burden on coal-fired power plants. They are especially attractive in urban environments because they require practically no fuel storage facilities (compressed gas arrives in pipes to the power plant), no particle removal system, and no scrubber for SO_2 removal. There is also no solid waste to dispose of. However,

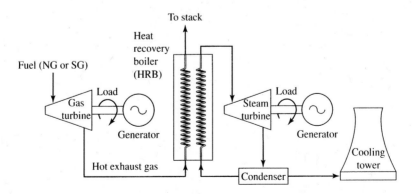

Figure 5.14 Gas turbine combined cycle power plant, schematic.

combined cycle power plants may require a NO_x control system because in the gas turbine copious quantities of thermal NO_x may be created. Of course, a condenser and cooling tower are also required for the steam portion of the combined cycle.

5.3.2 Coal Gasification Combined Cycle

A combined cycle power plant can be fueled by coal, but then the coal needs first to be gasified. The gasified coal (syngas) propels the topping cycle gas turbine. A power plant using coal gasification and combined cycle is called an integrated gasification combined cycle plant (IGCC).

Various coal gasification methods were developed already in the nineteenth century for providing piped gas for home heating, cooking, and lighting. This was called *city gas* and was used before natural gas became widely available in the second half of the twentieth century. In the world wars, Germany used gasified coal as a fuel for automobiles, trucks, and military vehicles.

Coal can be gasified to low-, medium- and high-heating value syngas. The processes differ depending upon whether air or pure oxygen is used for gasification and whether the product gas is rich or devoid of CO_2. For an IGCC plant, high-heating value syngas is preferred.

The process starts with crushing the coal. If the coal has a tendency for caking, preoxidation my be necessary to render its surface more porous. The crushed coal is fed to a retort, where, in the presence of catalysts, it is exposed to pure oxygen and steam:

$$3C + O_2 + H_2O \rightarrow 3CO + H_2 \tag{5.10}$$

Pure oxygen (99%+) is supplied from a special *air separation unit* constructed at the plant site. The resulting gaseous mixture contains a host of higher-molecular-weight organic compounds, as well as hydrogen sulfide. The next step is called *quenching,* in which the heavy oils and tar are removed from the mixture by condensation and aqueous solution scrubbing. This is followed by removal of H_2S by the above-described Claus or similar process.

The syngas resulting from reaction (5.10) has a relatively low heating value of 250–500 Btu per standard cubic foot (sfc), which equals 9.1–18.2 MJ m^{-3}. This gas would not be suitable as a fuel for an IGCC. To increase the heating value, the carbon monoxide and hydrogen is passed over a catalyst at about 400 °C to form principally methane, a process called *methanation:*

$$3H_2 + CO \rightarrow CH_4 + H_2O \tag{5.11}$$

The resulting syngas has a heating value of 950–1000 Btu/scf (36–38 MJ m^{-3}), similar to methane. For this syngas, a conventional gas turbine can be used. An alternative is the production of hydrogen in the *water gas shift reaction:*

$$CO + H_2O \rightarrow CO_2 + H_2 \tag{5.12}$$

The resulting hydrogen can be separated from CO_2 by membrane separation and used in high-efficiency H_2/O_2 fuel cells (see Section 3.12).

The estimated thermal efficiency of a coal gasification combined cycle plant is on the order of 40–45%, including the energy required for air separation and coal gasification. This efficiency exceeds that of a pulverized coal-fired steam plant with emission controls, having an efficiency of 36–38%. However, the capital and operating costs of such a plant would be considerably higher than that of a pulverized coal-fired steam plant. At present, these costs are not offset by the higher thermal efficiency.

In the United States, the Electric Power Research Institute together with a consortium of private utilities built a demonstration IGCC power plant including an air separation unit in Barstow, California, called the Coolwater Plant. It had a net electric power output of 105 MW, but because not all components were optimized and integrated, its thermal efficiency was only 31%. The air emissions met the stringent requirements of the California Air Resources Board. It operated satisfactorily for 5 years (1984–1989), but was moth-balled thereafter because of the high operating costs. However, the Coolwater Plant demonstrated that IGCC is feasible and can be compliant with strict emission standards.

5.3.3 Cogeneration

Cogeneration is the term applied to systems that provide both electrical power and useful heat from the burning of fuel. In industrial or commercial installations the heat may be used for space heating or material processing. The incentive for cogeneration is primarily financial in that the cost of supplying electricity and heat via a cogeneration scheme might be less than supplying them separately, such as purchasing electric power from a supplier while generating heat from an in-plant furnace or boiler. Whether or not a cogeneration system reduces the amount of fuel needed to supply the electricity and heat depends upon the details of the cogeneration system. Also, emission controls may be more efficient and cheaper on a large-scale electric power plant than on a smaller-scale cogeneration plant—unless the latter is fueled by natural gas, which inherently produces less pollutant emissions. In 2000, about 12% of U.S. electric generation capacity was in cogeneration facilities.

When a heat engine drives an electric generator to produce electricity, it also provides a stream of hot exhaust gas. Where the exhaust gas is warm enough to be used for process or space heat, some of the exhaust gas enthalpy may be extracted to satisfy the heat requirement in a cogeneration plant. If Q_{fuel} is the rate of fuel heat consumption needed to generate electric power P_{el} and process heat Q_{proc} in a cogeneration plant, then

$$Q_{fuel} = P_{el} + Q_{ex} \tag{5.13}$$

$$P_{el} = \eta_{th} Q_{fuel} \tag{5.14}$$

$$Q_{proc} = \eta_{xch} Q_{ex} = \eta_{xch}(1 - \eta_{th}) Q_{fuel} \tag{5.15}$$

where Q_{ex} is the enthalpy flux of the exhaust gas, η_{th} is the thermal efficiency of the electrical generation process, and $\eta_{xch} \leq 1$ is the fraction of the exhaust stream enthalpy that is delivered as heat for processing. The split of fuel heat Q_{fuel} between electric power, $\eta_{th} Q_{fuel}$, and process heat, $\eta_{xch}(1 - \eta_{th}) Q_{fuel}$, depends upon the thermal efficiency η_{th} of the heat engine and the heat exchanger effectiveness η_{xch}. The latter depends principally on the temperature at which process heat is delivered compared with the temperature of the exhaust gas—being greatest when the difference is large, and least when it is small. When process heat is needed at high temperatures, the usable process heat from a cogeneration system may be too small compared to the electric power generated to justify this complex system, and it may be more economical and fuel efficient to buy electricity from an efficient central power plant and generate process heat in an efficient in-plant boiler or furnace.

Cogeneration is most useful when process or space heat is required at a low temperature. Then Q_{ex}, instead of being dumped in a condenser and cooling tower, is heat exchanged with the device that needs the relatively low temperature heat, say for drying or space heating. When a central power station (or incinerator) is located in a densely populated or commercial downtown area, Q_{ex} may be piped directly into the buildings for space and water heating. This is called *district heating*.

5.3.4   Fuel Cell

The functioning of a fuel cell was described in Section 3.12. A fuel cell is not a heat engine. In a fuel cell, some of the chemical energy of the fuel is directly converted into electrical energy, with the rest appearing as heat rejected to the environment. Its theoretical thermal efficiency in terms of electrical energy generated versus fuel chemical energy input can be close to 100% when producing a low output of power. However, because of parasitic heat losses (e.g., ohmic resistance) and because of the power requirements of auxiliary equipment (e.g., pumps, fans), current fuel cells using natural gas or hydrogen and air (instead of oxygen) have a much lesser thermal efficiency when operating at maximum power, in the 45–50% range. Furthermore, if hydrogen is used as a fuel, it has to be generated separately in some fashion, which requires energy. For example, hydrogen can be generated from water by electrolysis. But, the splitting of water into hydrogen and oxygen requires about 18 MJ of electric energy per kilogram of water, more than is generated in the fuel cell by the electrolytic hydrogen. Fuel cell hydrogen is usually generated by the reforming of methane (see Section 3.14.1).

5.4   CONCLUSION

Fossil-fueled power plants consume 55.5% of the world's fossil energy. Worldwide about two-thirds of the electrical energy is generated by fossil energy, 80% of which is coal. Almost all fossil-fueled power plants work on the principle of heat engines where 25–40% of the fossil energy input is converted into electrical energy, the rest is wasted in the form of heat rejection to a cold reservoir and parasitic heat losses. Furthermore, fossil-fueled power plants, especially coal-fired ones, produce a host of pollutants, including particulate matter, sulfur and nitrogen oxides, and toxic organic and inorganic byproducts of combustion. The emissions of these pollutants must be controlled in order to safeguard public health and the environment. Emission control devices subtract from the thermal efficiency of power plants and add to the electricity generating cost. Power plants also produce large amounts of solid and liquid waste, consume significant amounts of fresh water, and last, but not least, emit large quantities of CO_2, a greenhouse gas that contributes to global warming.

However, life in an urban-industrial society cannot be imagined without electricity. We only can strive toward increasing the thermal efficiency of power plants, improving their emission control technologies, and gradually replacing them with fuel cells and solar, wind, and other renewable energy power plants. Eventually, with the depletion of fossil fuels and the looming crisis of global warming, nuclear-fueled power plants also may have to be utilized in greater proportion than is currently the case.

PROBLEMS

Problem 5.1

A coal-fired power plant (PP) has a rated power of 1000 MW(el). It works at 100% capacity (base-loaded) at a thermal efficiency of 35%. It burns coal having a heating value of 30 MJ/kg and a sulfur content of 2% by weight, a nitrogen content of 1% by weight, and a mineral (ash) content of 10% by weight.

(a) How much electricity does the power plant produce per year (kWh/y)?

(b) How much coal does the PP consume per year (metric tons per year, ton/y)?

(c) How much SO_2 does the PP emit per year (ton/y)? (Note that SO_2 has double the molecular weight of S.)

(d) How much NO_2 does the PP emit per year (ton/y), assuming "thermal" NO_2 formation doubles the rate of "fuel" NO_2 emissions? (Note that NO_2 has 3.28 times the molecular weight of N.)

(e) How much fly ash does the PP emit per year (ton/y)? Assume all mineral content is emitted as fly ash.

Tabulate the results of (a), (b), (c), (d), and (e).

Problem 5.2

Using the sulfur and nitrogen content of coal as in the previous problem, calculate the emission rate of SO_2 and NO_2 per unit heat input (lb SO_2/MBtu and lb NO_2/MBtu), assuming again that total NO_2 emissions are twice the fuel NO_2 emissions. Compare these values to the 1970 U.S. emission standards for coal-fired large boilers: 1.2 lb SO_2/MBtu and 0.7 lb NO_2/MBtu.

Problem 5.3

A 1000-MW(el) power plant, of 35% thermal efficiency and 90% average capacity factor, uses coal with a heating value of 30 MJ/kg. The combustible part of coal has the chemical formula CH. The coal is burned with 20% excess air. Calculate the volumetric flow rate of the flue gas Q in $m^3\ s^{-1}$ at STP.

Problem 5.4

How much air (m^3/s at 1 atm) needs to be pumped through the boiler of the plant in Problem 5.3 in order to burn the coal (chemical formula CH) with 20% excess air?

Problem 5.5

The coal in the power plant of Problem 5.3 also contains 10% by weight mineral matter that is converted in the boiler to fly ash. Calculate the mass concentration of the fly ash in the flue gas (g/m^3).

Problem 5.6

The collection efficiency η of particles in an electrostatic precipitator is given by the Deutsch equation

$$\eta = 1 - \exp(-Aw/Q)$$

where A is the area of collection plates, w is the drift velocity ($= 0.05\, d_p$ m/s, where d_p is the particle diameter in μm), and Q is the volumetric flow rate. (a) Calculate the collection efficiency for $d_p = 0.1, 0.3, 1$, and $3\ \mu$m, using $A = 75,500$ m^2 and Q from Problem 5.3. Tabulate and plot these results. (b) Calculate the plate area necessary for 99% removal efficiency of 1-μm-diameter particles.

Problem 5.7

The plant with the parameters given in Problem 5.1 has an electrostatic precipitator with the removal efficiencies and weight distribution in the given size ranges (shown in Table P5.7). Determine the rate of fly-ash emission (g/s).

TABLE P5.7

Particle size, μm	0–5	5–10	10–20	20–40	$\geq$40
Removal efficiency, %	70	92.5	96	99	100
Weight, %	14	17	21	23	25

Problem 5.8

A coal has a heating value of 12,000 Btu/lb and the following molecular composition: $C_{100}H_{100}S_1N_{0.5}$. It is burnt in air, without excess oxygen.

(a) Calculate the emission rate of SO_2 and NO_2 in lb/MBtu. Once more assume that the total emission rate of NO_2 is twice the emission rate of fuel NO_2. Compare this with the 1970 U.S. emission standards for large coal-fired boilers.
(b) Calculate the mole fraction and volume fraction (ppmV) of SO_2 and NO_2 in the flue gas.

Problem 5.9

The flue gas of the plant in Problem 5.8 is to be treated with flue gas desulfurization (FGD) using a limestone ($CaCO_3$) wet scrubber to remove the SO_2. Assume that 2 moles of $CaCO_3$ are necessary for every mole of SO_2. How much limestone is consumed per ton of coal?

Problem 5.10

The flue gas of the plant in Problem 5.8 is to be treated with flue gas denitrification (FGN) using urea ($CO(NH_2)_2$) injection directly into the boiler. Assume that a molar ratio NO_2:$CO(NH_2)_2 = 1{:}1$ is necessary. How much urea is consumed per ton of coal?

Problem 5.11

A 1000-MWel pulverized coal steam plant operating at 35% thermal efficiency rejects one-third of the coal heating value (30 MJ/kg) to the once-through seawater cooling system. At the discharge point the seawater can only be 5 °C warmer than at the intake point. How much seawater needs to be pumped through the cooling system (m³/s)?

BIBLIOGRAPHY

El-Wakil, M. M., 1984. *Powerplant Technology*. New York: McGraw-Hill.

Decher, R., 1994. *Energy Conversion*. New York: Oxford University Press.

Weston, K. C., 1992. *Energy Conversion*. St. Paul: West Publishing Co.

Cooper, C. D., and F. C. Alley, 1986. *Air Pollution Control*. Prospects Heights: Waveland Press.

Takeshita, M., 1995. *Air Pollution Control Costs for Coal-Fired Power Stations*. Coal Research Publication IEAPER/17. London: International Energy Association.

Nuclear-Fueled Power Plants

CHAPTER

6

6.1   INTRODUCTION

The U.S. developers of commercial nuclear power plants in the late 1950s and 1960s promised to produce vast amounts of electrical energy, electricity that would be "too cheap to meter." The plants would not cause air pollution and other detrimental environmental effects. Furthermore, countries, including the United States, that import part or all of their fossil fuel needed for electric power would not be subject to fossil energy import embargoes and price escalations.

In 1990, nuclear power plants supplied more than 20% of the electricity in the United States, and in 2001 more than 100 nuclear power plants are in operation there. In France, more than 50 plants are operating, supplying more than three-quarters of its electricity. Japan has about 40 plants, supplying more than one-third of its electricity. Altogether, more than 400 nuclear-fueled power plants operate in the world in 2001, supplying about 17% of the global electricity consumption.

However, in the last two decades of the twentieth century, nuclear power plants fell into disfavor. The Three Mile Island nuclear power plant accident in the United States in 1979, the Chernobyl power plant accident in the former USSR in 1986, and the recent nuclear fuel processing plant accident in Tokaimura in Japan in 1999 all raised grave concerns in the public eye toward further electricity generation in nuclear power plants. Also, the issue of the disposal of the high-level radioactive waste that keeps accumulating at the plants has not been resolved in the United States and worldwide.

In the United States, economics is also a factor affecting the maintenance of existing nuclear power plants and building new ones. The complexity of nuclear power plants is staggering, making electricity production costs in existing plants equal to or greater than that in fossil-fueled plants. At present, the capital investment of a new nuclear power plant is two- to several-fold higher than a pulverized-coal-fired or natural-gas-fired combined cycle plant, including the capital investment in emission control equipment that fossil-fueled power plants require. Also, fossil fuel is relatively cheap at present, ranging from $2 to $5 per MBtu, so that fuel cost is not a deterrent to its use. In some countries, notably France and Japan, which lack fossil fuel resources, energy security arguments appear to predominate over safety concerns or economic factors, so that nuclear power plants continue to supply an increasing fraction of the electricity demand.

In the future the situation may change. The global fossil energy resources are finite. We have seen in Section 2.7.6 that at the current consumption rate—let alone if consumption will increase with population and economic growth—fluid fuels (oil and gas) will be depleted within a century. While coal resources may last longer, the environmental effects of coal use, notably the greenhouse effect, militate against wider use of coal. Renewable energy may play an increasing role in marginal energy use, but it is doubtful that renewable energy will replace large-scale centralized fossil fueled or nuclear power plants supplying the base load for urban-industrial areas.

Nuclear energy resources are far more abundant than fossil fuel resources. It is estimated that high-grade uranium ores could provide the present mix of reactors for about 50 years, but utilization of lower-grade ores (with concomitant increase of refined uranium fuel price) would last for many centuries. Utilization of thorium ores and fast breeder reactors could extend nuclear energy resources to millennia.[1] Thus, it is possible that worldwide nuclear power plants will again win public favor and become economically competitive with other energy sources.

In this chapter we describe the fundamentals of nuclear energy, its application for electricity generation, the nuclear fuel cycle, and the problems associated with nuclear power plants in regard to their safety, nuclear weapons proliferation, and radioactive waste disposal.

6.2 NUCLEAR ENERGY

Nuclear energy is derived from the binding force (the "strong" force) that holds the nucleons[2] of the atomic nucleus together. The binding force per nucleon is greatest for elements in the middle of the periodic table and is smallest for the lighter and heavier elements. When lighter nuclei *fuse* together, energy is released[3]; when heavier nuclei undergo *fission*, energy is also released. When a nucleus of ^{235}U (an isotope of uranium) is bombarded with a neutron, it splits into many fission products with the release of two to three times as many neutrons as were absorbed. For example, one of the fission reactions is the splitting of ^{235}U into ^{144}Ba and ^{89}Kr, with the release of 3 neutrons plus 177 MeV of energy[4,5]:

$$^{235}\text{U} + \text{n} \rightarrow {}^{144}\text{Ba} + {}^{89}\text{Kr} + 3\text{n} + 177\text{MeV} \tag{6.1}$$

where n stands for a neutron.

---

[1]"Nuclear Electricity," 1999. Uranium Information Centre, Melbourne, Australia.

[2]For our purposes we shall consider as nucleons only positively charged protons and chargeless neutrons. Other particles have been observed in nuclear disintegration experiments, but they are not germane to our discussion.

[3]We shall discuss fusion at the end of this chapter.

[4]Electron volt (eV) = energy gained by an electron when accelerated through an electrical potential difference of 1 volt. 1 eV = 1.602E(−19) J; 1 MeV = 1.602E(−13) J.

[5]The energy released in a fission reaction can be calculated by means of the famous Einstein equation that ties energy to mass, $E = mc^2$. The masses on the left-hand side of equation (6.1) are ^{235}U = 235.04394 amu (1 atomic mass unit = 1.66E(−27) kg), n = 1.00867 amu; on the right-hand side ^{144}Ba = 143.92 amu, ^{89}Kr = 88.9166 amu, 3n = 3.026 amu. If we subtract the sum of the masses on the right-hand side from the sum of the masses on the left-hand side, there is a "mass deficit," $\Delta m = 0.19$ amu. This mass deficit is converted into energy $\Delta E = \Delta mc^2 = 0.19$ amu $\times$ 1.66E(−27) kg $\times$ [3E(8)]^2 m^2 s^{-2} = 2.84E(−11) J = 177 MeV. (1 amu deficit is equivalent to 931.5 MeV.) The fission of ^{235}U produces 2.84E(−11) J $\times$ 6.023E(23) atoms mole^{-1} ÷ 0.235 kg mol^{-1} = 7.3E(13) J kg^{-1} energy. In comparison, the combustion of carbon produces 3.3E(7) J kg^{-1}, about 2 million times less energy per unit weight than a fission reaction.

Most of the fission products are radioactive (see Section 6.3 below). Because more than one neutron is released in the fission reaction, a chain reaction[6] develops, with an increasing rate of release of energy. The greater portion (about 80%) of the released energy is contained in the kinetic energy of the fission products, which manifests itself as sensible heat. A part of the remaining energy is immediately released in the form of γ and β rays and neutrons from the excited fission products. The rest of the fission energy is contained in delayed radioactivity of the fission products.

At the same time that ^{235}U splits into fission products with the release of 2–3 neutrons, a part of the neutrons can be absorbed by the more abundant ^{238}U in the fuel, converting it in a series of reactions to an isotope of plutonium, ^{239}Pu:

$$^{238}\text{U} + \text{n} \rightarrow {}^{239}\text{U} + \gamma \rightarrow {}^{239}\text{Np} + \beta \rightarrow {}^{239}\text{Pu} + \beta \tag{6.2}$$

This reaction series is accompanied by γ and β radiation.

^{239}Pu is a *fissile*[7] element that can sustain a chain reaction (see breeder reactors in Section 6.4.4). Furthermore, because plutonium is a different element than uranium, it can be extracted chemically from the spent fuel and then used as fresh fuel for reloading a reactor. The extracted plutonium can also be used in atomic bombs. Thus, reprocessing of spent fuel carries the risk of nuclear weapons proliferation. Reprocessing of spent fuel for the purpose of extracting plutonium is currently banned in the United States.[8]

6.3 RADIOACTIVITY

Radioactivity is the spontaneous decay of certain nuclei, usually the less stable isotopes of the elements, both natural and man-made, which is accompanied by the release of very energetic radiation. After the emission of radiation, an isotope of the element (or even a new element) is formed, which is usually more stable than the original element. It is important to note that radiation emanates directly from the nucleus, not the atom as a whole. This is an important distinction, because X-ray radiation, although equally damaging as radioactivity, emanates from the inner electronic shells of the atom, not the nucleus.

In radioactive decay, there are three types of radiation: α, β, and γ. Only the latter is a form of electromagnetic radiation; the two former are emissions of very high energy particles. All three are called ionizing radiation because they create ions as their energy is absorbed by matter through

---

[6] A chain reaction propagates exponentially due to the release of more than one reactant (in this case a neutron) per step than the one that initiated it.

[7] A *fissile* nucleus is one that can split after absorption of a thermal neutron. Examples are ^{235}U, ^{239}Pu, and ^{233}U. A *fertile* nucleus is one that can convert into a fissile nucleus after absorption of a fast neutron. Examples of fertile nuclei are ^{238}U and ^{232}Th.

[8] It has to be emphasized that while the chain reaction in a nuclear power plant has to be carefully controlled, a nuclear reactor cannot explode like an atomic bomb. Atomic bombs contain highly enriched ^{235}U (95%+) or ^{239}Pu, in contrast to nuclear reactors that contain a maximum of 3–4% enriched ^{235}U or ^{239}Pu. An uncontrolled chain reaction in the reactor can lead to a "meltdown" and possible dispersal of radioactive fission products, as happened, for example, at Chernobyl. But there is never any danger of an accident at a nuclear power plant resulting in an uncontrolled chain reaction with the explosive force of an atomic bomb.

which they travel. Ionizing radiation can travel only so far into matter, depending upon its energy and character.

In *alpha radiation,* a whole nucleus of a helium atom, containing two protons and two neutrons, is emitted. Because two protons of the original nucleus are lost (and consequently, two electrons must be lost from the electronic orbitals in order to maintain electric neutrality), the daughter isotope moves two elements *backward* from the parent element in the periodic table. For example, the isotope ^{239}Pu disintegrates into the isotope ^{235}U with the emission of an α particle:

$$^{239}\text{Pu} \rightarrow \, ^{235}\text{U} + \alpha(^{4}\text{He}) \tag{6.3}$$

Another example is the disintegration of an isotope of radon ^{222}Rn (which is a gas at normal temperatures) into polonium ^{218}Po. The latter emits another α particle with the formation of a stable isotope of lead ^{214}Pb:

$$^{222}\text{Rn} \rightarrow \, ^{218}\text{Po} + \alpha(^{4}\text{He}) \tag{6.4}$$

$$^{218}\text{Po} \rightarrow \, ^{214}\text{Pb} + \alpha(^{4}\text{He}) \tag{6.5}$$

Because α particles are relatively heavy, their penetration depth into matter is very small, on the order of a millimeter. A sheet of paper or a layer of dead skin on a person can stop α radiation.

In *beta radiation,* an electron is emitted. This electron does not stem from the electronic orbitals surrounding the nucleus, but from the nucleus itself: a neutron converts into a proton with the emission of an electron. In that conversion, a parent isotope converts into a daughter isotope, which is an element *forward* in the periodic table because a proton has been added to the nucleus. An example is the decay of strontium ^{90}Sr into yttrium ^{90}Y:

$$^{90}\text{Sr} \rightarrow \, ^{90}\text{Y} + \beta(^{0}\text{e}) \tag{6.6}$$

Because ^{90}Sr is one of the products of uranium fission, this isotope is a major source of radiation from the spent fuel of a nuclear reactor. The emitted electron is relatively light; therefore it can penetrate deeper into matter, on the order of centimeters. To shield against β radiation, a plate of metal, such as lead, is necessary.

Generally, the emission of β radiation is followed by the emission of *gamma radiation.* Because the number of protons or neutrons does not change in that radiation, the radiating isotope does not change its position in the periodic table. Gamma radiation is essentially the emission of very short-wave, and hence energetic, electromagnetic radiation. An example is the emission of γ radiation from an isotope of cobalt ^{60}Co:

$$^{60}\text{Co} \rightarrow \, ^{60}\text{Co} + \gamma \tag{6.7}$$

The γ radiation from ^{60}Co finds medicinal use, such as destroying cancer cells, and hence provides a therapy for certain types of cancer. Because γ rays carry no mass, its penetration into matter is very deep, on the order of meters, and very heavy shielding is necessary to protect against it.

The depth of penetration of α, β, and γ radiation increases with the energy of the radiation. For a 1-MeV energy, the penetration depths in water or tissue are 1E(−3), 0.5, and 33 cm, respectively, and 1E(−2), 7, and 55 cm for 10-MeV energy. Their penetration depths in air are about a thousand times greater than in water.

In addition to the above three types of radiation, there is the emission of neutrons. Usually, neutrons are emitted as a consequence of a nuclear fission. We already mentioned the fission of ^{235}U with the emission of three neutrons [equation (6.1)]. These neutrons can enter other nuclei of ^{235}U, which leads to a chain reaction. Also, neutron irradiation of other nuclei can lead to their destabilization, with the successive emission of β or γ radiation.[9] Radioactivity accompanies the whole nuclear power plant fuel cycle, from mining of uranium ore, through uranium extraction, isotope enrichment, fuel preparation, fuel loading, reactor operations, accidents, plant decommissioning, and, last but not least, spent fuel disposal.

6.3.1 Decay Rates and Half-Lives

The decay rate of an ensemble of radioactive nuclei, to which the diminishing intensity of radioactivity is proportional, is governed by the law of exponential decline:

$$-dN/dt = kN \qquad (6.8)$$

where N is the number of decaying nuclei, or their mass, present at time t, and k is the decay rate in units of t^{-1}. Integration yields

$$N = N_0 \exp(-kt) \qquad (6.9)$$

where N_0 is the number of nuclei, or their mass, at the start of counting time. The time after which the number of decaying nuclei is halved is called the half-life $t_{1/2}$:

$$t_{1/2} \equiv \frac{\ln 2}{k} \qquad (6.10)$$

Some radioactive nuclei decay very fast, and their half-lives are measured in seconds; others decay slowly, and their half-lives can be days, years, or even centuries.

The decay rate has great importance in regard to radioactive waste disposal. For example, spent fuel of a nuclear power plant contains many radioactive isotopes, such as strontium-90 (half-life 28.1 y), cesium-137 (half-life 30 y), and iodine-129 (half-life 15.7 million years). Strontium-90 and cesium-137 will decay to small amounts in hundreds of years, but iodine-129 will stay around practically forever.[10] Table 6.1 lists some radioactive isotopes that play a role in the fuel cycle of nuclear power plants, along with their radiation and half-lives.

6.3.2 Units and Dosage

The level of radioactivity of a sample of substance is measured by the number of disintegrations per second. The SI unit of radioactivity is the becquerel (Bq), which is one disintegration per second (see Table A.1). A more practical unit of measurement of radioactivity is the curie (Ci),

[9]The irradiation by neutrons is called neutron activation. It is used, for example, for identification and quantitation of elements in environmental samples. Research reactors at universities and other institutions are used to irradiate samples inserted into their reactors for neutron activation analysis.
[10]With a half-life of 28.1 years, ^{90}Sr will decay to 1% of its initial value in 187 years. ^{137}Cs with a half-life of 30.2 years will decay to 1% in 201 years.

TABLE 6.1 Some Isotopes in the Nuclear Fuel Cycle, with Half-Lives and Radiation

Isotope	$t_{1/2}$	Activity
Krypton-87	76 min	β
Tritium (^{3}H)	12.3 y	β
Strontium-90	28.1 y	β
Cesium-137	30.2 y	β
Xenon-135	9.2 h	β and γ
Barium-139	82.9 min	β and γ
Radium-223	11.4 d	α and γ
Radium-226	1600 y	α and γ
Thorium-232	1.4E(10) y	α and γ
Thorium-233	22.1 min	β
Uranium-233	1.65E(5) y	α and γ
Uranium-235	7.1E(8) y	α and γ
Uranium-238	4.5E(9) y	α and γ
Neptunium-239	2.35 d	β and γ
Plutonium-239	2.44E(4) y	α and γ

which is defined as 3.7E(10) Bq, or 3.7E(10) disintegrations per second. The radioactivity of one gram of radium-233 is 1 Ci, and that of cobalt-60 is 1 kCi.[11] But for mixtures of radioactive isotopes, such as found in ore samples or spent reactor fuel, the radioactive level measured in curies cannot tell us the amount or composition of the radioactive components, only their total disintegration rate.

Because exposure of humans to α, β, and γ radiation can be harmful, we need practical units of measurement of exposure to them. The SI unit of absorbed dose of radiation is the gray (Gy), which equals one joule of absorbed energy per kilogram of matter penetrated by the radiation. Another commonly used unit of absorbed dose is the rad, which equals 1E(-2) Gy.[12]

The absorbed energy is not entirely satisfactory as a measure of the harmfulness to humans of ionizing radiation, because other qualities of the radiation are also important. To take these into account, a different unit, the sievert (Sv), is used to measure what is called the dose equivalent. Like the gray, it has the dimensions of J/kg. The Sievert takes into account the quality of the absorbed radiation. An equivalent dose of 1 Sv is received when the actual dose of radiation (measured in grays), after being multiplied by the dimensionless factors Q (the so-called quality factor) and N (the product of any other multiplying factors), is 1 J/kg. Q depends on the nature of radiation and has a value of 1 for X rays, γ rays, and β particles; 10 for neutrons; and 20 for α particles. N is a factor that takes into account the distribution of energy throughout the dose. An alternative unit of dose equivalent is the rem, defined as 1E(-2) Sv.

---

[11]The specific radioactivity level of a sample is obtained by dividing the radioactivity level by the mass or volume of the sample (e.g., Ci/kg or Ci/L).

[12]In the treatment of cancer with radiation, the absorbed dose needed to kill cancer cells is of the order of 100 Gy.

An accumulation of absorbed α, β, and γ radiation over time is called a radiation dosage. For example, the average person in the United States receives in one year a dosage of about 360 millirems (3.6 mSv), of which 200 is from radon-86, 27 from cosmic rays, 28 from rocks and soil, 40 from radioactive isotopes in the body, 39 from X rays, 14 from nuclear medicine, and 10 from consumer products and other minor sources.

The average person on earth receives about 2.2 mSv y^{-1}. A short-term dose of 1 Sv causes temporary radiation sickness; 10 Sv is fatal. After the Chernobyl accident in the former Soviet Union, the average dose received by people living in the affected areas surrounding the plant over a 10-year period, 1986–1995, was 6–60 mSv. The 28 radiation fatalities at Chernobyl appear to have received more than 5 Sv in a few days; those suffering acute radiation sickness averaged 3–4 Sv.

6.3.2.1 Biological Effects of Radiation

The greatest risk to humans of nuclear power plant operations is associated with radioactivity, also called ionizing radiation, because of the creation of ions left by the passage of α, β, and γ rays and neutrons. Radioactivity affects humans and animals, causing somatic and genetic effects.[13] Somatic effects can be acute when an organism is subjected to large doses of radiation, or chronic when the exposure is at low levels, but over protracted periods. Acute effects include vomiting, hemorrhage, increased susceptibility to infection, burns, hair loss, blood changes, and, ultimately, death. Chronic effects, which usually manifest themselves over many years, include eye cataracts and the induction of various types of cancer, such as leukemia, thyroid cancer, skin cancer, and breast cancer. Genetic effects may become apparent in later generations but not in the exposed person. These effects are due to mutations in the genetic material—for example, chromosome abnormalities or changes in the individual's genes that make up the chromosomes.

6.3.2.2 Radiation Protection Standards

The prescription of radiation protection standards is an onerous and controversial task. In the United States, this task was vested in the Committee on the Biological Effects of Ionizing Radiation (BEIR) of the National Academy of Sciences, and internationally it was vested in the United Nations Scientific Committee on the Effects of Atomic Radiation (UNSCEAR). The task is difficult because direct evidence on biological effects of radiation comes (unfortunately) from high-level exposures, such as received by the population of Hiroshima and Nagasaki during atomic bombing and by the workers at the Chernobyl and Tokaimura accidents. Lower-level exposure data can only be obtained from animal studies, then extrapolated to humans. However, even in animals, the effects of low-level exposure can only be established statistically, observing a large cohort of animals over lengthy periods and over many generations.

In setting radiation standards, the following three assumptions are made:

(a) There is no threshold dose below which radiation has no effect.

(b) The incidence of any delayed somatic effect is directly proportional to the total dose received.

(c) There is no dose-rate effect.

---

[13]Somatic effects pertain to all cells in the body; genetic effects pertain to egg and sperm cells.

Basically, these assumptions constitute the linear-no-threshold (LNT) hypothesis. The assumptions mean that even the slightest radiation could lead to delayed somatic or genetic effects, and the occurrence of delayed effects does not depend on the stochastic nature of radiation, or whether the given dose is received over a short or extended time.

Based on the LNT hypothesis, the Nuclear Regulatory Commission in the United States set a standard of exposure for workers in nuclear power plants to 50 millisievert per year (50 mSv y^{-1} = 5 rem y^{-1}). For the general population—that is, any person in the region outside the plant boundary—the standard is 1 mSv y^{-1} = 100 mrem y^{-1}.

6.4 NUCLEAR REACTORS

A nuclear reactor in a nuclear fueled power plant is a pressure vessel enclosing the nuclear fuel that undergoes a chain reaction, generating heat which is transferred to a fluid, usually water, that is pumped through the vessel. The heated fluid can be steam, which then flows through a turbine generating electric power; or it can be hot water, a gas, or liquid metal that generates steam in a heat exchanger, the steam then flowing through a turbine.

The first controlled nuclear reactor was built and demonstrated by Enrico Fermi in 1942. It was constructed under the bleachers of the stadium at the University of Chicago. The reactor had dimensions 9 m wide, 9.5 m long, and 6 m high. It contained about 52 tons of natural uranium and about 1350 tons of graphite as a moderator, and cadmium rods were used as a control device. The reactor produced an output of only 200 W and lasted only a few minutes. Fermi's "pile" ushered in the nuclear age.

The first commercial scale nuclear power plant of 180-MW capacity went into operation in 1956 at Calder Hall, England. In the United States a 60-MW station started operating in 1957 at Shippingport, Pennsylvania. Before that, an experimental breeder reactor that produced electricity was demonstrated in 1951 near Detroit, Michigan. The first nuclear-powered submarine, the *Nautilus,* was launched in 1954. Submarine reactors produce steam that drives a turbine, which in turn propels the submarine.

We have discussed before that the energy of a nuclear reactor is derived from splitting a fissile heavy nucleus, such as ^{235}U or ^{239}Pu. In a nuclear reactor of a power plant, the splitting of the nucleus and sustaining of the ensuing chain reaction has to proceed in a controlled fashion.

The basic ingredients of a nuclear reactor are *fuel rods,* a *moderator, control rods,* and a *coolant.* A schematic of a reactor is depicted in Figure 6.1.

The *fuel rods* contain the fissile isotopes ^{235}U and/or ^{239}Pu. Natural uranium contains about 99.3% ^{238}U and 0.7% ^{235}U. The concentration of the fissile isotope ^{235}U in natural uranium is not enough to sustain a chain reaction in most power plant reactors; therefore, this isotope needs to be "enriched" to 3–4% (for enrichment processes, see Section 6.5).[14] The fuel rods contain metallic uranium, solid uranium dioxide (UO_2), or a mix of uranium dioxide and plutonium oxide, called MOX, fabricated into ceramic pellets. The pellets are loaded into zircalloy or stainless steel tubes, about 1-cm diameter and up to 4 m long.

---

[14]Some power plant reactors use natural uranium containing 0.7% ^{235}U as the fuel with different moderator and coolant combinations.

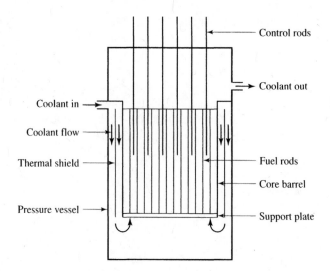

Figure 6.1 Schematic of nuclear reactor with water as coolant/moderator.

Moderators are used to slow the energetic neutrons that evolve from the fission reaction, yielding low-energy neutrons, also called thermal neutrons. This increases the probability for the neutrons to be absorbed in another fissile nucleus, so that the chain reaction can be propagated. Moderators contain atoms or molecules whose nuclei have high neutron scattering and low neutron absorption characteristics. Typical moderators are light water (H_2O), heavy water (D_2O), graphite (C), and beryllium (Be). The light or heavy water moderators circulate around the fuel rods. Graphite or beryllium moderators constitute a block into which fuel rods are inserted. For example, the original Fermi "pile" consisted of a graphite block into which metallic uranium fuel was inserted. The Chernobyl-type reactors also use graphite as a moderator.

Control rods contain elements whose nuclei have a high probability of absorbing thermal neutrons, so that they are not available for further splitting of fissile nuclei. In the presence of control rods, the chain reaction is controlled or stopped altogether. Typical control rods are made of boron (B) or cadmium (Cd).

The chain reaction inside the reactor is governed by the *neutron economy coefficient k*. Under a steady state, the number of thermal electrons is invariant with time, $dn/dt = 0$, and $k = 1$. The reactor is then in a *critical* condition. When $k < 1$, the reactor is *subcritical;* when $k > 1$, it is *supercritical*. A nuclear reactor becomes critical when control rods are lifted out of the core of the reactor to a degree where more than one neutron released by the fission of a fissile nucleus survives without being absorbed by the control rods. The position of the control rods determines the power output of the reactor. Monitoring the critical condition in a nuclear reactor while varying the output is quite complicated. Generally, nuclear power plants are run at full load, providing the base load of a grid. Running the plant at full load is also more economical.

Once the remaining fuel in the rods cannot sustain the rated capacity of the plant, even with complete withdrawal of the control rods, the fuel rods need to be replaced. This occurs every 2–3 years.

Heat must be constantly removed from the reactor. Heat is generated not only by the fission reaction, but also by the radioactive decay of the fission products. Heat is removed by a *coolant,* which can be boiling water, pressurized water, a molten metal (e.g., liquid sodium), or a gas (e.g., helium or CO_2). The accident at the Three Mile Island power plant near Harrisburg, Pennsylvania, in 1979 occurred because after shutdown (full insertion of the control rods), the reactor was completely drained of its coolant, so that the residual radioactivity in the fuel rods caused a meltdown of the reactor.

The heat removed by the coolant, in the form of steam or pressurized hot water, is used in conventional thermodynamic cycles to produce mechanical and electrical energy. In addition to the control rods position, the mass flow of the coolant also determines the plant's power output.

6.4.1 Boiling Water Reactor (BWR)

About one-quarter of the U.S. nuclear power plants are of the BWR kind. The schematic of a BWR is depicted in Figure 6.2; the reactor assembly itself is depicted in Figure 6.3. Most BWR's use 3–4% enriched ^{235}U as the fuel. Light water serves as both the coolant and moderator. As the control rods are withdrawn, the chain reaction starts and the coolant–moderator water boils. The saturated steam has a temperature of about 300 °C and pressure of 7 MPa. After separation of condensate in the steam separator, the steam drives the turbine that drives the generator. After expansion in the turbine, the steam is condensed in the condenser and returned via the feed water pump to the core.

The direct cycle has the advantage of simplicity and relatively high thermal efficiency, because the steam generated in the core directly drives a turbine without further heat exchangers. The thermal efficiency of a BWR is on the order of 33%, calculated on the basis of the inherent energy of the nuclear fuel that has been consumed in power production.

In a BWR there is no need for an extra moderator, because the coolant light water slows down the fast neutrons to thermal velocities that can engage in further fission reactions. Another advantage of a BWR is the fact that it is self-controlling. When the chain reaction becomes too intense, the coolant water boils faster. Because of its low density, steam has little or no moderating

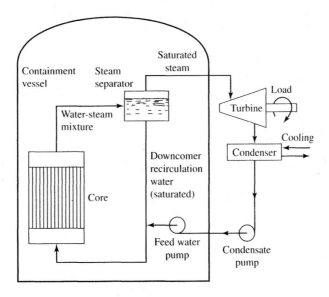

Figure 6.2 Schematic of a boiling water reactor (BWR) power plant.

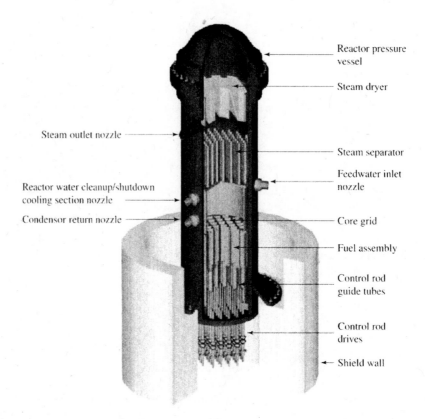

Figure 6.3 Cutaway of a boiling water reactor.

propensity. Thus, the reduction of the liquid content of the coolant water automatically slows the chain reaction.

In a BWR, the core is enclosed in the primary containment vessel made of steel and surrounded by reinforced concrete. A secondary containment vessel made of reinforced concrete (the dome-shaped building visible at nuclear power plants) contains the steam separator and the spent fuel storage pool. The steam turbine, condenser, and electric generator are located outside of the secondary containment vessel. Even though the coolant water is demineralized, some radioactive material may leach from the core into the coolant water and subsequently be transferred by the steam to the steam turbines. Furthermore, the coolant water may contain traces of mildly radioactive isotopes of hydrogen (tritium [^{3}H]) and nitrogen (^{16}N and ^{17}N). The steam remaining in contact with the turbine and other equipment loses its radioactivity quite fast, and with proper precaution no significant exposure is presented to workers in the power plant or the population outside the plant.

6.4.2 Pressurized Water Reactor (PWR)

The majority of nuclear power plants in the United States and worldwide are of the PWR type. A schematic of a PWR is depicted in Figure 6.4. In the primary loop surrounding the reactor core, the coolant water is kept at a high pressure of about 15 MPa, so that the water does not boil into

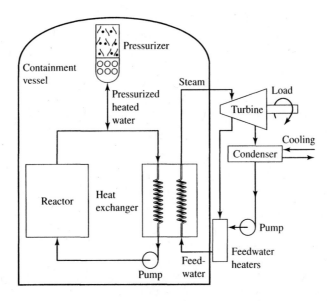

Figure 6.4 Schematic of a pressurized water reactor (PWR) power plant.

steam even though the temperature is in the 340–350 °C range. At that temperature, the pressure of the water exceeds the vapor pressure, so there is only liquid phase. The hot, nonboiling water is pumped into a heat exchanger, which is located together with the reactor core inside a heavy steel and reinforced concrete containment vessel. In the heat exchanger, feedwater is boiling into steam at about 7 MPa in a secondary loop. The steam drives a turbine that drives the electric generator in a conventional Rankine cycle. The steam turbine and condenser are located outside of the containment vessel.

One advantage of the PWR is that because of the single phase of the coolant water, the moderating capacity of the (light) water can be precisely adjusted, unlike in a BWR, where the coolant is in two phases, liquid and vapor. Usually, boron in the form of boric acid is added to the coolant to increase the moderating capacity. In such a fashion, fewer control rods are necessary to maintain the reactor at the design capacity. The other advantage is that the steam which is generated in the heat exchanger never comes into direct contact with the coolant water. Thus, any radioactivity that may be present in the coolant is confined to the primary loop inside the containment vessel. Because of the heat exchanger the overall thermal efficiency of a PWR is somewhat lower than that of a BWR, on the order of 30%.

A special type of PWR is the Canadian Deuterium Uranium (CANDU) reactor. This type of reactor is also used in Argentina, India, Pakistan, and Korea. CANDU uses natural uranium fuels (without enrichment) and heavy water (D_2O) as the moderator. Heavy water absorbs practically no neutrons; thus its neutron economy is superior to that of light water. On the other hand, its moderating capacity is less than that of light water; therefore neutrons have to travel twice as far to be slowed down (become thermal) as in light water. The fissile material is ^{235}U as in enriched uranium reactors, but some of the fertile ^{238}U is converted into fissile ^{239}Pu, which can participate in the chain reaction or can be extracted for fuel recycling or for weapons production. Instead of uranium enrichment facilities, the CANDU-type reactor requires heavy water production stills. Depending on which facilities a country possesses, on economic incentives of a particular government, and on the wish of a country to produce weapons grade plutonium, it may be more advantageous to operate a natural uranium/heavy water reactor than to operate an enriched uranium/light water reactor. In the United States, because of the availability of enriched uranium, no heavy water power plants are operating.

6.4.3   Gas-Cooled Reactor (GCR)

A reactor type that was developed in particular in Great Britain is the *gas-cooled reactor*. In fact, the first commercial power plant put into operation, in 1956 at Calder Hall, was a GCR. Generally, these reactors are fueled by natural or enriched uranium, either metallic or ceramic uranium oxide. The moderator is graphite; and as the name implies, the coolant is a gas, normally CO_2, but helium can also be used. Because of the lower heat transfer capacity of gases as compared to liquids, the contact surfaces and flow passages in the reactor must be larger than those in liquid-cooled reactors. In order to obtain a reasonable thermal efficiency, GCRs are run at higher temperatures than PWRs or BWRs. This necessitates cladding and piping materials that can withstand the higher temperatures. Some GCRs are using enriched uranium to boost the thermal efficiency.

An example of a high-temperature gas (CO_2)-cooled reactor is operating at Hinkley Point in the United Kingdom. Its net output is 1250 MW. The fuel is UO_2 with 2.6% ^{235}U. The CO_2 leaves the reactor at 655 °C and 4.3 MPa. The CO_2 is pumped to a heat exchanger where steam is generated at 540 °C and 17 MPa. The plant achieves a thermal efficiency of close to 42%.

In the United States a 40-MW high-temperature gas-cooled power plant was constructed and operated near Philadelphia, Pennsylvania. It is now decommissioned. Another 330-MW plant was operated near Platteville, Colorado. While providing interesting experience, the plant had many engineering problems and is by now also decommissioned. Research and development on GCRs is continuing, and their revival may occur in the future.

6.4.4   Breeder Reactor (BR)

In a *breeder reactor*, fissile nuclei are produced from fertile nuclei. The principal breeder mechanism is the conversion of ^{238}U to ^{239}Pu as shown in equation (6.2). The intermediary ^{239}U has a half-life of 23 minutes, then converts to neptunium ^{239}Np with a half-life 2.4 days, which in turn decays to ^{239}Pu, with a half-life of 24,000 years. The ^{239}Pu formed, while a fissile nucleus, does not participate to a significant extent in the chain reaction, but accumulates in the spent fuel from which it is later extracted and reused.

Unlike ^{235}U, which efficiently undergoes fission with slow thermal electrons whose energy is in the tenths of eV range, ^{238}U captures efficiently fast neutrons in the MeV range. To obtain this wide spectrum of neutron energies, a coolant/moderator other than light or heavy water is required. The preferred coolant is liquid sodium, and such a reactor is called the liquid metal fast breeder reactor (LMFBR). The sodium nucleus has a larger mass than does hydrogen or deuterium; therefore a neutron colliding with a sodium nucleus bounces off with nearly its original momentum, whereas a neutron colliding with a hydrogen nucleus (e.g., a proton) imparts nearly half of its momentum to hydrogen.

The efficiency of fuel utilization in a breeder reactor is expressed as the ratio of the number of fissile nuclei formed to the number destroyed. This ratio is called breeding ratio (BR):

$$BR = \frac{\text{Number of fissile nuclei produced}}{\text{Number of fissile nuclei destroyed}} \qquad (6.11)$$

When BR is greater than 1, breeding occurs. Note that both in the numerator and denominator of equation (6.11) fissile nuclei include not only ^{235}U, but ^{233}U and initially present ^{239}Pu.

Breeder reactors could also use thorium as a fuel. ^{232}Th is a fertile nucleus that can be converted into fissile ^{233}U by the following reaction sequence:

$$^{232}Th + n + \gamma \rightarrow {}^{233}Th + \beta \rightarrow {}^{233}Pa + \beta \rightarrow {}^{233}U \qquad (6.12)$$

Here ^{233}Pa is an isotope of element 91, protactinium. Because worldwide thorium ores have about an equal abundance as uranium ores, the use of thorium would extend the nuclear fuel resources by about a factor of two. So far, thorium-based breeder reactors are only in the planning and development phase.

A major problem with breeder reactors is the need to use liquid sodium as a coolant. (A helium-cooled breeder reactor was designed by Gulf General Atomic Corporation in the 1960s, but was not constructed.) In addition to the appropriate moderating characteristics of sodium, it also has an excellent heat transfer capacity. Sodium melts at 90 °C and boils at 882 °C. This allows the reactor to run hotter, and consequently a higher thermal efficiency is obtained. But sodium is a nasty chemical. It burns spontaneously in air, and it reacts violently with water. Furthermore, ^{23}Na (the natural isotope) can absorb a neutron to convert first into ^{24}Na, and then into the stable ^{24}Mg. The intermediary ^{24}Na emits very energetic β and γ radiation. Therefore, more than one sodium loop is required, a separate loop for the sodium coolant circulation and another loop for the water/steam cycle. A schematic of a LMFBR with two sodium loops is presented in Figure 6.5. The reactor core, sodium loops, heat exchangers, and steam loops are all located in a containment vessel; the steam turbine, condenser, and the rest of the generating plant are outside of the vessel.

Several fast breeder reactors were operated in the United States, United Kingdom, France, Germany, India, Japan, and Russia. The largest (400-MW) breeder reactor in the United States operated from 1980 to 1994 at Clinch River, Tennessee, but was terminated because of concern about nuclear proliferation. Breeder reactors still in operation are the 250-MW Phenix in France (the 1240-MW Superphenix was shut down in 1998), the 40-MW FBTR in India, the 100-MW Joyo in Japan, the 135-MW BN 350 formerly in USSR but now in Kazakhstan, and the 10-MW BR 10, the 12-MW BOR 60, and the 600-MW BN 600 in Russia.

Most breeder reactors, with the exception of those in Japan, were operated for the primary purpose of producing weapons-grade plutonium. This is certainly one of the drawbacks of breeder-type power plants. While they produce more fissile fuel than they consume in the process of power production, the fissile fuel, plutonium, can be fairly easily extracted to make atomic bombs. Strict international supervision will be necessary if, in the future, breeder-type power plants will commonly be used for electric power production.

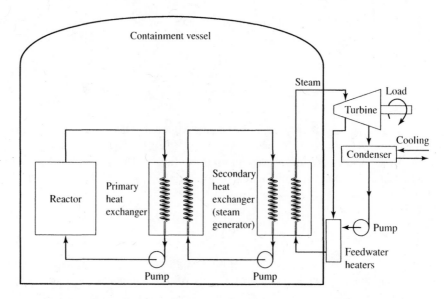

Figure 6.5 Schematic of a liquid metal fast breeder reactor (LMFBR) power plant.

6.5 NUCLEAR FUEL CYCLE

The nuclear fuel cycle starts from mining uranium (or thorium) ore, through extraction of the useful uranium concentrate, gasification to UF_6, enrichment of ^{235}U, conversion to metallic uranium or oxide of uranium, fuel rod fabrication, loading of a reactor, retrieval of spent fuel, reprocessing of spent fuel, and finally fuel waste disposal. A block diagram of the nuclear fuel cycle is presented in Figure 6.6.

6.5.1 Mining and Refining

Uranium ores containing variable concentrations of uranium are found in many parts of the world. Rich ores may contain up to 2% uranium, medium-grade ores 0.5–1%, and low-grade less than 0.5%. In the United States, ores are found in Wyoming, Texas, Colorado, New Mexico, and Utah. Large deposits are found in Australia, Kazakhstan, Canada, South Africa, Namibia, Brazil, and Russia.

The most economic way of extracting the ore is from open surface mines. Because uranium deposits are always associated with decay products (daughters) of uranium, such as radium and radon, these ores can be radioactive, and workers' protection must start at the mining phase. Open pit mines are well-ventilated, so most radiation, especially that associated with radon, which is a gas, escapes into the atmosphere. However, masks must be worn to prevent inhalation of mining dust which can contain radioactive elements.

The ore is crushed and ground. The ground ore is leached with sulfuric acid. Uranium, together with some other metals, dissolves. Uranium oxide with the approximate composition U_3O_8, called yellow cake, is precipitated, dried, and packed into 200-liter drums for shipment. The radiation from these drums is negligible. However, the solids remaining after leaching with acid may contain

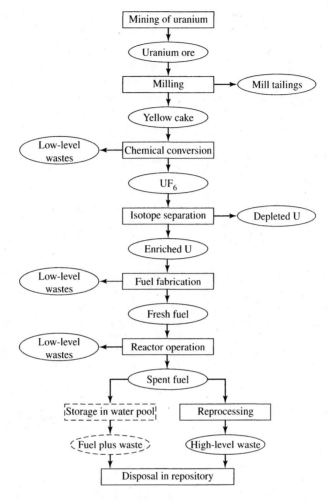

Figure 6.6 Nuclear fuel cycle.

radioactive isotopes of radium, bismuth, and lead. These solids are pumped as a slurry to the tailings heap, also called tailings dam. The tailings must be covered with clay or other impenetrable material to protect humans and animals from radiation exposure.

6.5.2 Gasification and Enrichment

The U_3O_8 concentrate is shipped from the mines to the enrichment facilities. The concentrate has the normal isotope distribution, about 99.3% ^{238}U and 0.7% ^{235}U. With the exception of CANDU-type heavy-water-moderated reactors, all other reactors need uranium enriched with ^{235}U.

For enrichment, a gaseous uranium compound is necessary. The concentrate is treated with hydrogen fluoride gas and the uranium oxide is converted to uranium hexafluoride UF_6. This is a white solid that sublimates to a vapor at a pressure of 1 atmosphere at 56 °C. The gaseous UF_6 is enriched by one of the following processes.

Membranes with small pores allow *gaseous diffusion* to occur. The rate of diffusion is a function of pressure and temperature and is also a function of the molecular diameter of the diffusing gas. $^{235}UF_6$ has a slightly smaller diameter than $^{238}UF_6$. Thus, the lighter isotope diffuses somewhat faster through the pores than the heavier isotope. In one passage, the enrichment factor is very small, but by forcing the gases to pass many membranes ("stages"), almost any degree of enrichment can be achieved. For light-water-moderated nuclear reactors, the desired enrichment is 3–4%. If the enrichment factor is a fraction of a percent per stage, hundreds to thousands of stages are required for the desired enrichment level. This requires large and expensive facilities with huge electricity consumption for the vacuum pumps and the blowers that force the gases through the membranes. Therefore, gaseous diffusion facilities are usually located near a supply of abundant and cheap electricity. In the United States, gaseous diffusion plants are located in Oak Ridge, Tennessee, and Hanford, Washington, where large hydroelectric dams are located nearby.

In the gaseous diffusion process about 85% of the uranium feed is rejected as depleted uranium. Currently, these "tails" are stockpiled for eventual future use in fast breeder reactors.

The gaseous diffusion facilities were established during the second world war for the production of weapons-grade highly enriched uranium. The relatively low enriched uranium used for power plant reactors is a byproduct of the wartime effort. Critics of nuclear power plants often charge that were enriched uranium not available from diffusion plants constructed and operated by governments, nuclear electricity would be far more expensive, or not available at all.

Recently, in the United States and Europe, enrichment plants have been built that work on the principle of a *gaseous centrifuge*. When the uranium gases are spun very fast in a centrifuge, the heavier $^{238}UF_6$ spreads toward the edges of the centrifuge while the lighter $^{235}UF_6$ concentrates toward the center. Here, the enrichment is dependent on the rate of revolutions and the time spent in the centrifuge. Gaseous centrifuges are less energy-consuming than diffusion plants and require lower capital investment.

Development is in progress in various countries on *laser enrichment* of uranium. In this process, metallic uranium is vaporized in an oven. A stream of vaporized uranium atoms exits the oven port. A laser beam with a very narrow wavelength band is shone unto the atomic stream to differentially excite to a higher electronic state only ^{235}U, but not the heavier isotope. The excited atom is then ionized with ultraviolet light. The ionized ^{235}U is collected on a negatively charged plate.

After enrichment, the UF_6 gas or metallic uranium is converted to uranium dioxide UO_2. This is a ceramic-like material that is fabricated into pellets. The pellets are loaded into fuel rods. A 1000-MW power plant needs about 75 metric tons of uranium dioxide pellets per load.

6.5.3   Spent Fuel Reprocessing and Temporary Waste Storage

In boiling and pressurized water reactors the fuel stays in the reactor for 2–3 years, generating electricity. After that period, the level of fission products and other neutron absorbers has built up, and the fission reaction has slowed down, with a concomitant decline in steam and electricity production. At that time, the fuel rods have to be replaced with fresh ones. In CANDU-type reactors, fuel rods have to be replaced every 18 months. The retrieved fuel rods emit a high level of radiation because of the radioactive fission products and other neutron-activated isotopes that have accumulated in the spent fuel rods. Generally, the extracted fuel rods are stored in the containment vessel of the power plant, in steel and concrete-walled water pools or dry casks.

Once the radioactive level of the spent fuel has declined sufficiently to be handled by remote control and proper shielding, it should be taken away to a permanent disposal site. Unfortunately, in the United States this is not happening because a permanent disposal site has not yet been approved. Except for the earlier reprocessing of spent fuel (now banned in the United States), every spent fuel rod is still stored in the water pool or dry cask at the power plants where they have been withdrawn from the reactor. At some power plants the storage space is running out, and their reactors may have to be shut down.

In other countries, after due decline of radioactivity, the spent fuel is reprocessed. About 96% of the original uranium in the fuel is still present, although it contains less than 1% ^{235}U. Another 1% of the uranium has been converted to ^{239}Pu. The spent rods are chopped up and leached in acid. Uranium and plutonium dissolve and are separated chemically from the rest of dissolved elements. The recovered uranium is sent back to the enrichment facilities. The recovered plutonium is mixed with natural uranium and made into fresh fuel, called mixed oxide (MOX). France reprocesses about 2000 tons per year of spent fuel, United Kingdom 2700 tons, Russia 400 tons, India 200 tons, and Japan 90 tons, but Japan sends spent fuel abroad for reprocessing. These amounts do not include plutonium reprocessing from defense establishment reactors for weapons fabrication.

The liquid wastes generated in reprocessing plants are stored temporarily in cooled stainless steel tanks surrounded by reinforced concrete. After a cooling period, the liquid wastes are *calcined* (evaporated to dry powder) and *vitrified* (encased in molten glass). The molten glass is poured into stainless steel canisters. In the United Kingdom, France, Belgium, and Sweden the canisters are stored in deep silos, pending permanent disposal.

6.5.4   Permanent Waste Disposal

Perhaps the largest problem facing nuclear power plants is the permanent disposal of spent fuel, or the waste remaining after extracting the still useful fuel from the spent fuel. The level of radioactivity of the spent fuel declines about tenfold every hundred years. After about 1000 years the level reaches that of the original ore from whence the fuel (uranium) was extracted.[15]

The only practical way of disposing of the waste would be in stable geologic formations known not to suffer from periodic earthquakes and where the water table is either absent or very deep beneath the formation. For example, deep salt formations have these characteristics; otherwise the salt would have leached out long ago.

In the United States a formation at Yucca Mountain, Nevada, has been selected for permanent disposal. Current plans call for the repository facility to be excavated from volcanic tuff at a depth of about 300 meters beneath Yucca Crest, which is 300 meters above the local water table. When filled, the repository is projected to store 70,000 tons of spent fuel and 8000 tons of high-level military waste. Many studies have been undertaken to model the fate of the disposed waste. Most studies concluded that the waste would be undisturbed for periods of 10 thousands to millions of years. Yet some uncertainties remain in these assessments that need to be clarified before official approval of Yucca Mountain as the permanent spent fuel disposal site in the United States. Current plans are for a 2010 opening of this site.[16]

---

[15]"The Environmental and Ethical Basis of Geological Disposal of Long Lived Radioactive Waste," 1995. OECD/Nuclear Energy Agency.

[16]See a series of articles by J. F. Ahearne, K. D. Crowley, W. E. Kastenberg, L. J. Gratton, and D. W. North in *Physics Today*, **50**, 22–62, June 1997.

The lack of permanent waste disposal systems and the unfavorable economics has deterred privately and publicly financed electricity producers from building new nuclear power plants in the United States. No other country has yet solved the permanent disposal problem of radioactive waste from nuclear power plants.

6.6 FUSION

As noted in Section 6.2, a large amount of energy is evolved when light nuclei fuse together. For example, the following fusion processes are accompanied by energy evolvement:

$$^2D + {}^3T \rightarrow {}^4He + n + 17.6 \text{ MeV} \tag{6.13}$$

$$^2D + {}^2D \rightarrow {}^3He + n + 4 \text{ MeV} \tag{6.14}$$

$$^2D + {}^3He \rightarrow {}^4He + p + 18.3 \text{ MeV} \tag{6.15}$$

The sum of the masses of the nuclei that fuse (the left-hand side of the equations) is not exactly the sum of the masses of the fused nucleus plus the mass of the ejected neutron or proton (the right-hand side of the equations). The "mass deficit" appears as the evolved energy. The ejected neutrons or protons collide with surrounding matter so that their kinetic energy is converted into sensible heat. Such fusion reactions power the sun and other self-luminous stars, as well as thermonuclear bombs, also called hydrogen bombs.

It would be desirable to perform fusion reactions under controlled conditions, so that the evolved heat energy can be transferred to a coolant working fluid, which in turn would drive a turbo-machinery. The advantages of fusion-based power plants are threefold: (a) The "raw" material or fuel available for fusion reactors is almost unlimited, because deuterium is a natural isotope of hydrogen to the extent of 1 deuterium atom in 6500 hydrogen atoms.[17] Tritium is not found in nature, but can be manufactured from an isotope of lithium in the following reaction:

$$^6Li + n \rightarrow {}^3T + {}^4He + 4.8 \text{ MeV} \tag{6.16}$$

(b) The fusion reactions would produce a minimal amount of radioactivity. Some radioactive isotopes may be created due to absorption of neutrons in materials surrounding the fusion reactor. Also, tritium is mildly radioactive, emitting low-energy β rays with a half-life of about 12 years. (c) There is no spent fuel waste from which ingredients could be extracted for fabricating fission nuclear weapons.

The difficulty of achieving a controlled fusion is in overcoming the electrical repulsion force of the positively charged nuclei. To overcome the repulsive force, the colliding nuclei must have a kinetic energy comparable to a temperature of tens of million degrees. At such temperatures, atoms are completely dissociated into positively charged nuclei and free electrons, the so-called *plasma* state. For the release of significant amounts of energy, many nuclei must collide. Hence the plasma needs to be confined to a small volume at a high pressure.

---

[17]Of course, separating deuterium from hydrogen requires considerable amounts of energy.

Attempts of controlled fusion processes have been conducted since the 1950s. So far only limited success has been achieved; here, success means that an equal or greater amount of energy is released as was consumed in the fusion experiment, the so-called "break-even" point. Optimistic estimates predict that commercial power plants based on fusion will be operative in the next 40–50 years. Pessimistic estimates claim that fusion-type power plants will never be practical, or they will be too expensive compared to power plants based on fission reactors or renewable energy, let alone fossil energy.

6.6.1 Magnetic Confinement

Most approaches to plasma confinement and inducement to fusion rely on magnetic field confinement. The magnetic field is created inside cylindrical coils in which a current flows. The cylindrical coils form a circle, so that the magnetic field has the shape of a toroid (doughnut). The plasma particles travel in helical revolutions along the magnetic field lines.

The first toroidal magnetic field reactor was constructed in the former USSR, hence the acronym *Tokamak,* short in Russian for toroidal magnetic chamber. Further confinement of the plasma is provided by an additional current flowing in the plasma itself. The plasma current is induced by transformer action from external coils. The plasma is heated by a combination of the resistive dissipation from the current flowing in the plasma and from external sources, such as radio-frequency waves. Also, energetic particles may be injected into the plasma, such as high-velocity ions from an accelerator.

Tokamak-type fusion machines are operating in Russia, Europe, Japan, and United States. In 1993, the Tokamak reactor at the Princeton Plasma Physics Laboratory using the deuterium–tritium fusion achieved a nominal temperature of 100 million degrees and a power of 5 million watts for about 4 seconds. Plans are underway to build a $1.2 billion International Thermonuclear Experimental Reactor, jointly funded by the United States, Japan, Russia, and several European countries. This facility will use deuterium–tritium as a fuel and superconducting magnets for confinement.

6.6.2 Laser Fusion

Laser-induced fusion apparatuses are operating at the Lawrence Livermore Laboratory, California; Los Alamos Laboratories, New Mexico; and Rochester University, New York.

A fuel (e.g., a mixture of deuterium and tritium) is contained in a small (1-mm) sphere made of glass or steel. The mixture is irradiated with several high-intensity laser beams. The sphere is compressed and heated by implosion. The glass or steel outer layer evaporates. The temperature of the content gases increases to tens of millions degrees, and the pressure rises to thousands of atmospheres. Laser beams of megajoules of energy are employed in very short pulses, a billionth of a second long. The power at the center of the sphere reaches 1E(15) W. The initial beam is from a neodymium-glass laser radiating at 1 μm. A crystal is used to generate higher harmonics in the ultraviolet range. Unfortunately, the laser system operates at about 1% efficiency, so that 99% of the input energy does not result in light emission. Higher-efficiency lasers pulsing at a higher rate will be necessary for break-even power production, let alone for a commercial power plant.

Plans are underway to build the National Ignition Facility at the Lawrence Livermore Laboratory, California, employing a deuterium–tritium fuel and 192 laser beams.

6.7 SUMMARY

More than 400 nuclear power plants are currently operating in the world, supplying about 17% of the global electricity demand of more than 2E(15) kWh per year. While complicated, the technology is well-developed, and the power plants operate relatively reliably.

Most of the current power plants use a reactor of the pressurized water type (PWR) in which the fuel is uranium enriched to 3–4% uranium-235; the pressurized light water serves both as the coolant and the moderator, and cadmium or boron control rods serve to control the power output or to shut down the reactor. The boiling water reactors (BWR) use the same fuel and control rods, but the coolant/moderator is in two phases: liquid water and steam. In some BWR the moderator is graphite. Another type of reactor is used in the CANDU power plants, where the fuel is natural or slightly enriched uranium, the coolant/moderator is heavy water, and the control rods are similar to the other reactors. Breeder reactors have been primarily used to produce weapons-grade plutonium. They use natural uranium and a coolant/moderator that usually is liquid sodium. Most of the breeder reactors have been phased out since sufficient plutonium for weapons production has been stockpiled, and there is no immediate scarcity of natural uranium.

Further advances in developing novel nuclear power plants are possible that will increase their thermal efficiency, use less expensive fuels, such as natural uranium and thorium, and, foremost, ensure their absolute safety.

Despite persisting concerns about nuclear power plant safety, only one acknowledged serious accident occurred—that at Chernobyl in 1986—involving 28 mortalities and a yet unknown number of latent morbidities. The Three Mile Island accident in 1979 is not known to have caused excess mortality or morbidity. The Tokaimura accident in 1999 was not associated directly with power plants, but with reprocessing spent fuel.

Nevertheless, the future of nuclear power plants is very much uncertain. In part the uncertainty stems from the public's perception that nuclear power plants are inherently unsafe, notwithstanding the contrary records and statistics. The other factor is economics. Although governments directly or indirectly subsidized nuclear power plants by funding research, development, and the infrastructure upon which the plants were founded, the production cost of electricity in existing plants is either equal to or greater than that in fossil-fueled plants. New plants may become more expensive than existing ones because of the increased capital investment needed for added safety features. Production costs may also increase because a larger number of technical personnel will be required to ensure safe operation of the new plant and also because of the imbedded cost of safe disposal of spent fuel and decommissioning the plant after reaching retirement.

Realistically, the only conditions under which nuclear power plants will increase their share of global electricity supply are as follows: (a) fossil fuels will become scarce and/or expensive; (b) international or national policies to prevent the threat of global warming will militate against electricity production by fossil-fueled power plants; (c) renewable energies cannot satisfy the increasing electricity demand; (d) national policies of governments that wish to be independent of foreign fossil fuel supplies; (e) national policies of governments that wish to develop nuclear weapons, the ingredients of which (i.e., plutonium) come from reprocessed spent fuel of power plants.

In any case, if nuclear power plants are to increase their share of electricity supply, the risk of radiation accidents must be brought to near zero, and the problem of long-term disposal of radioactive waste must be solved.

PROBLEMS

Problem 6.1

(a) Calculate the mass deficit (Δm) in atomic mass units (amu) of the following fission reaction. (Use literature values for the exact masses of the isotopes and neutrons.)

$$^{235}U + n \rightarrow {}^{139}Xe + {}^{95}Sr + 2n$$

(b) Calculate the energy (MeV) released per one fission. (c) Calculate the energy released per kilogram of ^{235}U, and compare it to the energy released in the combustion of 1 kg of carbon.

Problem 6.2

(a) Calculate the mass deficit (Δm) in atomic mass units (amu) of the following fusion reaction.

$$^{2}D + {}^{3}T \rightarrow {}^{4}He + n$$

(b) Calculate the energy released (MeV) per one fusion. (c) Calculate the energy released per kilogram of deuterium.

Problem 6.3

In a nuclear accident there is a release of ^{90}Sr that emits γ rays with a half-life of 28.1 y. Suppose 1 μg was absorbed by a newly born child. How much would remain in the person's body after 18 and 70 years if none is lost metabolically?

Problem 6.4

The isotope ^{223}Ra has a half-life of 11.4 days. ^{223}Ra decays at a rate of 1 Ci per gram of radium isotope. (Ci = Curie = 3.7E(10) disintegrations/sec.) What will be the decay rate (Ci) of the 1-gram sample after 10, 100, and 1000 days?

Problem 6.5

The isotope ^{129}I has a half-life of 15.7 years. In a nuclear power plant accident, 1 kg of the iodine isotope is dispersed into the surroundings of the plant. How much of the iodine isotope will remain in the surroundings after 1, 10, and 100 years?

Problem 6.6

Define the following terms and give examples as appropriate:

    fissile nucleus

    fertile nucleus

    chain reaction

neutron economy, sub- and supercriticality

fuel rod

moderator

control rod

coolant

Problem 6.7

For each of the following, create a schematic diagram and describe and list its advantages and disadvantages:

    (a) Boiling water reactor

    (b) Pressurized water reactor

    (c) Breeder reactor

Problem 6.8

Create a block diagram of, and describe, the nuclear fuel cycle. List the risks at each step.

Problem 6.9

Why is there no danger of a nuclear power plant reactor exploding like an atomic bomb?

BIBLIOGRAPHY

Bodansky, D., 1996. *Nuclear Energy—Principles, Practices and Prospects*. New York: American Institute of Physics.

El-Wakil, M. M., 1982. *Nuclear Energy Conversion*. La Grange Park: American Nuclear Society.

El-Wakil, M. M., 1984. *Powerplant Technology*. New York: McGraw-Hill.

Glasstone, S., and A. Sesonske, 1994. *Nuclear Reactor Engineering*. New York: Chapman and Hall.

Hinrichs, R. A., 1996. *Energy*. Orlando: Harcourt Brace College Publishers.

Mataré, H. F., 1989. *Energy Facts and Future*. Boca Raton: CRC Press.

Murray, R. L., 1989. *Understanding Radioactive Waste*. Columbus: Battelle Press.

Tester, J. W., D. O. Wood, and N. A. Ferrari, Eds., 1991. *Energy and the Environment in the 21st Century*. Cambridge: MIT Press.

Turner, James E., 1986. *Atoms, Radiation, and Radiation Protection*. New York: Pergamon Press.

Renewable Energy

7.1 INTRODUCTION

Renewable energy systems draw energy from the ambient environment rather than from the consumption of mineral fuels (coal, oil, gas, and nuclear). The ultimate source of most renewable energy production is the sun, whose total radiant energy flux intercepted by the earth provides a much greater source of power than can be captured by practical renewable energy schemes.[1] Despite the vast amounts of energy that are potentially available from renewable sources, collecting and utilizing that energy in an economical and effective manner is far from an easy task.

There are many reasons for the growth of interest in renewable energy systems. Such systems are independent of fuel supply and price variabilities and are thereby economically less risky. Renewable energy resources are more uniformly distributed geographically than are fossil fuels, providing indigenous energy resources for most fuel-poor nations. Some forms of renewable energy operate efficiently in small units and may be located close to consumers, reducing energy transmission costs. In the United States in recent years, governmental regulation of the electric utility system has provided some economic incentive for the construction of renewable energy systems as a small part of the mix of electric power sources. Because renewable energy sources have lesser environmental effects than conventional energy sources, especially in regard to the emission of toxic and greenhouse gases to the atmosphere, they are expected to become an important source of energy in future years as environmental controls become more stringent. But the major obstacle to greater use of renewable energy is its generally higher cost compared to conventional sources.

The renewable technologies that currently are employed or show promise of becoming practical are: (a) hydropower, (b) biomass power, (c) geothermal power, (d) wind power, (e) solar heating and thermal electric power, (f) photovoltaic power, (g) ocean tidal power, (h) ocean wave power, and (i) ocean thermal electric power. As shown in Figures 2.2 and 2.4, renewable energy accounts for 4.4 % of total U.S. 1996 energy production and 8.2% of world energy production in 1997, almost all of it in the form of hydropower, but the amount of non-hydro renewable power is growing rapidly.

Some renewable energy systems produce only electrical power, which has a higher economic value than heat. Among these are hydro, wind, photovoltaic, tidal, and ocean wave power. Nevertheless, biomass, geothermal, and solar systems, which can deliver both electric power and heat, are equally important renewable energy sources. Table 7.1 lists the 1997 U.S. renewable energy pro-

---

[1]Exceptions are tidal power, which derives its energy from the gravitational energy of the earth–moon–sun system, and geothermal power, whose source is nuclear reactions in the earth's interior.

TABLE 7.1 1997 U.S. Renewable Energy Production[a]

Type	Electrical (10^9 kWh/y)	Thermal (10^9 kWh/y)	Electrical Capacity (GW)	Capacity Factor (%)
Hydroelectric	359	—	79.8	51.4
Biomass	57.8	614	10.7	61.7
Geothermal	14.6	47.2	2.85	15.6
Wind	3.38	—	1.62	23.8
Solar	0.89	18.8	0.33	30.4
Total	435.7	680	95.3	52.2

[a] Data from USDOE, 1998. DOE/EIA-0603(98) and DOE/EIA-0484(98).

duction by type of source. Both electric and thermal contributions, where appropriate, are listed. Of the electrical power component, hydropower accounted for 83%, biomass 13%, geothermal 3%, wind 0.8%, and solar 0.2%. Together, renewable electricity accounted for 13% of U.S. 1997 electricity production. (On a worldwide basis, renewable electricity is a greater proportion, 21.6%.) Also listed in Table 7.1 for each type is the installed electrical capacity and capacity factor (ratio of the actual annual electrical energy produced to that which would be possible if the source ran at its installed capacity continuously for a full year).[2]

Renewable energy is characterized by several important traits:

- Ubiquity of renewable energy sources.
- Low intensity of energy fluxes being captured, compared to conventional systems.
- Random, intermittent nature of renewable energy fluxes.

Some form of renewable energy is available everywhere on earth. Averaged over a year, the entire earth's surface (including the polar regions) intercepts some sunlight, making solar insolation universally available; although low-latitude, dry (clear) climates have the most, whereas the polar regions have the least, solar flux at ground level. Solar insolation induces circulation in the earth's atmosphere, giving rise to wind and ocean waves, along with precipitation into elevated drainage basins, providing secondary sources of solar energy. Biomass crops and forests grow in tropical and temperate lands having sufficient precipitation. Ocean tides are noticeable along continental margins. Temperature gradients beneath the earth's surface exist everywhere, providing the possibility of geothermal energy. But it is generally true that there is considerable geographic variability of each form of energy, so that some locations are more favorable for its development than others.

Fossil- and nuclear-fueled power systems operate under conditions of intense energy transfer within their components, typically of the order of 1E(5) W/m^2, so that they produce large amounts of power per unit volume. In contrast, renewable energy fluxes are much smaller, although quite different among the various sources, as listed in Table 7.2, requiring larger structures to reap the same amount of power. As a consequence, the capital cost per unit of average

---

[2]For the United States in 1997, the electrical power produced by both utilities and industrial sources amounted to 3340E(9) kWh/y, the installed electrical capacity was 683.2 GW, and the average capacity factor was 55.8%.

TABLE 7.2 Average Energy Flux in Renewable Energy Systems

Source	Area	Heat (W/m^2)	Work (W/m^2)
Solar	Collector	150	20
Photovoltaic	Cell		30
Hydropower	Drainage basin		0.01
Wind	Turbine disk		40
Geothermal	Field	0.1	0.02
Biomass	Field	0.5	0.1
Ocean tidal	Tidal pond		1
Ocean thermal	Surface area		
Ocean wave	Frontal area		10,000

power output is higher for renewable than for conventional power plants, but the cost of fuel for the conventional plant may more than offset its lower capital cost, making renewable energy cheaper.

Energy from sunlight is only available about half of the time, and it reaches its higher intensities in the several hours on either side of local noon. On cloudy and partly cloudy days, there is a considerable reduction in solar insolation compared with clear days. Electric power generated from sunlight must be stored or integrated into a distribution network with other power sources to be useful in an industrial society. Similar considerations apply to wind, wave, and tidal power systems, which are also intermittent sources. On the other hand, hydropower, biomass, and geothermal systems have storage capabilities that permit them to deliver power when it is needed, as do conventional systems.

In this chapter we explain the physical basis for each of the renewable energy sources listed above and also discuss the technology used to collect and utilize that energy. In some cases it is possible to identify costs and performance for systems in use. We also discuss environmental effects which, while greatly reduced compared to traditional energy sources, can be significant.

7.2 HYDROPOWER

Before steam engines were developed, mechanical power generated by river water flowing past water wheels was the major source of power for industrial mills. These mills had to be located near river falls so that water could be diverted from an impoundment upstream of the falls and fed to a water wheel or turbine discharging to a lower level downstream of the falls. The need for mechanical power and sites for industrial facilities soon outgrew the availability of hydropower, leading to the introduction of steam engines and, eventually, steam electric power plants that distribute their power via electric lines to consumers far removed from the power plant site. Nevertheless, today hydropower continues to be an important source of electric power generation, supplying 10.7% of U.S. and 19% of world electricity production.

In the United States in 1980, the installed hydropower capacity totaled about 63 GW generated in 1384 plants distributed throughout the country, as listed in Table 7.3, operating at an average capacity factor of 51%. It is estimated that the total potential capacity in the United States is about

TABLE 7.3 Hydropower Development in the United States in 1980[a]

Region	Number of Plants	Installed Capacity (GW)
Pacific Northwest	160	28.1
California	173	7.5
South Atlantic and Gulf	119	5.6
Great Lakes	203	3.9
Tennessee	47	3.7
Other	682	14.5
Total	1384	63.3

[a]Data from Gulliver, John S., and Roger E. A. Arndt, 1980. *Hydropower Engineering Handbook.* New York: McGraw-Hill.

three times this amount. Most of the U.S. hydropower plants are small, with the average power being 46 MW. The largest, Grand Coulee, located on the Columbia river, can generate 7.5 GW. Worldwide there are seven plants generating more than 5 GW located in Venezuela (10.6), United States (7.5), Brazil/Paraguay (7.4), former Soviet Union (6.4 and 6.0), and Canada (5.3 and 5.2). China is presently constructing a dam at the Three Gorges section of the Yangtse river that will generate 18.2 GW.

An economical hydropower plant requires a reliable supply of water flow from an elevated source that can be discharged at a lower elevation nearby. Most sites require the construction of a dam at a location along a river where a large volume of water can be impounded behind the dam at a level higher than the river bed. A pipe connected to the upstream reservoir (called a *penstock*) conducts high-pressure water to the inlet of a hydroturbine located at a level at or below the downstream riverbed, into which the turbine discharge water is released, generating mechanical power to turn an electric generator. If the difference in height of the water between upstream and downstream of the dam is h (called the *head*) and the volume flow rate of water through the turbine is Q, then the maximum power the turbine can generate is $\rho g h Q$, where $\rho = 1\mathrm{E}(3)$ kg/m^3 is the mass density of water and $g = 9.8$ m/s^2 is the acceleration of gravity. For a 1000-MW hydropower plant utilizing a head of 10 m, a flow of more than $1\mathrm{E}(4)$ m^3/s would be required. Only the largest of rivers can supply flows of this magnitude.

River water flow depends upon precipitation in the drainage basin upstream of a dam site. Only a fraction of the precipitation reaches the river course, with the remainder being lost to evaporation into the atmosphere and recharging of the underground water aquifer. On a seasonal or annual basis, precipitation and river flow is variable, so that it is desirable to impound a volume behind the hydropower plant dam that can be used in times of low river inflow. In locations where a substantial portion of annual precipitation is in the form of snow, its sudden melting in the spring will produce a surge in flow that may exceed the capacity of the hydroplant to utilize or store. It is usually not possible or economical to utilize the entire annual river flow to produce power.

In mountainous terrain it may be possible to locate sites having a large head, of the order of several hundreds of meters, but with only moderate or low flow rates. Such sites may prove economical because of the large power output per unit of water flow. On the other hand, low-head "run of the river" plants that store no water behind their dams but can utilize whatever flow is available during most of the year have also proved economical.

Figure 7.1 The rotating component of a hydroturbine: low-head Kaplan turbine (left), medium-head Francis turbine (center), and high-head Pelton turbine (right). (By permission of VA TECH HYDRO.)

The technology of hydropower is well developed. Hydroturbines are rotating machines that convert the flow of high-pressure water to mechanical power. For low-head, high-volume rate flows the Kaplan turbine is an axial flow device that somewhat resembles a ship's propeller in shape (see Figure 7.1). For higher heads, the Francis turbine is a radial inflow reaction turbine in which the flow is approximately the reverse of a centrifugal pump. For very small flow rates, the Pelton wheel utilizes a nozzle to accelerate the high-pressure water stream to a high-velocity jet that subsequently impinges on the peripheral blades of the turbine wheel. Powering a synchronous electric generator, hydroturbine installations can convert more than 80% of the ideal water power available to electric power.

Most hydropower plants require the construction of a dam and spillway, the latter to bypass excess water flow around the dam when the reservoir is full and the turbines are operating at full capacity. The civil works required for the dam, spillway, and power house can be a considerable fraction of the cost of the installation, so that the overall cost per unit of electrical power generation is higher than that for the hydroturbine and generator alone. A favorable site for a hydropower plant is one for which the cost of the civil works is not large compared to the power-producing equipment.

An example of a hydropower installation is shown in Figure 7.2. On the far left and right is an earthen dam that backs up the river flow, forming a higher-level pool extending miles upstream. In the center, on the left, is a spillway that passes excess river flow; and in the center, on the right, is a power house with five turbine-driven generators discharging water to a lower level downstream.

Given the unevenness of river flow and the finiteness of reservoir volume, hydropower plants do not operate at their rated capacity year-round. A typical capacity factor is 50%, which represents a compromise between the desire to utilize all the hydroenergy available in the river flow and the necessity to limit the capital costs of doing so.

Hydroplants have indefinitely long, useful lives. The civil works and power-generating machinery are very robust, and operating expenses are very low. Many large hydropower plants have been constructed by national governments, and their cost has been paid in part from government revenues as well as by electricity consumers. The construction of the dam may provide benefits other than electric power, such as water for irrigation or flood control, that reduce the costs allocated to electricity production. Nevertheless, there are hydropower sites that can be developed that are economically competitive with fossil fuel power plants.

Figure 7.2 The Noxon Rapids dam on the Clark Fork River in Montana generates 466 MW. (By permission of George Perks/Avista Corp.)

7.2.1 Environmental Effects

Hydropower plants can have severe environmental effects. Where a reservoir is formed behind a dam, aquatic and terrestrial ecosystems are greatly altered. Downstream of the dam, the riverine flow is altered, interfering with the ecosystems that have adjusted to the natural variable river flow pattern. The reservoir interrupts the natural siltation flow in the river and its contribution to alluvial deposits downstream. The flooding of land that previously served for agriculture and human habitation may be a significant social and economic loss. The construction of dams interferes with the migration of anadramous fish and adversely affects their populations, even where fish ladders are employed. Currently, removal of U.S. hydropower dams in the Pacific northwest and Maine have been recommended to revive endangered salmon fisheries.

7.3 BIOMASS

Until the onset of the industrial revolution led to the exploitation of fossil fuels, wood was the principal fuel available to humans, being used for cooking and space heating. In many developing nations today, wood supplies up to half of the energy consumption, which is very low on a per capita basis. Wood crops, a renewable resource, replace themselves every 50–100 years. The amount of non-food-producing land available for energy crops, like wood, is insufficient to supplant entirely current fossil fuel consumption. Nevertheless, a variety of agricultural and silvicultural crops or their byproducts can contribute to energy supplies and thereby replace some fossil fuel consumption.

Another form of biomass that can be converted to a useful fuel is animal waste. Digesters can generate methane from farm animal or human wastes, with the residue of this process being suitable for crop fertilizing. Organic matter in municipal waste landfills generates methane in an uncontrolled process that can supply low heating value gas.

In the United States in 1997, biomass provided 29% of renewable energy and 1.25% of total energy. It also contributed 13% of renewable electric power and 1.7% of all electric power.

Organic matter in terrestrial plants and soil is, among other things, a temporary storage system for solar energy. The conversion of atmospheric CO_2 and H_2O to organic matter by photosynthetic reactions in plants stores the energy of visible light from the sun in the form of chemical energy of the organic matter. The latter may be utilized in the same manner as fossil fuels, releasing CO_2 back to the atmosphere. Thus no net emissions of CO_2 to the atmosphere result from this cycle. However, the efficiency of conversion of solar radiation to biomass energy is very low, less than 1 %, with the terrestrial average energy conversion rate amounting to 0.5 W/m^2.

The food energy stored in agricultural crops like grains is only a fraction of that in the entire crop mass, so that agricultural crop residues are potential sources of biomass energy. It has been estimated that crop and forest residues in the United States have a heating value equal to 12% of fossil fuel consumption, but only about one-fifth of this is readily usable.[3] The energy required to collect, store, and utilize this residue further reduces the amount of energy available from it to replace fossil fuels. Furthermore, crop residues are often used to build soil mass and fertility by composting, during which some, but not all, of the residue carbon is released to the atmosphere as CO_2. Using residues as fuel reduces the amount of carbon uptake in soil.

The overall photosynthetic process by which water and carbon dioxide are combined to form carbohydrate molecules in plants may be summarized as

$$nCO_2 + mH_2O \rightarrow C_n(H_2O)_m + nO_2 \qquad (7.1)$$

The first step of this overall process is the photosynthetic one, in which solar radiation provides the energy needed to start the process that ends in the production of carbohydrates, such as sugar, starch, or cellulose. Subsequent steps dissipate some of that energy, storing the rest in chemical form. The process of conversion of solar energy into biomass is called primary production. All living systems depend ultimately upon this process to maintain their viability.

Agricultural crops supply humans with food and materials (e.g., lumber, paper, textiles) as well as energy. On an energy basis, the price to consumers of these commodities exceeds that of fossil fuels, often by a considerable amount. To compete economically with traditional agricultural commodities for scarce agricultural resources, biomass energy supplies must use otherwise under-valued byproducts and waste streams or efficient, low capitalization facilities that convert primary biomass to higher value fuels.

The principal processes that utilize the energy content of primary biomass are as follows:

- Combustion
- Gasification
- Pyrolysis
- Fermentation
- Anaerobic digestion

Woody plants and grasses can be burned directly in stoves, furnaces, or boilers to provide cooking, space, or process heat and electric power. Alternatively, biomass may be converted to a gaseous fuel composed of H_2 and CO in a thermal process that conserves most of its heating value, with the fuel being combustible in boilers and furnaces. Pyrolysis, the thermal decomposition of biomass,

---

[3]D. Pimentel et al., 1981. *Science* **212**, 1110.

produces a combination of solid, liquid, and gaseous products that are combustible. Fermentation and distillation of carbohydrates produces ethanol (C_2H_5OH), a valuable liquid fuel that is commonly blended with gasoline for motor vehicle use. Anaerobic digestion produces a gaseous mixture of CO_2 and CH_4. The gaseous fuels can be upgraded to more desirable forms, albeit at some loss of energy.

The processing of biomass to a form that is more readily used as a replacement for fossil fuel inevitably results in a loss of some of its heating value and generates a production cost of the conversion. To compete economically with fossil fuels, which have small processing costs, biomass-derived fuels must utilize low cost forms of primary biomass.

Except for wood, which may be harvested year-round, biomass energy crops, like food crops, must be stored after harvesting to provide a steady supply of feed stock for the remainder of the year. The economic cost of storage and maintaining a year's inventory adds to the price differential between biomass and fossil fuels, which have much smaller inventories.

During the 1970s, when international oil prices skyrocketed, many agricultural-based, biomass-derived liquid fuel production schemes were investigated. Of particular interest was the energy gain factor, the ratio of the energy value of the fuel produced (usually ethanol) to the energy consumed in producing, harvesting, and processing the biomass. Not only must this ratio exceed unity, so that there is a net output of energy from the biomass conversion process, but it must be sufficiently greater that other, non-energy costs of production can be covered by the sales revenues generated.[4] Such requirements could only be met by utilizing a high-energy crop like sugar cane and incorporating the energy supply of the crop residue into the fuel synthesis process. Currently, these energy and economic constraints make fuels derived from biomass food crops, such as grains or sugar cane, noncompetitive with fossil fuels at current world market prices.

As an example, in the United States, corn is currently employed as a feedstock to manufacture ethanol for use as an additive in motor vehicle fuel. In the fermentation and distillation process that converts corn starch to ethanol, only 65% of the grain heating value is preserved in the ethanol output from the process.[5] In addition, fossil fuel is consumed in the production and harvesting of corn and the production of ethanol. As a consequence, use of ethanol as a vehicle fuel additive results in only a 50–60% reduction in fossil fuel energy use and a 35–46% reduction in greenhouse gas emissions, compared to the use of ordinary vehicle fuel.[6] Thus the energy gain factor for corn-generated ethanol is 2–2.5.[7]

The use of wood harvesting residues to generate process heat and electric power in paper mills, eliminating their use of fossil fuels, is now common in the paper industry. The residues are collected along with the wood used to feed the pulping operation. While the residues supply some or all of the mill's energy requirements, there is generally no significant surplus that might otherwise generate electric power for other purposes. (The same is true for wood harvested for lumber.) In contrast

---

[4]For fossil fuels, the energy consumed in production, processing, and distribution is less than 10% of the energy of the end product.

[5]Probstein, Ronald F., and R. Edwin Hicks, 1990. *Synthetic Fuels.* Cambridge: pH Press.

[6]Wang, M., C. Saricks, and M. Wu, 1999. *J. Air Waste Manage. Assoc.* **49,** 756–772.

[7]The heating value of a bushel of corn is about 400 MJ. At an average price of $3 per bushel, this energy costs about $8 per million Btu. Average prices of coal and natural gas in the 1990s were about $1.25 and $3 per million Btu.

Figure 7.3 A gasifier at a power plant in Burlington, Vermont, processes about 200 tons per day of wood chips and waste, generating a gaseous fuel that provides 8 MW of electric power from a steam plant. (By permission of DOE/NREL-PIX.)

to this use of an otherwise valueless byproduct, electric power plants fueled by wood harvested for fuel only purposes have been constructed. The harvesting and transportation costs of the wood fuel for these plants, added to the market value of the wood component as pulp feed or lumber (in contrast to the residue component), generates a costly fuel and makes these plants uneconomical compared with fossil fuel power plants.[8]

Figure 7.3 shows a gasifier that converts wood waste to gaseous fuel for use in an electric power plant. By converting from a solid to a gaseous form, the fuel can be used in a combined cycle power plant, doubling the thermal efficiency of about 20% obtained in conventional wood burning power plants.

7.3.1 Environmental Effects

The use of crops or their residues to supply fuel to replace fossil fuel creates environmental impacts similar to those associated with agriculture and silviculture: consumption of manufactured fertilizers, spreading of pesticides and herbicides, soil erosion, consumption of irrigation water, and interference with natural ecosystems. Air emissions from the combustion of biomass fuels are not always less than those from the fossil fuels they replace.

---

[8]Expressed in 1999 dollars, the wood-fueled plants produce electric power at a cost of 8–12 cents per kWh compared to 4–6 cents per kWh for current fossil fuel plants.

7.4 GEOTHERMAL ENERGY

Fossil fuel deposits, which form the bulk of the world's current energy supply, are the residue of biomass formed millions of years ago, in which solar energy was stored in living matter. Made accessible by industrial technology, fossil fuel is the most easily and cheaply exploitable form of energy. But the energy stored in the hot interior of the earth is vastly greater in magnitude and is potentially exploitable, yet is almost entirely inaccessible. This energy is what remains from the gravitational collapse of the interplanetary material from which the earth was formed.

The earth's interior consists of a core of mostly molten material at a temperature of about 4000 °C, extending to a little more than half an earth radius and surrounded by a mantle of deformable material. The outer edge of the mantle is covered by a crust of solid material, of thickness between 5 and 35 km. Within the crust there is an outward flow of heat from the earth's interior of approximately 50 mW/m^2 which is accompanied by a temperature gradient of about 30 K/km.[9] No practical use can be made of such a feeble flow of heat unless it can be substantially amplified. By drilling wells to depths of 5–10 km and pumping from them water or steam heated to 200–300 °C, enough heat may be extracted to generate hundreds of megawatts of electrical power in a single geothermal plant.

The most economical sites at which to develop geothermal energy are those where the subsurface temperatures are highest and underground water and steam deposits are closest to the surface. Such sites are found mostly at the borders of the earth's tectonic plates, near active or recently inactive volcanos, hot springs, or geysers. Favorable sites of this type occupy only a small fraction of the earth's land area.

The principal countries that have installed geothermal electrical power generation systems are listed in Table 7.4, together with the amounts of installed electrical power. Of the world total

TABLE 7.4 Installed Electrical and Thermal Power of Geothermal Systems in 1993[a]

Country	Electrical (MW)	Thermal (MW)
United States	2,594	463
Phillipines	888	—
Mexico	752	8
Italy	637	360
New Zealand	285	258
Japan	270	3,321
Indonesia	144	—
El Salvador	105	—
Other	240	6,802
Total	5,915	11,204

[a]Data from Dickson, Mary H., and Mario Fanelli, Eds., 1995. *Geothermal Energy.* Chichester: John Wiley & Sons.

---

[9]This heat flux is tiny compared with the average solar energy flux to the earth's surface of about 500 W/m^2. It is the latter that determines the earth's surface temperature.

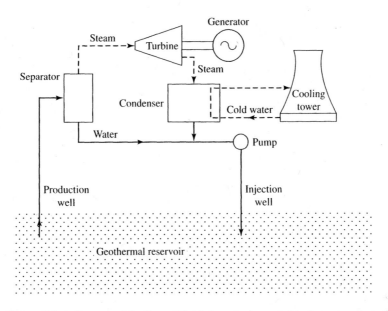

Figure 7.4 A diagram of the flow of fluids in a geothermal electric power plant.

of nearly 6000 MW, the United States accounts for 44%, most of which is located in northern California. In 1977, the average capacity factor for U.S. geothermally generated electricity was 15.6%.

Geothermal heat is recovered in many locations where the reservoir temperature is insufficient for economical power generation but where the low-temperature heat may be used for space heating or industrial processing. The heat recovery capacity of this type of geothermal energy supply is listed in the last column of Table 7.4. Assuming an average thermal efficiency of 20% for geothermal electric power generation, the figures in Table 7.4 suggest that nearly three-quarters of the world's geothermal heat recovery is used to generate electricity.

A typical geothermal electric power plant is sketched in Figure 7.4. A production well supplies pressurized warm water, a water and steam mixture, or only steam, depending upon the temperature and pressure conditions in the underground deposit. This fluid is fed to a separator, where steam is separated from liquid. The liquid water is pumped back into the ground via an injection well, while the steam is fed to a low-pressure turbine that generates electrical power. The turbine exhaust steam is condensed in a condenser, where cool water from a cooling tower removes heat from the exhaust stream. The condensed steam is reinjected into the ground. These systems typically supply steam at 160–180 °C and a pressure of 0.6–0.9 MPa and have thermal efficiencies of 20–25%. More elaborate systems than that shown in Figure 7.4, which improve the thermal efficiency, have been constructed, albeit at higher capital cost.[10]

A geothermal plant transfers heat from a hot underground reservoir to the earth's surface, where it can be profitably utilized to produce electrical power or for other heating purposes. If it is to be a source of renewable energy, this heat must not be removed at a faster rate than it can be replenished by conduction from deeper in the earth. In the regions where geothermal resources have

---

[10]See Dickson, Mary H., and Mario Fanelli, Eds., 1995. *Geothermal Energy.* Chichester: John Wiley & Sons.

been developed, this heat flux is of the order of 0.3 W/m^2 = 3E(5) W/km^2. A 100-MW power plant would require a geothermal field of 1700 km^2 to ensure a steady source of hot fluid. Most current geothermal plants remove heat at more than 10 times the replenishable rate so as to reduce the cost of collecting and reinjecting the hot fluid. Consequently, the underground reservoirs for these plants is slowly being cooled down, but this cooling is not significant over the useful economic life of the plants.

The problem of developing an economical geothermal resource is similar to that for developing a petroleum or natural gas resource. A reservoir of underground water of sufficiently high temperature must be located at depths that are accessible by drilling wells. It must be sufficiently large to supply the required heat for a period of 40 years or more. The resistance to underground fluid flow to and from the supply and injection wells must be low enough to produce the volume flow of hot fluid needed for the plant without requiring excessive pumping power. For these reasons it is never certain that a promising resource can be developed until sufficient exploratory drilling and testing has been conducted, which adds to the costs of geothermal plants.

An alternative geothermal system, known as "hot dry rock," would utilize impermeable rock formations lacking underground water, a more common geological condition that would greatly increase the potential availability of geothermal energy in proximity to prospective users. The plan is to fracture the formation surrounding the producing and injection wells by hydraulic pressurization or explosives. This would provide sufficient permeability to enable the pumping of water between the wells, heating it by contact with the hot rock. So far, exploratory attempts to develop these systems have failed, for various reasons related to the establishment of a satisfactory production potential. It appears that the cost of creating a permeable, water-filled underground resource equivalent to the natural ones that have already been or can be exploited is too great, using current technology and geological development practices. The economics of these systems is presently unfavorable to their development.

One recent development that utilizes the heat storage capacity of an underground aquifer is called the geothermal heat pump. In this system, water from a well supplies heat to a heat pump that then releases a greater amount of heat to a building space during cold weather. Because the well water is warmer than the ambient winter air, the heat pump produces much more heat per unit of electrical work required to run it than would be the case if the heat pump used cold winter ambient air as its heat source.[11] In climates where air conditioning is needed in summertime, this same system can be reversed to deposit the heat rejected by the air conditioner to the well water, which is more saving of power than rejecting it to the hot summer ambient air. This system is practical only for central air conditioning and heat pump systems and then involves the additional expense of well drilling. Depending upon the price of electricity, the savings in power may repay the additional investment.

7.4.1   Environmental Effects

Geothermal fluids contain dissolved solids and gases that must be disposed of safely. The solids are returned to the underground reservoir in the reinjected fluid. (Without reinjection, the discharge of geothermal water to surface waters would adversely affect water quality and possibly produce subsidence in the geothermal field.) Of the dissolved gases, some of which must be removed from

---

[11]Near-surface underground temperature in temperate climates is about 10 °C.

the condenser by steam ejectors (an energy loss), hydrogen sulfide, sulfur dioxide, and radon, all of which are toxic to humans, must be safely vented.

7.5 SOLAR ENERGY

A prodigious source of radiant energy, the sun emits electromagnetic radiation whose energy flux per unit area, called *irradiance,* decreases as the square of the distance from the sun center. At the mean sun–earth distance of 1.495E(8) km, the solar irradiance is 1367 W/m^2.[12] The sun's light is nearly, but not exactly, parallel; the sun's disc, viewed from the earth's mean orbit, subtends a plane angle of 9.3E($-$3) radian = 31.0 minutes and a solid angle of 5.40E($-$6) steradian. It is this energy stream from the sun that maintains the earth at a livable temperature, far above the cosmic background temperature of 2.7 K that would exist if the sun's irradiance were zero.

Not all of this solar radiation reaches the earth's surface. Some is reflected from the earth's atmospheric gases and clouds; some is absorbed by air molecules (principally oxygen, carbon dioxide, water vapor, and ozone), clouds, and atmospheric dust; some is scattered by air molecules and dust. As a consequence, only a fraction of the solar irradiance impinging on the earth's atmosphere (called the *extraterrestrial irradiance*) actually reaches ground level. Of that fraction, some retains its solar direction, called *beam irradiance,* while the remainder, called *diffuse irradiance,* has been scattered through large angles, approaching the ground from directions quite different than that of the sun.[13] Of the sunlight reaching the ground, some is absorbed and the rest reflected upward, with the latter undergoing scattering and absorption in the atmosphere on its way out to extraterrestrial space.[14]

The solar radiation approaching the earth comprises a wide band of wavelengths, but most (94%) of the radiant energy lies between 0.3 and 2 μm in wavelength. The distribution of the extraterrestrial solar irradiance among its various wave lengths is shown as a dashed line in Figure 7.5. The main components of the solar spectrum are the ultraviolet ($<$ 0.38 μm), the visible (0.38 μm to 0.78 μm), and the infrared ($>$ 0.78 μm). The energy content of these portions are 6%, 48%, and 46% of the total irradiance, respectively. The energy of an individual photon, the quantum packet of electromagnetic radiation, is proportional to its frequency and hence inversely proportional to its wavelength. Ultraviolet photons, having shorter wavelengths and hence higher individual energies than visible or infrared photons, are potentially more damaging to living organisms, even though their aggregate share of solar energy is much smaller.

---

[12]Because the earth's orbit around the sun is elliptical rather than circular, the solar irradiance varies by $\pm$3.3% = $\pm$45.1 W/m^2 during the year, being a maximum on December 26 and a minimum on July 1, when the earth is closest and furthest from the sun, respectively.

[13]Atmospheric molecules scatter blue light more than red light, making a clear sky uniformly blue and sunsets red.

[14]A small portion of absorbed incoming sunlight causes photochemical reactions in the atmosphere, mainly ozone formation in the stratosphere and smog formation in the lower troposphere. Also, some of the ground-level incident radiation absorbed by plants results in photosynthesis. Nevertheless, the solar energy invested in chemical change in the atmosphere and plants is tiny compared to the heating of the atmosphere, hydrosphere, and geosphere by absorbed sunlight.

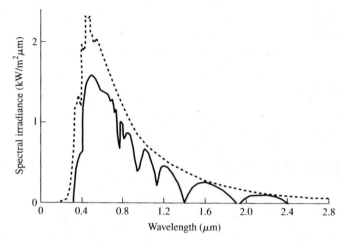

Figure 7.5 The spectral distribution of the solar beam irradiance as a function of wavelength. The dashed line is extraterrestrial irradiance, and the solid line is clear-sky ground-level irradiance.

The portion of the incoming beam irradiance that is absorbed or scattered by the atmosphere depends upon the wavelength of the incoming light. Scattering predominates for short wavelengths (blue), and molecular absorption predominates for long wavelengths (red). The net beam irradiance at ground level—that is, what remains after scattering and absorption—is shown as a solid line in Figure 7.5. The sharp dips in spectral irradiance in the red and infrared regions are caused by molecular absorption by carbon dioxide, water vapor, and ozone (which are greenhouse gases), while the ultraviolet absorption is caused by diatomic oxygen and ozone. Under the best of conditions, for a clear sky and the sun overhead, about 80% of the extraterrestrial beam irradiance reaches the earth's surface.

The distinction between beam and diffuse radiation is important to the design and functioning of solar energy collection systems. If the collector is a flat surface exposed to the sky, it will collect both beam and diffuse radiation. But if an optical focusing system is used to intensify the solar radiation impinging on the collector, only beam radiation is collected, which is a lesser amount than the total of beam and diffuse. The proportions of beam and diffuse radiation depend upon the amount of cloud cover, with a clear sky providing mostly beam radiation and a heavy cloud cover resulting in entirely diffuse radiation.[15] Summed over daytime periods, this relationship is exhibited in Figure 7.6. The abscissa is the ratio of daily total solar radiation on a horizontal flat plate to the extraterrestrial value and is called the *clearness index,* while the ordinate is the ratio of the daily total diffuse radiation to daily total radiation. As the sky becomes less clear, Figure 7.6 shows that both the beam and diffuse radiation decrease, but the former decreases proportionately faster. Focusing systems collect little energy on cloudy days.

The amount of solar energy falling upon a solar collector depends upon its orientation with respect to the sun's direction, as viewed from the earth's surface. The solar direction varies with the local hour of the day and the day of the year. Fixed collectors remain in the same position year-round. Common types of fixed collectors are horizontal (e.g., solar pond), tilted at an angle

---

[15]If one can't see the sun through the clouds and no shadows form, the light is all diffuse.

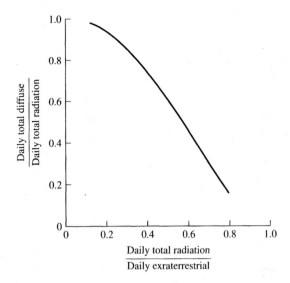

Figure 7.6 The ratio of daily diffuse to total radiation as a function of the ratio of daily total to extraterrestrial radiation, showing that cloudiness increases the diffuse portion of total radiation.

above the horizontal, and vertical (windows). On the other hand, focusing collectors may move in the vertical and azimuthal directions to direct the collector aperture to be normal to the solar direction. Each type of collector receives a different amount of solar energy.

Table 7.5 lists the clear-day solar irradiance at 40°N latitude[16] incident upon four common types of collectors: a horizontal surface (first column); a south-facing flat plate tilted 40° above the horizontal (second column); a vertical south-facing surface (third column); and a focusing collector that is oriented both vertically and horizontally to collect direct (or beam) irradiance (fourth column). The daily total of energy incident upon these collectors is listed in the upper part of the table for four days of the year: the spring and fall equinoxes (21 March and 21 September) and the summer and winter solstices (21 June and 21 December). On the equinoxial days the sun's position in the sky is essentially identical, while the solstices mark the annual extremes in solar irradiance. The lower half of the table lists the maximum hourly solar irradiance, which for the collectors selected occurs at local noon. The annual average value of these items is also listed in the table.

Several important conclusions about clear-sky solar irradiance may be drawn from the entries in Table 7.5. For the horizontal collector, the summer daily total is more than three times the winter value.[17] By tilting the collector to 40°, the difference between summer and winter is much reduced, to a ratio of 1.5:1, so that the daily total is more nearly uniform year-round. For a vertical

[16]This is the median latitude for the lower 48 U.S. states, which generally lie between 30° and 50°N latitude. A lower latitude will experience a greater solar irradiance and a lesser variability between summer and winter, and vice versa for a higher latitude.

[17]The daily irradiance on a horizontal surface maintains the surface atmospheric temperature level. The difference between summer and winter irradiance accounts in part for the atmospheric temperature difference between these seasons.

TABLE 7.5 Clear-Sky Irradiance at 40 °N Latitude

Date	Horizontal (MJ/m^2)	40° Tilt (MJ/m^2)	Vertical (MJ/m^2)	Direct Normal (MJ/m^2)
	Daily Total			
21 March	21.02	26.44	16.84	33.09
21 June	30.05	25.24	6.92	36.08
21 September	20.29	25.28	16.08	30.73
21 December	8.87	18.54	18.68	22.44
Annual Average	20.06	23.88	14.63	30.58
	Maximum Hourly			
21 March	810	1027	656	968
21 June	958	911	309	879
21 September	785	987	630	914
21 December	451	867	829	898
Annual Average	751	948	606	915

collector, such as a window, the winter daily total irradiance is much greater than that in summer, making a south-facing window a good source of winter space heat and a small source of summer air-conditioning load. The focusing collector's daily irradiance is about 30% higher than the tilted flat plate, which may be hardly enough in itself to justify the complication of moving the collector to track the sun. On an annual average basis, there is only a factor of two difference between the greatest and least values of the daily totals for the various collectors of Table 7.5. Similar distinctions apply to the maximum noontime hourly irradiance, listed in the lower half of Table 7.5. Under most circumstances, these values lie within the range 800–1000 W/m^2, or 60–75% of the extraterrestrial irradiance of 1367 W/m^2.

The values of Table 7.5 apply only to clear-sky irradiance. The effect of clouds and atmospheric dust will reduce the available irradiance to lesser values. Figure 7.6 illustrates how cloudiness, in reducing the irradiance on a horizontal surface, has a greater effect on beam irradiance than does diffuse, or scattered, light. The average value of the clearness index, the ratio of daily total irradiance to the extraterrestrial value, varies with geographic location and season of the year. As a consequence, seasonal or annual average irradiances can be as low as half of the values shown in Table 7.5.

As will be seen below, only part of the incident solar energy flux can be collected by a well-designed collector system. While a collector can absorb a high fraction of the solar irradiance, it also loses heat to the ambient atmosphere, the more so as it attempts to store the collected energy at a temperature higher than atmospheric. Practical solar collectors may average between 5 and 10 MJ/m^2 of collected energy per day at typical U.S. locations, or 2 to 4 GJ/m^2 per year. If this energy were to be valued monetarily as equal to that for the heating value of fossil fuel, which currently is in the range of 3–5 \$/GJ, then the dollar value of the collected energy would be 6–20 \$/m^{2}y. On the other hand, if it were valued as equal to the retail price of electric energy, which is about 30 \$/GJ, then the annual value of the collected solar energy would be 60–120 \$/GJ. It is for this reason that solar collectors are economically advantageous if they replace electricity for the heating of domestic hot water.

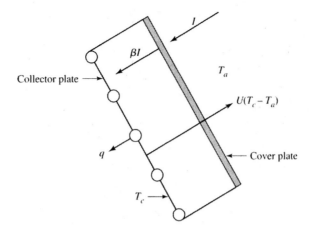

Figure 7.7 A flat plate collector absorbs incident sunlight on a thermally conducting collector plate that is cooled by a flow of liquid delivering the absorbed heat to a storage tank. A glass cover plate reduces the heat loss to the ambient environment.

7.5.1 The Flat Plate Collector

The most common and economical form of solar thermal energy collector is the flat plate collector, illustrated in Figure 7.7. It consists of a thermally conducting collector plate equipped with passages through which a heat transfer fluid passes, transferring heat from the collector plate to a fluid storage tank. The collector plate surface facing the incoming sunlight is treated to absorb as much of that sunlight as possible. A transparent cover (usually glass) is placed parallel to the collector plate, forming an enclosed space that reduces the heat loss to the surrounding atmosphere, whose temperature T_a is less than that of the collector plate, T_c.

Of the incoming solar irradiance I that falls on the cover plate, some is reflected and absorbed by the cover plate, and only a portion of what is transmitted through the cover plate is absorbed by the collector plate. Overall, if the fraction of the incoming solar irradiance I that is absorbed by the collector plate is β, then the solar heat flux to the plate is βI.[18] The warm collector plate will lose heat to the surrounding atmosphere by heat conduction, convection and radiation at a rate that is proportional to the temperature difference $T_c - T_a$ between the plate and the ambient atmosphere. The proportionality constant $\mathcal{U}$, called the overall heat transfer coefficient, depends in part upon the heat transfer properties of air and the radiant heat transfer properties of the collector and cover plates. As a consequence, the net unit heat flux q that is collected in the storage system is the difference between the absorbed irradiance and the loss to the atmosphere:

$$q = \beta I - \mathcal{U}(T_c - T_a) \tag{7.2}$$

Note that the higher the temperature of collection, the smaller the amount of the heat collected,

---

[18]The value of β depends upon the angle of incidence of the incoming solar radiation, the refractive and transmissive properties of the glass cover plate (see Figure 7.7), and the absorptive properties of the collector surface for solar wavelengths. For the best designs, about 80% of the incident solar radiation is absorbed by the collector plate.

there being a maximum collector temperature $(T_c)_{max}$ at which no heat is collected:

$$(T_c)_{max} = T_a + \frac{\beta I}{\mathcal{U}} \tag{7.3}$$

The fraction of the incident irradiance that is collected, q/I, is called the collector efficiency η:

$$\eta \equiv \frac{q}{I} = \beta - \frac{\mathcal{U}(T_c - T_a)}{I} \tag{7.4}$$

The collection efficiency η is a linearly decreasing function of the ratio $\mathcal{U}(T_c - T_a)/I$, and the heat collection rate q decreases with increasing collector temperature. The maximum collector efficiency is β when heat is collected at ambient temperature. The efficiency falls to zero when the collector temperature reaches $(T_c)_{max}$. Either of these limits of operation has no practical value because either no heat is collected or it is collected at ambient temperature, which has no useful function. Practical collector systems will function at intermediate conditions where $\eta < \beta$ and $T_c < (T_c)_{max}$.

Flat plate collectors operate at lower efficiency in winter than in summer. When we use typical collector values of $\beta = 0.8$ and $\mathcal{U} = 5$ W/m^{2}K, winter and summer clear-day noontime values of $I = 870$ W/m^2, $T_a = 70\,°$F (21 $°$C), $I = 910$ W/m^2, and $T_a = 32\,°$F (0 $°$C), and a collector temperature of $T_c = 140\,°$F (60 $°$C) suitable for domestic water heating, the corresponding winter and summer collector efficiencies are $\eta_w = 45.5\%$ and $\eta_s = 58.6\%$. At other hours, where I is less than these noontime values, the collector efficiencies will be less. This will also be the case when the sky is cloudy. Year-round collector efficiencies are likely to be in the range of 30–50%.

Because of the vagaries of solar irradiance from day to day, a solar collector, no matter how big, can never completely satisfy the demand for year-round heat for domestic hot water or space heating, and a backup supply must be available for satisfactory operation. The relationship between annual heat collection and collector area is sketched in Figure 7.8(a). A very small collector accumulates only a small amount of heat, whereas an oversized one collects enough heat to meet the daily demand on all but a few very cold, cloudy days, when auxiliary heat must be supplied. The capital cost of a collector system is plotted in Figure 7.8(b), showing how the cost increases with collector area, but is finite for small areas because the storage, piping, and control system needed constitute

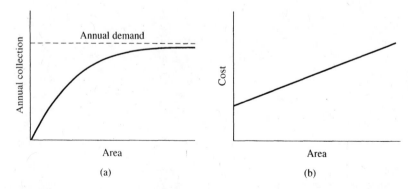

Figure 7.8 The characteristics of solar heat collection systems as a function of collector area: (a) annual heat collection and (b) capital cost.

an irreducible cost. An optimum design would minimize the ratio of cost to heat collection and would have an area intermediate between a small and large value because these extremes have either small heat collection or large cost, making the ratio larger than the optimum.

7.5.2 Focusing Collectors

The purpose of employing focusing solar collector systems is to increase the intensity of the solar radiation falling on the collector, thereby making it possible to collect solar energy at a higher temperature and with a smaller collector area than for a simple flat plate system. The factor by which the solar irradiance is increased is called the concentration ratio, CR.

The principle of light concentration is sketched in Figure 7.9. A curved parabolic mirror whose axis points in the sun's direction will form an image of the sun at its focal point, a distance F from the mirror called the focal length. The image dimension D_i is equal to the product of the focal length L and the small plane angle $\alpha = 9.3E(-3)$ radian that the sun subtends when viewed from the earth. The ratio of mirror dimension D_m to image dimension D_i is thus found to be

$$\frac{D_m}{D_i} = \frac{D_m}{\alpha F} = 107.5 \frac{D_m}{F} \qquad (7.5)$$

For an ordinary mirror of circular shape that forms an image as does a camera (called spherical), the concentration ratio is equal to the ratio of the area of the mirror to that of the image, or $(D_m/D_i)^2$. For a cylindrical mirror, which focuses light only in one dimension, the concentration ratio is (D_m/D_i). Thus

$$CR = 1.156E(4) \left(\frac{D_m}{F}\right)^2 \qquad (spherical)$$

$$= 1.075E(2) \left(\frac{D_m}{F}\right) \qquad (cylindrical) \qquad (7.6)$$

In camera lenses, the ratio of focal length to lens diameter is called the *f number*. The smaller its value, the greater is the lens' light gathering power, or concentration ratio, permitting film exposure in dim lighting conditions. For focusing solar collectors, it is not practical to construct mirrors with F/D_m less than about 2.

The collectors for focusing systems, which operate at higher temperatures, are of different design than those for flat plate systems. Nevertheless, their collection efficiency and maximum

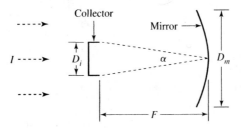

Figure 7.9 A focusing mirror (D_m) concentrates the solar irradiance I (dashed lines) on a smaller collector (D_i) located at the focal point of the mirror.

temperature obey the same relationships of that of equations (7.4) and (7.3), but with the collector irradiance increased by the factor (CR):

$$\eta \equiv \frac{q}{I} = \beta - \frac{\mathcal{U}(T_c - T_a)}{I(CR)} \qquad (7.7)$$

$$(T_c)_{max} = T_a + \frac{\beta I(CR)}{\mathcal{U}} \qquad (7.8)$$

in which CR is to be taken from equation (7.6), and the absorbed light fraction β and heat transfer coefficient $\mathcal{U}$ are those for the focusing system collector. It should also be noted that I is the solar beam irradiance and does not include the diffuse light, which cannot be focused by the mirror. But given the large values of the concentration ratio CR that are available to focusing systems, they can operate at high efficiency even with high collector temperatures.

Focusing systems collect solar energy at a sufficiently high temperature to use that energy in a heat engine cycle to generate electric power efficiently. Figure 7.10 shows a spherical focusing system consisting of individual mirrors attached to a movable frame whose optical axis tracks the sun's position in the sky. The mirrors focus on a Sterling cycle heat engine that generates 25 kW of electrical power, with about 24% of the solar beam irradiance falling on the mirrors. An alternative spherical system is shown in Figure 7.11 where individual mirrors arrayed about a central focusing axis each track the sun's motion so as to focus the sun's image on a collector at the top of a central tower. In this system, two megawatts of electrical power are generated in a Rankine cycle engine that is heated by hot fluid circulating through the collector at the tower's top. A cylindrical collector system can be seen in Figure 7.12, where the collector is a pipe through which fluid circulates. Nine California power plants using cylindrical collectors generate a total of 354 MW of electrical power.

Figure 7.10 A parabolic mirror system focuses sunlight on a Sterling cycle heat engine that produces electrical power. (By permission of DOE/NREL-PIX.)

Figure 7.11 A spherical mirror system consisting of individual focusing mirrors arrayed around a central tower that collects the focused light. (By permission of DOE/NREL-PIX.)

Figure 7.12 A cylindrical collector mirror focuses sunlight on a glass pipe containing fluid. (By permission of DOE/NREL-PIX.)

7.5.3 Photovoltaic Cells

Photovoltaic cells are solid-state devices that generate electric power when irradiated by solar light. They provide an alternative method of electric power production to that of solar thermal systems generating electric power by a heat engine supplied with heat from a solar collector. Like solar thermal systems, only a fraction of the solar irradiance incident upon a solar cell can be converted to electric power. Nevertheless, their mechanical and electrical simplicity, negligible operating cost, and ability to produce power in any quantity makes the use of photovoltaic power systems very attractive. A major obstacle to widespread use of photovoltaic cells is their current high cost per unit of power output.

The process whereby the energy flux in solar radiation is converted to electrical power is quantum mechanical in origin. The energy of sunlight is incorporated in packets of electromagnetic radiation called photons, each of which possesses an energy of amount hc/λ, where h is Planck's constant, c is the speed of light, and λ is the wavelength of light. Sunlight contains photons of all wavelengths and energies, but most of the solar energy content lies in the wavelength range of 0.3–2.5 μm (see Figure 7.5). If a photon can deliver all its energy to an electron in a semiconductor, for example, it can move the electron to a higher energy state of electric potential magnitude equal to $hc/e\lambda$, where e is the electron charge. For solar photons, the corresponding electric potential range is 0.5–2.5 V. If the energetic electron can flow through an electric circuit while experiencing this potential change, it can deliver electrical energy to an external load.

The components of a photovoltaic cell are sketched in Figure 7.13. The cell consists of a *base* layer of a *p*-type semiconductor, about 250 μm in thickness, joined to an extremely thin *emitter* layer of an *n*-type semiconductor about 0.5 μm thick that is exposed to the solar irradiance. Solar radiation is absorbed mostly in the thin region close to the junction of the two materials. Current generated by the cell is collected at positive and negative electrodes attached to the exterior of

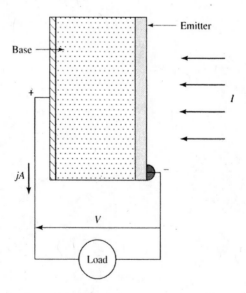

Figure 7.13 The elements of a photovoltaic cell of area A exposed to a solar irradiance I and supplying a total current jA to an external load at a potential difference V. The current density in the cell is j A/m^2.

the p-type and n-type semiconductor layers, respectively. The external circuit that receives the electrical power generated by the cell is connected to these electrodes.

The n-type and p-type semiconductor materials consist of a pure semiconductor, such as silicon, doped with a small amount of another element. In the n-type layer, silicon is doped with P or As, elements that readily give up an electron from the outer shell and are thereby called *donors*. In the p-type layer, silicon is doped with B or Ga, which accept an electron into their outer shell and are called *receptors*. The mobile electrons in the n-type layer and the mobile "holes" in the p-type layer provide charge carriers that permit an electric current to flow through the cell. At the junction of the two layers, there exists an electric potential difference between them required to maintain thermodynamic equilibrium between the charge carriers in either layer.

When sunlight falls upon the cell, some photons penetrate to the region of the interface and can create there an electron–hole pair, provided that the photon energy equals or exceeds the gap energy E_g needed to move an electron from the valence band to the conduction band; that is, provided that the wavelength λ is less than hc/E_g. The electron and hole then move to the negative and positive electrodes respectively and provide a current that moves through the external circuit from the positive to the negative electrodes with an accompanying electric potential drop, both sustained by the flow of photons into the cell.

In these processes only a fraction of the solar energy flux is utilized to create electrical power to feed into the external circuit. Long-wavelength photons ($\lambda > hc/E_g$) have insufficient energy to create an electron–hole pair, so their absorption merely heats the cell. Short-wavelength photons ($\lambda < hc/E_g$) have more energy than needed, with the excess ($hc/\lambda - E_g$) appearing as heating, not electrical power. Typically, only about half of the solar irradiance is available to produce electrical power. Of this amount, only a fraction eventually results in electric power flowing to the external circuit because of various additional internal losses in the cell.

When exposed to sunlight, a photovoltaic cell generates electric current and a cell potential difference, depending upon the solar irradiance level and the electrical characteristics of the load in the external circuit. In the limit where the external circuit is not closed, no current can flow but an open-circuit voltage V_{oc} is generated that increases with solar irradiance, but not proportionately so, as sketched in Figure 7.14(a). At the opposite limit of an external short circuit, where the cell voltage is zero, a short-circuit current (current density j_{ss} A/m^2) flows through the cell in an amount proportional to the solar irradiance [see Figure 7.14(a)]. In both these extremes, no electrical power is generated because either the current or voltage is zero. For intermediate cases where the cell delivers electrical power to a load, the cell voltage and current density, V and j, are each less than the limiting values of V_{oc} and j_{ss}. For a given value of the irradiance, the relationship between V and j, sketched in Figure 7.14(b), depends upon the electrical characteristics of the load. The power output is the product Vj, which reaches a maximum $(Vj)_{max}$ at a point intermediate between the open- and short-circuit conditions. The maximum power per unit area that the cell can deliver, $(Vj)_{max}$, is generally about 60–80% of the product $V_{oc}j_{ss}$.

The efficiency η of a photovoltaic cell is the ratio of the electrical power generated per unit area (Vj) to the solar irradiance (I) impinging on the cell surface:

$$\eta \equiv \frac{Vj}{I} \qquad (7.9)$$

The efficiency increases with increasing irradiance and, for any level of irradiance, is a maximum at the maximum of output power. Photovoltaic cell efficiencies are rated for a solar irradiance of

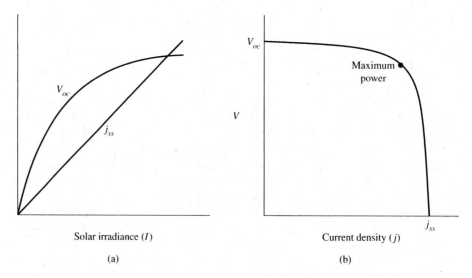

Figure 7.14 (a) The open-circuit voltage V_{oc} and short-circuit current density j_{ss} of a photovoltaic cell as functions of solar irradiance. (b) The voltage versus current density characteristics of a photovoltaic cell for a fixed value of solar irradiance.

1 kW/m^2 under the conditions for maximum power. Typical efficiencies are in the range of 15–25 %, with $V_{oc} \sim 0.5$ V and $j_{ss} \sim 300$–400 A/m^2.

The photovoltaic cell is but one component of a practical system for converting sunlight directly to useful electric power. Many individual cells must be wired together to produce output power at desirable voltage and current levels, and the array of cells must be mounted in a manner that optimizes the solar irradiance input. Photovoltaic power is direct-current (DC) power and must be converted to alternating-current (AC) form for most uses. These other constituents of a practical photovoltaic power supply add to its capital cost, although at the present time the basic cell cost is so high that it constitutes by far the major component of system cost.

7.6 WIND POWER

The use of wind to provide mechanical power for grinding grain and propelling ships and boats dates from ancient times, three or four millennia ago. Substantial improvement in wind technology within the past thousand years, especially for transoceanic vessels, made possible the migration of populations to western and southern continents and the initiation of intercontinental trade. The technology of sailing vessels had reached a high level in the mid- to late nineteenth century when it was suddenly displaced by steel-hulled, fossil-fuel-powered steamships that greatly transformed intercontinental travel and trade.

At one time, wind-powered mills for grinding grain and pumping water were common in western Europe, numbering some 10,000 by the twelfth century. By present standards their design was rudimentary and their use, like that of sailing vessels, was displaced by the advent of industrial power in the nineteenth century.

The modern wind turbine is a development of the last few decades that utilizes the latest technology in design and manufacture. Currently used exclusively to produce electrical power, wind turbines are usually tied into electrical transmission and distribution system, although some turbines are used to power remote installations. Despite the many ingenious forms of wind turbines that have been developed, the predominant type used today is the horizontal axis machine, mounted on a support tower, that is free to rotate so as to align the axis with the wind direction. The turbine rotor, invariably a three-bladed propeller-like structure, drives an electric power generator through a speed increasing gear. Because the turbine–generator must be able to rotate with respect to its supporting tower, electrical power is transmitted through slip rings to the electrical transmission system. Because the turbomachinery is mounted at the top of a tower 50 or more meters in height, with the wind turbine rotating and the whole power unit swiveling into the wind, a reliable and strong supporting structure that will withstand the highest expected storm winds is required.

A wind turbine more closely resembles an airplane propeller than it does a steam or gas turbine rotor. The wind turbine blades are long and slender; the tip of the blade moves at a speed much greater than the wind speed. An airplane propeller is designed to produce a large thrust while the wind turbine must produce power, which is the product of the torque that the wind applies to the turbine rotor times its rotational speed. Nevertheless, the wind turbine blade shape is quite similar to that of an airplane propeller.

The source of power from the wind is the flow of air through the wind turbine. If V is the wind speed, each unit mass of air possesses a kinetic energy of amount $V^2/2$. If $A = \pi R^2$ is the area subtended by the rotating blades of length R, then the mass flow rate of air through an area A of the undisturbed wind stream is $\rho V A$, where ρ is the air density. The maximum rate at which the wind kinetic energy could be supplied by the wind flow passing through the turbine is the product of the mass flow rate, $\rho V A$, and the kinetic energy per unit mass, $V^2/2$, for a value of $\rho V^3 A/2$. In practice, the wind turbine power available from aerodynamically perfect design is less than this value because the action of the turbine modifies the surrounding wind flow, reducing the mass flow rate below the value of $\rho V A$.

To illustrate this effect, Figure 7.15 shows the ideal wind flow past a turbine. Because the turbine extracts some of the kinetic energy of the wind flow, the wind speed is reduced in the

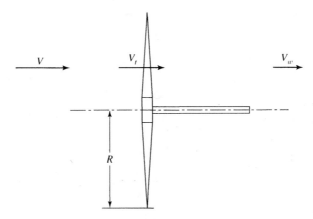

Figure 7.15 The flow of wind through a wind turbine whose blade radius is R is slowed at the turbine disk and is slowed even further in the wake region down stream of the turbine.

vicinity of the turbine to a value V_t at the turbine and an even lower value of V_w downstream of the turbine, called the wake region. The turbine power $\mathcal{P}$ is then the product of the mass flow rate through the turbine, $\rho V_t A$, and the reduction of kinetic energy of the wind, $V^2/2 - V_w^2/2$, or

$$\mathcal{P} = \frac{1}{2}\rho V_t(V^2 - V_w^2)A \tag{7.10}$$

As the air slows down both upstream and downstream of the wind turbine, it undergoes a pressure rise that is very small compared to atmospheric pressure. This provides a pressure drop across the wind turbine of amount $\rho(V^2 - V_w^2)/2$ and a corresponding axial thrust force of $\rho(V^2 - V_w^2)A/2$. But this force must also equal the reduction in momentum of the wind flow, $\rho V_t A(V - V_w)$. It then follows that V_t is the average of V and V_w:

$$V_t = \frac{1}{2}(V + V_w) \tag{7.11}$$

The power $\mathcal{P}$ can now be expressed in terms of V and V_w as

$$\mathcal{P} = \frac{1}{2}\rho V^3 A\left[\frac{(1 + V_w/V)^2(1 - V_w/V)}{2}\right] \tag{7.12}$$

The factor in brackets has a maximum value of 16/27 when $V_w = V/3$, in which case 8/9 of the wind's kinetic energy has been removed by the wind turbine. As a consequence, the wind turbine power cannot exceed the limit:

$$\mathcal{P} \le \frac{16}{27}\left[\frac{1}{2}\rho V^3 A\right] \tag{7.13}$$

The foregoing considerations do not reveal the detailed mechanism whereby the wind flow exerts a torque on the wind turbine rotor in the direction of its rotation, thereby generating mechanical power. Figure 7.16 depicts the amount and direction of the wind flow relative to a section of the turbine blade at a radius r from the turbine axis. In the tangential direction, the velocity

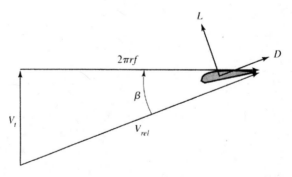

Figure 7.16 The motion of the wind relative to a turbine blade consists of an axial speed V_t and a tangential speed $2\pi rf$ that generates a lift force L and drag force D.

component is $2\pi rf$, where f is the rotational frequency of the turbine shaft.[19] In the direction of the turbine axis the wind speed is V_t. The net relative speed V_{rel} is thereby $\sqrt{V_t^2 + (2\pi rf)^2}$, and this velocity lies at an angle $\beta = \arctan(V_t/2\pi rf)$ to the tangential direction, as shown in Figure 7.16. The airfoil-shaped section of the turbine blade, pitched at a small angle of attack to this direction, develops a lift force L and drag force D that are, respectively, perpendicular and parallel to the direction of the relative velocity. The net force exerted on the blade section in the direction of its motion is $L \sin\beta - D\cos\beta$, and the rate at which this force does work on the turbine, which is the power developed by the wind, is thereby $2\pi fr(L\sin\beta - D\cos\beta)$.

There are two important deductions that can be made from this analysis. The first is that it is necessary that the airfoil lift L should greatly exceed its drag D, which can be secured by utilizing efficient airfoil shapes in forming the turbine blades. The second is that the turbine power is enhanced by turbine tangential speeds that are large compared to the wind speed, so as to come as close as possible to the power limit of Equation (7.13). Optimum designs incorporate a speed ratio (turbine tip speed/wind speed) of about five to seven.

The performance of a wind turbine may be described in terms of a dimensionless power coefficient C_p defined as

$$C_p \equiv \frac{\mathcal{P}}{\frac{1}{2}\rho V^3 A} \tag{7.14}$$

where A is the turbine rotor disk area. The power coefficient cannot exceed the ideal value of 16/27, but is otherwise a function of the speed ratio, as sketched in Figure 7.17, reaching a maximum at the design speed ratio. At lower speed ratios the airfoil angle of attack is too great and the aerodynamic drag D increases greatly, reducing turbine power. At higher speed ratios, the airfoil angle of attack decreases, reducing the lift L and also the turbine power.

An economical wind turbine design must take into account several factors. The power available from the turbine and that delivered by the electric generator it drives is matched at the design wind

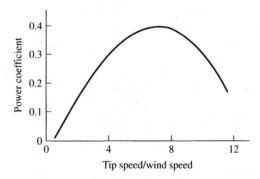

Figure 7.17 The power coefficient of a wind turbine is a function of the ratio of the tip speed to wind speed. Its maximum value is less than the theoretical value of $16/27 = 0.593$.

---

[19]Strictly speaking, the relative speed in the tangential direction is somewhat less than $2\pi rf$ because the blade tangential force induces a small amount of tangential flow. This is generally negligible for efficient wind turbine designs.

UNITED STATES ANNUAL AVERAGE WIND POWER

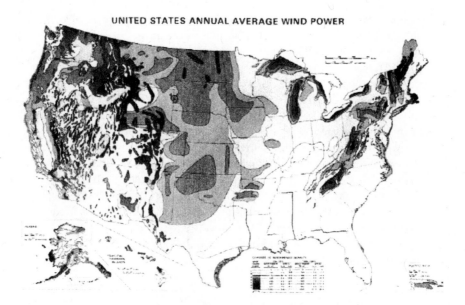

Figure 7.18 A U.S. map showing areas graded by the annual average wind energy flux $\rho V^3/2$, with darker areas indicating higher wind energy flux. (Data from Elliot, D. L., C. G. Holladay, W. R. Borchet, H. P. Foote, and W. F. Sandusky, 1987. *Wind Energy Resource Atlas of the United States.* DOE/CH 10093-4. Golden: Solar Energy Research Institute.)

speed V_d. When the wind speed is higher than V_d, the turbine can deliver much more power (since $\mathcal{P}$ varies approximately as V^3) but the electric generator cannot absorb this additional power without overheating and burning out, so the rotor pitch angle is adjusted to deliver only the design power output to the generator. On the other hand, at wind speeds less than V_d, much less power can be be generated, by a factor of about (V/V_d^3), and the electrical output is reduced below the design value. The economical optimum design is one that maximizes the ratio of the average electrical power output to the capital cost of the wind turbine installation, given the wind speeds available at the site. For efficient designs used at desirable sites, the ratio of time-averaged power output to the rated output, called the *capacity factor*, is about 25–30%.

A desirable site for a wind power installation is one having a high average wind speed. Sites are commonly rated by the average value of the energy flux per unit area, $\rho V^3/2$.[20] High values of wind energy flux require smaller, less costly turbine rotors for a given electrical power output. Figure 7.18 identifies U.S. regions with high wind energy levels. These are located mostly at higher elevations and in mountainous regions, as well as in coastal areas.

At any one site, wind speeds are quite variable in magnitude and duration. Daytime wind speeds are higher than nighttime levels because of daytime solar heating of the atmosphere. Average winter season wind speeds exceed those of the summer. While the time-averaged annual wind speed $\overline{V}$ approximately measures the availability of wind energy, a more detailed knowledge of wind speed statistics is needed for an accurate assessment of a site's wind energy potential and the selection of

---

[20]Site energy fluxes are classified in one of five levels: class 1 (0–100 W/m²), class 2 (100–200 W/m²), ..., class 5 (more than 400 W/m²).

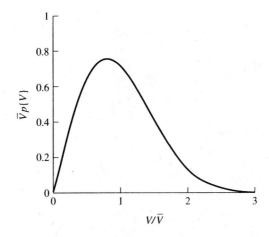

Figure 7.19 A probability distribution $p\{V\}$ of wind speed at a site for which the average speed is $\overline{V}$ shows low values at low and high speeds and a maximum near the average speed.

a suitable wind turbine design. A commonly used form of the probability distribution function of wind speeds is

$$p\{V\} = \frac{\pi}{2} \frac{V}{\overline{V}^2} \exp\left[-\frac{\pi}{4}\left(\frac{V}{\overline{V}}\right)^2 \right] \tag{7.15}$$

where $p\{V\}$ is the probability of a wind speed V, per unit of wind speed, at a site where $\overline{V}$ is the time-averaged wind speed.[21] In Figure 7.19, it can be seen that a very low or high wind speed, compared to the average, is very unlikely, and that the most probable wind speed is $0.8\,\overline{V}$. In addition, the average value of the cube of the wind speed, $\overline{V^3}$, to which the average wind power is proportional, can be calculated to be $\overline{V^3} = (6/\pi)\overline{V}^3 = 1.91\,\overline{V}^3$. Thus the average value of the wind energy flux $\rho\overline{V^3}/2$ is considerably more than its value at the average wind speed, $\rho(\overline{V})^3/2$.

At any site, wind speed and energy flux increase gradually with distance z above ground level. While it is theoretically advantageous to place a wind turbine on a high tower so as to take advantage of the higher energy flux available there, the greater cost of a higher tower may not be offset by a less costly turbine. Generally, tower height is proportional to turbine diameter, the turbine axis being about one to two diameters above ground.

Wind turbines need to be protected from damage by storm or hurricane level winds and, in northern climates, from icing conditions in winter months.

Wind turbines are customarily installed at adjacent sites called "wind farms," an example of which is shown in Figure 7.20. Currently, wind turbines are manufactured in the power range of a few hundred kilowatts to several megawatts, so that hundreds of them must be deployed to equal the output of a typical steam electric power plant. Maintenance of this many units at a remote or

---

[21]Equation (7.15) is called a Rayleigh distribution. It satisfies the integral conditions that $\int_0^\infty p\{V\}\,dV = 1$ and $\int_0^\infty V p\{V\}\,dV = \overline{V}$.

Figure 7.20 A row of wind turbines at a wind farm in Palm Springs, California (US). (By permission of DOE/NREL-PIX.)

not easily accessible site, in an unsheltered and sometimes hostile environment, presents practical difficulties that may lead lead to high operating costs.

Wind turbines often drive induction generators as the source of electrical power. Because induction generators operate at near synchronous speed when supplying 60-cycle power to an electric grid, the turbine speed is held constant over its load range, thereby operating at less-than-optimum efficiency at low and high speeds. A more efficient match between the wind turbine and electric generator is possible if the turbine and generator speed can be varied with the wind speed, but this requires conversion of the AC generator output to DC and then back to 60-cycle AC, which may be less economical despite the improved turbine efficiency.

7.6.1   Environmental Effects

There are several environmental drawbacks of wind energy systems. Wind turbines generate audible noise, somewhat akin to that of helicopters, but much less intense because of their much lower power levels. Nevertheless, wind turbines are unwelcome noisy neighbors in populated areas but have no adverse effect on livestock operations. They can, however, kill migrating birds that attempt to fly through the turbine. To some observers they provide visual blight, especially if located in otherwise undeveloped natural areas.

7.7   TIDAL POWER

The regular rise and fall of the ocean level at continental margins has been used in past centuries to produce mechanical power for grinding grain or sawing wood by damming water in coves or river mouths at high tidal levels and then releasing it to the sea through turbines or water wheels when the tide is at its lowest level. Intermittent tidal power proved practical only where the tidal range (the difference in elevation between high and low tides) exceeded a few meters. In recent years, tidal plants have been constructed at the La Rance estuary on the Brittany coast of France and at Annapolis Royal, Nova Scotia, in the Bay of Fundy, both of which generate electrical power.

The tidal rise and fall in the ocean is a consequence of the differential gravitational force exerted on opposite sides of the earth primarily by the moon, and secondarily by the sun, which gives rise to simultaneous higher ocean surface elevations on the earth's sides facing toward and away from the moon (or sun) and lower elevations halfway in between.[22] At a given point in the ocean, the period T of the lunar tide is one-half of the lunar day ($T = 12$ h, 24 min $= 4.464E(4)$ s); that of the sun is just 12 hours. The solar and lunar effects reinforce each other twice a month, at the time of a full moon and a new moon, giving rise to maximum, or *spring,* tides. Halfway between the spring tides, at the time of the first and last quarters of the moon's monthly cycle, the solar effect is least, resulting in the minimum, or *neap,* tides.[23]

The tidal range of the ocean at the continental margins is far from uniform worldwide. At a few locations, such as the Bay of Fundy and Cook Inlet in North America and the Bristol channel and Brittany coast in western Europe, resonance effects in the tidal motion lead to ranges as high as 10 m, which contrasts with an average oceanic range of 0.5 m. This large amplification of the oceanic tidal range is associated with a resonant motion of water movement across the shallow waters of the continental shelf into restricted bays or estuaries. About two dozen sites in North and South America, Europe, East and South Asia, and Australia have been identified as possessing this tidal amplification.

The principle of tidal power is illustrated in Figure 7.21, showing in elevation a cross section of a *tidal pool* separated from the ocean by a *dam* (or *barrage*). A *sluiceway* in the dam allows water to flow quickly between the sea and the tidal pool at appropriate times in the tidal cycle, as does also a turbine in a *power house*. If the tidal pool is filled to the high tide level and then isolated from the sea until the latter reaches its low tide level, a volume of water is impounded within the tidal pound of amount $\overline{A}H$, where $\overline{A}$ is the average surface area of the tidal pond and H is the tidal range. If the center of mass CM of this water is a distance h above the low tide level, work can be extracted from this volume at the time of low tide by allowing it to flow through a turbine until the

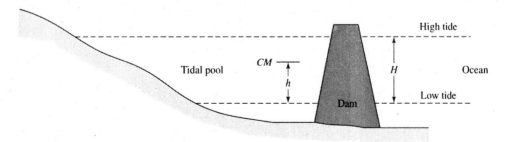

Figure 7.21 A cross section of a tidal power pool with dam.

---

[22]The solar effect is about 40% of that caused by the moon.

[23]There are additional perturbations of the tidal range associated with the position of the moon relative to the solar ecliptic plane and the variation of both the moon–earth and sun–earth distances.

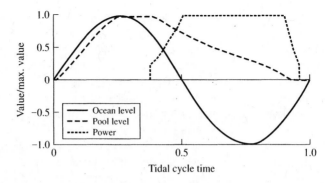

Figure 7.22 The pool and ocean levels, and power output, of a single effect tidal power plant during one tidal cycle. (Data from Fay, J., and M. Smachlo, 1983. *J. Energy,* **7**, 529.)

pool level reaches that of low tide. The maximum amount of this work is equal to the product of the mass of fluid $\rho \overline{A} H$, the acceleration of gravity g, and the average distance h by which it falls during outflow to the sea, for a total maximum energy of $\rho g \overline{A} H h$. If the emptied pool is then again closed off from the sea until high tide occurs, an additional amount of energy, $\rho g H \overline{A}(H - h)$, may be reaped for a total energy of

$$\text{Ideal tidal energy} = \rho g \overline{A} H^2 \tag{7.16}$$

and total average power of

$$\text{Ideal tidal power} = \frac{\rho g \overline{A} H^2}{T} \tag{7.17}$$

The ideal tidal power per unit of tidal pool surface area, $\rho g H^2 / T$, which equals 0.22 W/m² for $H = 1$ m, increases as the square of the tidal range, showing the importance of tidal range in an economical power plant design.

If power is generated both during inflow to and outflow from the tidal pool, the design is called a *double effect* plant. Because it is difficult and expensive to design a turbine and power house that operates in both flow directions, most tidal power plants operate in the outflow direction only, called a *single effect* plant.[24] Because h is usually more than half of H, more energy can be recovered on the outflow than the inflow to the pool.

For a typical single effect plant, Figure 7.22 shows the pool and sea surface levels and power output during one tidal cycle. Beginning at midtide, the sluice gates are opened and sea water flows into the pool, filling it to the high tide level in the first quarter period, at which point the sluice is closed. Shortly thereafter, the turbine inlet is opened and the turbine power output rises quickly to its rated power for nearly half of the tidal cycle. When the difference in water level between pool and ocean becomes small, near the end of the cycle, the turbine is closed down.

---

[24]The La Rance plant (see Table 7.6) was built to operate as a double effect plant, including the possibility of incorporating pumped storage, but now operates exclusively as a single effect outflow tidal plant for various practical reasons.

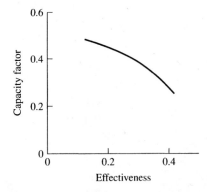

Figure 7.23 The turbine capacity factor η as a function of power plant effectiveness ϵ for single effect tidal power plants. (Data from Fay, J., and M. Smachlo, 1983. *J. Energy,* **7**, 529.)

TABLE 7.6 Tidal Power Plant Characteristics[a]

Site Location	Annapolis Royal	La Rance
Pool area $\overline{A}$ (km^2)	4.8	12.9
Tidal range H (m)	6.3	8.5
Rated turbine flow rate (m^3/s)	378	3240
Rated turbine head (m)	5.5	8
Rated electric power (MW)	17.8	240
Sluice gate flow area (m^2)	230	1530
Capacity factor, η	0.321	0.225
Effectiveness, ϵ	0.171	0.331

[a]Data from Fay, J., and M. Smachlo, 1983. *J. Energy,* **7**, 529.

There are two measures of performance of a tidal power plant. One is the *capacity factor η,* the ratio of the average turbine power output to the rated turbine output. A high capacity factor provides more annual revenue per unit of capital investment in the turbine/generator. A second measure is the plant *effectiveness ϵ,* the ratio of the average power output to the ideal power of equation (7.17). A high effectiveness allows the plant to utilize most of the tidal power available at the site. Unfortunately, it is not possible to obtain high values for both these measures simultaneously, because increasing one causes the other to decrease. Figure 7.23 shows the relationship of these parameters for single effect plants.[25]

The characteristics of two plants that are currently in operation are listed in Table 7.6. The smaller plant, at Annapolis Royal (Figure 7.24), has a higher capacity factor but lower effectiveness than the larger one at La Rance, in agreement with the trends of Figure 7.23. The average electrical power outputs per square kilometer of tidal pool area are 1.19 MW/km^2 and 4.19 MW/km^2, respectively, for the Annapolis Royal and La Rance plants.

---

[25]For the cycle shown in Figure 7.22, $\eta = 0.4$ and $\epsilon = 0.25$.

Figure 7.24 The tidal power plant at Annapolis Royal, Nova Scotia, Canada. In the foreground is the entrance to the power house from the tidal pool; in the background is the exit into the tidal waters of the Annapolis Basin.

The hydroturbine of a tidal power plant operates at a low head that is somewhat less than the tidal range H (see Figure 7.22). This requires an axial flow machine with a low flow velocity of about $\sqrt{gH}$. The turbine axis must be located below the lowest water level in the cycle so as to avoid cavitation, so the turbine diameter is generally less than H. The maximum power from a single turbine is of the order of $\rho(gH)^{3/2}H^2$, which works out to about 10–30 MW for practical plants. For large installations, multiple turbines are required.

7.7.1 Environmental Effects

The principal environmental effects of tidal power plants are a consequence of the changes to the tidal flow in the pool and, to a lesser extent, the ocean exterior to the pool. For a single effect plant, the flow into the pool is reduced by about half, decreasing the intertidal zone by about the same amount and reducing the average salinity of the pool waters when the pool is an estuary, both of which alter the nature of the original marine ecosystems. The plant also imposes an impediment to the movement of marine mammals and fish. The patterns of siltation, often a significant natural process at sites having large tidal ranges, is changed by the presence of a tidal plant. Ship navigation into the pool is prevented unless a lock is built into the dam. The unpredictable and possibly adverse changes to existing ecological systems, including fisheries and other wildlife, has weighed heavily in the assessment of proposals for new tidal power plants.

7.8 OCEAN WAVE POWER

The damage to ocean shorelines caused by storm-driven waves is ample evidence of the dynamic power released when waves impinge on a coastal shore. It would seem that harnessing the power

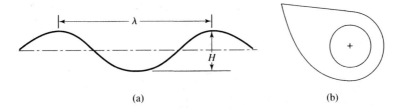

Figure 7.25 (a) The profile of an ocean wave of wavelength λ and wave height H. (b) A sketch of the Salter cam, a device for extracting power from a wave.

of waves could provide a source of electrical energy analogous to that of the wind. The technology for doing so, however, presents both economic and engineering challenges.

Ocean waves are formed on the ocean's surface by action of the wind. By friction and pressure forces exerted at the air–water interface, wind energy is transferred to the energy of gravity waves moving across the ocean surface. The wave energy moves with the wave, but the ocean water has no net motion in the direction of the wave propagation. By suitable mechanical devices, the energy in a wave may be converted to mechanical power.

Ocean waves involve periodic motion of the water at or close to the ocean surface that is sustained by gravity [see Figure 7.25(a)]. The cyclic frequency f of an ocean wave is related to its wavelength λ by

$$f = \sqrt{\frac{g}{2\pi\lambda}} \tag{7.18}$$

where g is the acceleration of gravity. The velocity c at which the wave pattern moves across the surface of the water, called the phase velocity, is the product of the wavelength times the frequency:

$$c = f\lambda = \sqrt{\frac{g\lambda}{2\pi}} = \frac{g}{2\pi f} \tag{7.19}$$

Long-wavelength waves move faster than those of shorter wavelength, leading to dispersion, or spreading of the energy in an ocean wave system.

The energy of an ocean wave consists of two parts: the kinetic energy of the moving water particles in the wave and the gravitational potential energy of the fluid displaced from its equilibrium position of a flat horizontal surface. It is this motion and displacement imparted to the ocean surface by the wind that constitutes the source of power that can be abstracted by a wave power system.

The motion of the water in an ocean wave is confined mostly to a layer of depth $\lambda/2\pi$ below the surface. If H is the wave height, measured from crest to trough as in Figure 7.25, then the velocity of a water particle in the wave is about Hf and the kinetic energy of the water in this layer, per unit of surface area of the wave, is the product of the water mass density ρ times the depth $\lambda/2\pi$ times the kinetic energy per unit mass, $(Hf)^2/2$, for a total of $\rho\lambda H^2 f^2/4\pi = \rho g H^2/8\pi^2$. Thus the average kinetic energy per unit surface area of a wave system is of the order of $\rho g H^2$ and is independent of the wavelength or frequency. The potential energy of the wave, per unit surface area, is approximately the product of the displaced mass per unit area ρH, its vertical displacement H, and the gravitational acceleration g, for a total of $\rho g H^2$. An exact calculation of the kinetic and potential energies per unit of wave surface area reveals that they are equal and their sum can be expressed as

$$\text{Wave energy/area} = \frac{\rho g H^2}{8} \tag{7.20}$$

The wave energy of equation (7.20) is propagated in the direction of motion of the wave, which is generally the direction of the wind, at the group velocity of an ocean wave. The latter equals half the phase velocity c, or $g/4\pi f$. The flux of wave energy, or power, per unit of distance normal to the direction of wave propagation, is the product of the energy per unit area and the group velocity:

$$\text{Wave power/length} = \frac{\rho g^2 H^2}{16(2\pi f)} = \frac{\rho g^{3/2} H^2 \lambda^{1/2}}{16(2\pi)^{1/2}} \tag{7.21}$$

Unlike the wave energy, the wave power depends upon the wavelength, albeit as the half power of the wavelength.

Depending upon the wind speed, the surface of the ocean is covered with a random collection of waves of different height and wavelength. As the speed V of the wind increases, both the mean square wave height $\overline{H^2}$ and the mean wavelength increase and the mean frequency $\overline{f}$ decreases. Oceanographers have measured these quantities as a function of wind speed and found them to be empirically related by the following formulas[26]:

$$\overline{H^2} = 2.19\text{E}(-2)\frac{V^4}{g^2} \tag{7.22}$$

$$2\pi \overline{f} = \frac{0.70\,g}{V} \tag{7.23}$$

Substituting these relations into equations (7.20) and (7.21) for wave energy and power, we find that

$$\text{Wave energy/area} = 2.74\text{E}(-3)\frac{\rho V^4}{g} \tag{7.24}$$

$$\text{Wave power/length} = 3.89\text{E}(-3)\frac{\rho V^5}{g} \tag{7.25}$$

For example, a wind speed of 20 knots (10.29 m/s) will generate an energy density of 3.13 kJ/m², a wave power of 45.8 kW/m, an average wave height of 1.6 m, and an average wave frequency of 0.106 Hz, which is a wave period of 9.4 s.

The dependence of wave power on wind speed, V^5, is steeper than that for wind power, V^3. As a consequence, most of the wave power available at a site is collectible only when the wind exceeds its average value, as can be seen in Figure 7.26, showing the probability distribution of wave power with wind speed, assuming the wind speed probability distribution of Figure 7.19. Figure 7.26 contrasts sharply with Figure 7.19 for the distribution of wind speed. The average wave power for the distribution of Figure 7.26 is $60/\pi^2 = 6.08$ times the power available at the mean wind speed $\overline{V}$.

---

[26]These relations hold under conditions where the wind duration is long enough over a sufficiently great unimpeded fetch of ocean surface.

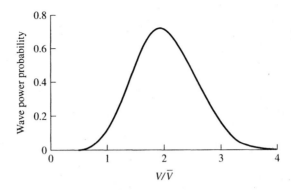

Figure 7.26 The probability of wave power per unit of relative wind speed $V/\overline{V}$, as a function of relative windspeed.

Most of the wave energy at a site is available only a small fraction of the time. Capacity factors for practical wave energy systems are inevitably small, in the range 15–20%. This is unfavorable to the economic viability of wave power and its suitability as a source of electrical power.

The most favorable sites for ocean wave power are the continental or island margins exposed to prevailing winds. For mid- to high latitudes these winds are westerly, while for tropical latitudes they blow from the east. At favorable locations, the annual average wave power available is about 50 kW/m. Because capacity factors are small for wave systems, the installed power at such locations would be about 250 kW/m.

Many ingenious mechanisms have been proposed, and some of them tested, to produce mechanical or electrical power from wave systems. They are of two types: (a) floating bodies that are anchored in place and (b) structures fixed to the sea bed. Mechanical power is produced by the relative displacement of parts of the structure or by the flow of fluid within it, caused by the impingement of waves on the device. The power that is produced is a fraction of the power available in the waves that are intercepted by the device.

One of the devices, which has been tested to confirm a high efficiency of wave power conversion, is illustrated in Figure 7.25(b). Called the Salter cam, the device consists of a horizontal cylindrical float, of asymmetrical cross section shown in Figure 7.25(b), that in still water floats with most of its volume below the sea surface. When waves impinge on the float, it rotates about a stationary circular spline in an oscillatory manner. An electrical power system applies a restraining torque to the float that absorbs most of the power in the incident wave system. The cam structure must have dimensions greater than the wave height, and be tuned to the average wave frequency, of the waves corresponding to its power capacity. In effect, the volume of the device will be proportional to its rated power.

An ocean wave system carries momentum as well as energy. If the wave energy is absorbed by a wave power system, a horizontal force per unit length will be exerted on the power system of amount equal to the energy per unit surface area, equations (7.20) and (7.24). A corresponding restraining force must be supplied by the anchoring gear that holds the power system in place.

The construction and operation of wave power systems immediately seaward of the ocean shorefront would have some adverse environmental impacts. They must be anchored to the sea bottom, disturbing some part of it. They may impede navigation and recreational use of the ocean front, and they definitely would have aesthetic drawbacks.

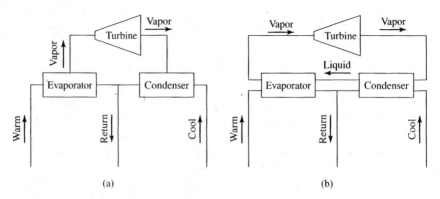

Figure 7.27 Diagrams of the fluid flow in an ocean thermal power plant. (a) An open cycle employs the water evaporated from warm sea water as the working fluid in a low pressure steam cycle. (b) A closed cycle uses ammonia as the working fluid, the latter being vaporized and condensed in heat exchangers supplied with warm and cool ocean streams.

7.9 OCEAN THERMAL POWER

The concept of generating mechanical power using warm surface water from a tropical ocean site was first advanced by Jacques d'Arsonval in 1881. The warm water would provide heat to a Rankine cycle heat engine while cool deep water would supply the cooling needed to condense the working fluid. In low-latitude regions of the ocean, within 20 degrees of the equator, water near the surface, where sunlight is absorbed, is warmer by about 20 K than water at depths greater than a kilometer. (The cool deep ocean water originates in the polar regions.) Commonly called *ocean thermal energy conversion* (OTEC), such a plant would necessarily have a small thermodynamic efficiency because it would operate with a temperature difference that is small compared to the absolute temperature of the heat source.[27] Although the heat that can be extracted from the ocean surface is very great, mechanical power is required to circulate the surface and deep water through the power plant, subtracting from the power generated by the Rankine cycle turbine. This and other practical difficulties have prevented OTEC plants from developing beyond the demonstration stage.

Two types of OTEC plants have been tried. The first, an open cycle, utilizes water evaporated from the warm stream under low pressure to supply a turbine discharging to an even lower pressure condenser cooled by a spray of cool water [see Figure 7.27(a)]. The second, a closed or hybrid cycle, employs heat exchangers that supply or remove heat from a closed Rankine cycle working fluid, usually ammonia, permitting a more economical and efficient turbine than that for the open cycle, but requiring a more expensive evaporator and condenser [see Figure 7.27(b)]. In both cycles, large volumes of warm and cool water must be pumped from the ocean through long, large-diameter ducts and then returned to the sea via a third duct. These systems must be constructed to minimize

---

[27]A Carnot cycle operating with a temperature difference of 20 K and supplied with heat from a warm source at 25 °C would have a thermodynamic efficiency of $20/(25 + 273) = 6.7\%$. A practical Rankine cycle would have even lower efficiency than the ideal Carnot cycle.

the amount of power required to circulate the fluid streams from and to the ocean via the power plant.

In 1979, a small closed-cycle demonstration plant mounted on a barge anchored off Keahole Point in Hawaii generated 52 kW of gross electrical power but only 15 kW of net power, with the difference being consumed in pumping losses. The heat exchangers were made from titanium plate. The cool water polyethylene pipe was 0.6 m in diameter and drew water from a depth of 823 m. Currently, experiments are continuing with a 50-kW closed-cycle plant located on shore at the Natural Energy Laboratory of Hawaii, with testing of heat exchangers that will be more economical.

In 1981, Japan demonstrated a shore-based closed-cycle plant utilizing a fluorocarbon working fluid. Cool water was drawn from a depth of 580 m. The gross and net powers from this plant were 100 kW and 31 kW, respectively.

In 1992, an open-cycle plant was tested at the Natural Energy Laboratory of Hawaii that generated 210 kW of gross electrical power and 40 kW of net electrical power. The turbine rotor for this plant weighed 7.5 tons.

There are several problems to be overcome if ocean thermal power systems are to become practical. Fouling of the warm water supply pipe and heat exchanger by marine organisms, and the corrosion of heat exchangers, needs to be eliminated. The mechanical power needed to pump fluids through the system must be kept well below the gross output of the turbine. All components of the system need to be constructed economically so that the capital cost per unit of net power output becomes competitive with other types of renewable energy systems.

7.10 CAPITAL COST OF RENEWABLE ELECTRIC POWER

Under the new regulatory regime being instituted in some states of the United States in 2000, and which perhaps will be extended subsequently to all of the states, the price of electricity paid by consumers is the sum of several parts. One part is the price charged by the electricity producer for electricity leaving the power plant. Other parts include the price for transmission by high-voltage power lines and distribution networks to the consumer's location. One might say that the electricity producer sells a product, electric energy (measured in kilowatt hours), while the transmission and distribution utility sells a service, transmitting the electric energy to the customer. Individual producers will compete to sell electric energy to consumers for the lowest price, but the transmission and distribution system will remain a public utility monopoly whose price of service will be regulated by public authorities. In a competitive market, electricity producers will price their product to cover the costs of production, plus a profit for privately owned producers.

Most of the currently installed renewable energy systems in the United States generate electric power (see Table 7.1). These systems collect naturally available forms of renewable energy and convert them to electrical power. The economic cost of doing so stems almost entirely from the capital investment required to produce and install the equipment that accomplishes this energy conversion. In contrast to fossil fuel power plants, where the cost of fuel and of operating and maintaining the plant may equal or exceed the annualized capital cost, renewable energy plants have small operating costs and, of course, no fuel cost. Because renewable energy plants have higher capital costs and lower utilization than fossil fuel plants, their cost of producing electricity is usually higher than that of fossil fuel plants.

TABLE 7.7 Capital Cost of Renewable Electric Power

Type	Capital Cost ($/kW)	Capacity Factor (%)	Capital Cost of Electricity (cents/kWh)
Hydropower	2000	50	6.8
Biomass[a]	1000	60	2.9
Geothermal	1500	15	17.1
Wind	1200	25	8.2
Photovoltaic	4000	30	22.8
Tidal	2000	30	11.4

[a] Wood-fueled steam plant.

To compare the renewable energy technologies with each other, and with fossil fuel plants, we computed the component of the cost of producing electricity that is allocated to financing the capital investment needed to construct the facility, for each type of renewable energy plant listed in Table 7.7. For each kilowatt of installed power, the annual cost of capital to be recovered by electricity sales is assumed to be 15% of the capital cost per kilowatt. When this is divided by the annual average output per kilowatt of installed power, which is the capacity factor times $24 \times 365 = 8760$ hours per year, the quotient is the capital cost component of the cost of electricity production at that plant, shown in the last column of Table 7.7.

For the renewable energy systems listed in Table 7.7 the capital cost component of electricity covers a large range. Part of this variability reflects the difference in unit capital cost, and the rest is a consequence of different capacity factors. Except for photovoltaic systems, these systems require turbine-generator machinery that constitutes a large share of the capital cost, which is in the range of 1000–2000 $/kW. The capacity factors, taken from Table 7.1, are mostly low, for reasons inherent in the variability of the renewable energy resource.[28] In contrast, the capital cost for modern efficient fossil fuel plants lie in the range of 500–1300 $/kW and the corresponding capital cost of electricity is 1.2–3.0 cents/kWh. Taking into account the costs of fuel and operations, these fossil fuel plants produce electricity at a cost of 4–6 cents/kWh. Only hydropower and wind energy plants come close to being competitive with new fossil fuel plants.[29]

It has been argued that research and production subsidies may so enlarge the market for renewable energy technologies that their capital costs will be reduced to the point that they become economically competitive with fossil fuel plants. In addition, if fossil fuel plants must reduce their CO_2 emissions by capturing and sequestering the CO_2, the extra cost of doing so may make renewable sources more economically attractive. In either case government intervention, either economic or regulatory, will be necessary to increase the market share for renewable electric power.

---

[28] The low historic value of 15% for geothermal power is a consequence of the market utilization for this power in California. Otherwise, the capacity factor should be in the range of 50%.

[29] The wood-fueled steam plant of Table 7.7 may have low capital cost, but the fuel cost is high because of low thermal efficiency, compared to fossil-fueled plants. To improve this efficiency, and thereby lower fuel cost, the fuel must be gasified and a combined cycle plant constructed, increasing the capital cost.

7.11 CONCLUSION

The renewable energy technologies that have been demonstrated to be technologically practical sources of power are hydro, biomass, geothermal, wind, solar thermal and thermal electric, photovoltaic, and ocean tidal. Of these, hydro and biomass comprise almost the entire renewable installed capacity in the United States in 1997, but the proportion of wind is increasing, especially worldwide. Neither ocean wave nor ocean thermal electric power has yet emerged from the development stage.

Hydro, biomass, and geothermal plants are able to supply electric power dependably on a daily and annual basis. Wind, solar thermal, photovoltaic, and ocean tidal power have diurnal rhythms and seasonal changes that do not necessarily match the demand for electric power. When linked to an integrated power transmission system, they displace fossil fuel consumption, reducing air pollutant and carbon emissions.

Renewable energy technologies are capital intensive, having capital costs per installed kilowatt that are two to eight times those of fossil fuel plants. Furthermore, these plants may have low capacity factors (the ratio of average to installed power), which increases the capital cost portion of the price of electricity they generate, compared with conventional plants. The economic competitiveness of renewable energy plants, compared to fossil fuel plants, is strongly dependent upon the costs of capital and fuel.

The adverse environmental effects of renewable energy plants are generally less than those of conventional power plants, but are by no means negligible. Hydro and tidal plants require extensive changes to the natural hydrologic system, and biomass plants emit air pollutants. Because of the low intensities of renewable energy, the size of a renewable power plant is larger than that of a conventional plant of equal power, requiring more land area.

PROBLEMS

Problem 7.1

A proposed hydroelectric power station would produce maximum power when the volume flow rate reached 100 m³/s at a head of 10 m. (a) If the turboelectric machinery operates with an efficiency of 85%, calculate the maximum electric power output of the facility. (b) If the annual average capacity factor of the plant is 65%, calculate the annual income from the sale of electric power if the selling price is $0.03/kWh. (c) If the capital cost of the power plant is $1000/kW, calculate the ratio of annual income to capital cost, expressing the result as %/y.

Problem 7.2

The annual average daily solar irradiance falling on agricultural land in the United States is about 10 MJ/m² per day. (a) If 0.1% of this is converted to biomass heating value, calculate the annual rate of biomass crop heating value that may be harvested per hectare of crop land. (b) If the biomass heating value is converted to electric power at an efficiency of 25%, calculate the annual average electric power generated per hectare of cropland. (c) If electric power is sold at 0.03 $/kWh, calculate the annual income, per hectare of land, from electricity sales of this biomass energy.

Problem 7.3

A large oak tree produces 2.2 tons of wood in 50 years of growth. The tree has a canopy of 10 m in diameter, and it collects solar energy for six months each year at an average rate of 177 W/m². What is the tree's efficiency for converting solar energy to wood heating value? (Assume a wood heating value of 20 MJ/kg.)

Problem 7.4

Draw a schematic diagram for a geothermal heat pump that works to supply space heat in winter and air conditioning in summer, and explain how it works.

Problem 7.5

The rate of heat q collected by a flat plate solar collector is given by equation (7.2). According to the limits of the second law of thermodynamics, the maximum electric power p_m that could be generated from this heat flux q would be lower by the factor $(1 - T_a/T_c)$, where T_c and T_a are the temperatures of the collector fluid and the atmosphere, respectively. (a) Derive an expression for the temperature ratio T_c/T_a that will maximize p_m, in terms of the collector parameters βI and $\mathcal{U}$, and the temperature T_a. (b) Calculate the numerical values of T_c/T_a, T_c, and p_m when $\beta = 0.8$, $I = 900$ W/m², $\mathcal{U} = 5$ W/m²K, and $T_a = 300$ K.

Problem 7.6

Calculate the collector surface area A required to heat 500 liters of water a day from 15 °C to 80 °C under conditions where the daily insolation on a slanted collector is 1.13 E(7) J/m², assuming 33% collector efficiency.

Problem 7.7

Explain why a solar flat plate collector used for domestic water heating can work even in subfreezing ambient temperatures.

Problem 7.8

A spherically focusing solar collector is being designed to generate electric power from a heat engine, similar to that shown in Figure 7.9. For this design, $\beta = 0.9$, the design solar irradiance is $I = 700$ W/m², and the concentration ratio is $CR = 2000$. Assuming that the heat loss rate from the collector, $\mathcal{U}(T_c - T_a)$, is equal to the black-body radiation from the collector, or σT_c^4 (where σ is the Stefan–Boltzmann constant [see Table A.3]), calculate the maximum collector temperature, $(T_c)_{max}$.

Problem 7.9

If incident solar radiation averages 700 W/m² for 8 hours, estimate the area A of 80% efficient heliostats needed to provide 10 MW of electrical power (as in Solar One). Assume that the efficiency of converting solar heat to electric energy is 35%.

Problem 7.10

Calculate the land area in km² that would be needed for a solar thermal power plant delivering 1000 MW of electrical power under the following conditions: The concentrating mirrors receive 700 W/m²; each concentrating mirror and its platform require a land area twice the area of the mirror itself; the efficiency of converting solar heat to electric energy is 35%.

Problem 7.11

An earth satellite photovoltaic power system is oriented toward the sun to intercept the solar irradiance of 1367 W/m². (a) If the photovoltaic cell efficiency is 17%, calculate the electrical power output per unit of surface area. (b) If 90% of the solar irradiance is absorbed by the cell, calculate the heat flow rate, per unit area, radiated to space that is needed to maintain the cell at a fixed temperature. (c) If this radiant heat flow is that of a black body (σT^4, where sigma is the Stefan–Boltzmann constant [see Table A.3] and T is the cell absolute temperature), calculate the cell temperature.

Problem 7.12

If solar irradiation is 700 W/m² and the efficiency of a photovoltaic cell is 10%, calculate the area (in m² and ft²) of a PV cell needed to run simultaneously—and without a grid or battery backup—a typical refrigerator, toaster, TV, and stereo set. Use electric power consumption data from your own appliances or from the literature.

Problem 7.13

A single-story retail store wishes to supply all its lighting requirement with batteries charged by photovoltaic cells. The PV cells will be mounted on the horizontal rooftop. The time-averaged lighting requirement is 10 W/m²; the annual average solar irradiance is 150 W/m²; the PV efficiency is 10%; the battery charging–discharging efficiency is 80%. What percentage of the roof area will the photocells occupy?

Problem 7.14

Calculate the power produced per m² of wind turbine disk area for a wind velocity of 20 miles per hour, assuming a power coefficient C_p of 0.5 and an air density of 1.2 kg/m³. Calculate the diameter of a wind turbine that would supply 1 MW of electrical power under these conditions.

Problem 7.15

Calculate the mechanical power produced by a wind turbine with rotor diameter 40 m when the wind speed is 8 m/s, assuming a power coefficient C_p of 0.3 and an air density of 1.2 kg/m³. How many households can this wind turbine supply, assuming an average power requirement of 3 kW/household? How many wind turbines would be needed for a city of 100,000 population with four people per household? Calculate the land area for this population, in m² and acres, needed for the wind turbine farm, assuming that the wind turbines should be placed 2.5 rotor diameters apart perpendicular to the wind, and 8 rotor diameters apart parallel to the wind. Place 10 windmills perpendicular to the wind, the rest parallel. Compare the wind farm area to that of a typical city of 100,000 (single-family homes, no high rises).

Problem 7.16

An inventor proposes to utilize a surplus ship's propellor, of diameter equal to 2 m, as an underwater "wind turbine" at a location where a tidal current is 2 m/s. Calculate the maximum power that this turbine could produce from the underwater flow.

Problem 7.17

For the Annapolis Royal tidal power plant, Table 7.6 lists the pool area $\bar{A} = 4.8$ km², tidal range $H = 6.3$ m, rated electric power $P_{el} = 17.8$ MW, and capacity factor $\eta = 0.321$. (a) Using equations (7.16) and (7.17) and these values, calculate the ideal tidal energy and ideal tidal power. [You may assume that the tidal period equals 4.46E(4) s.] (b) Calculate the average power and the effectiveness $\epsilon = $ (average power)/(ideal power), and compare it with the value of Table 7.6.

Problem 7.18

A wave energy system is being evaluated for a site where the design wind speed V is 20 knots. Using equations (7.22), (7.23), and (7.25), calculate the expected mean wave height $\sqrt{\bar{H^2}}$, mean wave frequency $\bar{f}$, and the wave power per unit length.

BIBLIOGRAPHY

Avery, William H., and Chih Wu, 1994. *Renewable Energy from the Ocean.* Oxford: Oxford University Press.

Armstead, H., and H. Cristopher, 1978. *Geothermal Energy.* London: E. F. N. Spon.

Baker, A. C., 1991. *Tidal Power.* London: Peter Peregrinus.

Boyle, Godfrey, Ed., 1996. *Renewable Energy. Power for a Sustainable Future.* Oxford: Oxford University Press.

Dickson, Mary H., and Mario Fanelli, Eds., 1995. *Geothermal Energy.* Chichester: John Wiley & Sons.

Duffie, John A., and William A. Beckman, 1991. *Solar Engineering of Thermal Processes,* 2nd edition. New York: John Wiley & Sons.

Fay, James A., 1994. *Introduction to Fluid Mechanics.* Cambridge: MIT Press.

Kreith, Frank, and Jan F. Kreider, 1978. *Principles of Solar Engineering.* Washington, D.C.: Hemisphere.

Lewis, Tony, 1985. *Wave Energy Evaluation for C.E.C.* London: Graham & Trotman.

Lunde, Peter J., 1980. *Solar Thermal Engineering.* New York: John Wiley & Sons.

Markvart, Thomas, Ed., 1994. *Solar Electricity.* New York: John Wiley & Sons.

McCormick, Michael, 1981. *Ocean Wave Energy Conversion.* New York: John Wiley & Sons.

Merrigan, Joseph A., 1982. *Sunlight to Electricity.* 2nd edition. Cambridge: MIT Press.

Probstein, Ronald F., and R. Edwin Hicks, 1990. *Synthetic Fuels.* Cambridge: pH Press.

Selzer, H., 1986. *Wind Energy. Potential of Wind Energy in the European Community. An Assessment Study. Solar Energy R&D in the European Community,* Series G, Volume 2. Dordrecht: D. Reidel.

Simeons, Charles, 1980. *Hydropower. The Use of Water as an Alternative Source of Energy.* Oxford: Pergamon Press.

Seymour, Richard J., 1992. *Ocean Energy Recovery: The State of the Art.* New York: American Society of Civil Engineers.

Warnick, C. C., 1984. *Hydropower Engineering.* Englewood Cliffs: Prentice-Hall.

Wortman, Andrze J., 1983. *Introduction to Wind Turbine Engineering.* Boston: Butterworth.

Transportation

8.1 INTRODUCTION

Near the end of the nineteenth century in the United States, a predominantly steam powered, coal-fueled transportation system had aided the transformation of what had been a mostly agricultural economy at the century's beginning into one dominated by industrial activity. Railroads, coastal ships, and river barges moved farm produce and forest products to urban consumers, mineral ores to refineries and steel mills, and coal to industrial, commercial and residential consumers. Except for the affluent who could afford horse-drawn carriages, most urban and rural workers walked from home to the place of employment. In the major cities, however, horse-drawn or electric-powered streetcars provided passenger transport for daily travel where residential areas spread beyond the urban core in response to population growth. Long-distance passenger travel was exclusively provided by trains and ships. Nevertheless, horse-drawn wagons were necessary for the distribution of food and goods within urban and rural regions alike, and animal power provided the source of much agricultural energy.

The development of the internal combustion engine (ICE) and its supply of liquid fossil fuel transformed both urban and rural communities in the twentieth century, most markedly by greatly expanding the kind and function of transportation and work vehicles. By the century's end, practically all adults had the use of an automobile for commuting to work or other personal daily travel. Passenger travel to and in central urban areas by bus and electric-powered rail (above ground, at ground level, or underground) complemented the need for personal travel where the density of travel could not be accommodated by automobiles alone. Most freight moved by truck, but bulk commodities were shipped by rail and river barge. The rapid development of commercial air travel, beginning at midcentury, expanded long-distance travel availability, superseding railroads, which had already lost market share to intercity buses and private automobiles. ICE powered tractors revolutionized agriculture, and rural farm population declined drastically. Work machines (bulldozer, chain saw, construction crane, etc.) greatly increased human productivity in tasks that previously consumed large amounts of hard human labor.

The growth of transportation in the twentieth century paralleled that of electric power, with both becoming major factors in the economies of developed nations at the century's end. In the United States in 1996, transportation and electricity accounted for about 25% and 45% of the total primary energy consumption, respectively. The availability of both electric power and transportation in modern industrial societies is an important factor in human productivity, although the efficiency of their use may well be improvable.

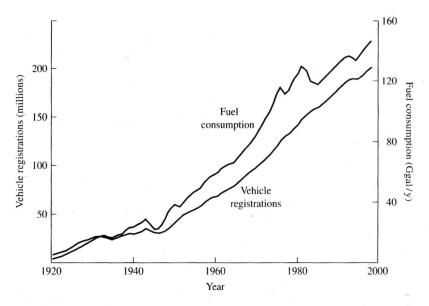

Figure 8.1 The number of registered highway vehicles and their annual fuel consumption in the United States since 1920. (Data from the National Highway Traffic Administration. 1999.)

The rapid growth in the number of U.S. highway transportation vehicles in the twentieth century is graphically depicted in Figure 8.1. The population of road vehicles in use has grown at a rate of about 3.4 million vehicles per year since 1950,[1] reaching 201.5 million in 1995, of which 64% are automobiles, 36% are trucks (light, medium, and heavy duty), and 0.3% are buses.[2] The number of these vehicles is closely approaching the number of eligible drivers, but the annual rate of increase in vehicles exceeds that of the U.S. population increase by one million per year.[3]

The growth in the number of highway vehicles, mostly privately owned, has been accompanied by a growth in highways and roads, publicly financed and maintained. In the United States there are currently 3.93 million miles of highways (79% rural, 21% urban), about 40 vehicles per highway mile. In the last decade, highway miles have increased at an annual rate of about 0.1% per year, much less than the 1.7% per year growth in the vehicle population. This slow growth in highway miles reflects both the difficulty of siting new roadways, especially in urban regions, and the scarcity of public funds for infrastructure improvement. The concentration of national vehicle ownership in urban areas (which accounts for only 21% of highway miles but 75% of the population), along

---

[1]This amounts to a current annual growth rate of 1.7%. The world vehicle population, about 600 million with production of about 50 million new vehicles per year, is growing at an annual rate of 2.2%.

[2]About 15.4 million new vehicles are sold each year, 12 million of which are needed to replace the aging members of the 201.5 million vehicles in use. The approximate vehicle life expectancy, equal to the ratio of vehicle population to new vehicle sales, is 13 years.

[3]The United States has the highest ratio of vehicle to human population of any nation. Most developing countries currently have vehicle/human ratios about the same as that of the United States in the 1920–1940 period.

TABLE 8.1 1995 U.S. Transportation Vehicle Use[a]

Type	Number (million)	Annual Miles (Gmile/y)	Miles per Vehicle (mile/y)	Miles per Gallon (mile/gal)
Passenger cars	132.2	1448	10,958	21.3
Light trucks, SUV	65.7	790	12,018	17.4
Trucks	6.72	178	26,500	6.2
Buses	0.69	6.4	9,312	6.6
Transit	0.12	3.6	30,900	4.7
Rail	1.22	30.4	24,900	8.7
Commercial air	0.0056	4.4	790,000	0.35

[a]Data from Bureau of Transportation Statistics, U.S. Department of Transportation, 1999.

with the faster growth of vehicle population than highway miles, has aggravated the problem of increasing highway congestion in U.S. cities.

Highway vehicle fuel consumption shows a growth proportionate to the vehicle population (see Figure 8.1), but annual variations are larger, reflecting the tempering effects of variations in economic activity and level of employment. In 1995, U.S. highway vehicle fuel consumption was 134.8 Ggal/y (510.2 GL/y), averaging out to 669 gallons per year (2532 L/y) per vehicle. Weighted by the fuel heating value, this annual consumption is 16.3 Quad/y = 0.54 TW.

A breakdown of the use of all transportation vehicles in the United States in 1995 is given in Table 8.1. Small passenger vehicles (cars, light duty trucks, sports utility vehicles, vans) account for 96% of all highway vehicles, 92% of highway vehicle miles, and 76% of highway fuel. These privately owned vehicles clearly dominate highway travel and energy use. Public passenger travel, by bus and transit on the ground and airplane above, accounts for only 0.6% of total transportation vehicle miles annually.

Small passenger vehicles are used only a small fraction of the time, mostly for short trips. Their average yearly mileage, about 12,000 miles per year, amounts to 33.7 miles per day, about an hour per day of operating time. The average one-way trip length is less than 10 miles, so vehicles do not move far from their home locations, on average.

In the United States, the manufacture, operation, and maintenance of highway vehicles and the refining and distribution of their fuels is pervasively regulated by both federal and state governments. Two of the regulatory objectives are of interest here: control of exhaust and evaporative emissions and vehicle fuel economy.[4] Other objectives include vehicle and passenger safety, operator competence, and owner fiscal liability. The regulation of emissions and fuel economy falls principally upon the vehicle manufacturer, to a lesser extent on the fuel supplier, and hardly at all upon the vehicle owner, whose principal responsibility is to maintain control equipment during the vehicle lifetime.

The problem of air pollutant emissions from transportation vehicles is primarily that associated with the private passenger vehicle. Their emissions, principally carbon monoxide (CO), oxides

---

[4]U.S. federal regulation of vehicle fuel economy was instituted after the oil crisis of the 1970s so as to reduce the dependence of the national economy on the supply of imported oil. At present, there is no legislative requirement that vehicle fuel economy should be regulated to reduce vehicular CO_2 emissions.

of nitrogen (NO_x), hydrocarbons (HC), sulfur dioxide (SO_2), and particulate matter (PM), are distributed geographically in proportion to vehicle usage, which is concentrated in urban regions. But the secondary pollutants photochemically formed from the direct emissions of NO_x and HC can reach elevated levels downwind from the vehicular sources and outside the urban region. As a consequence, all primary emissions are (or soon will be) regulated to ensure that primary and secondary air pollutant levels do not exceed harmful levels, either locally or regionally.

In this chapter we discuss the technology of the automobile as it affects the efficiency of fuel use and the emission of atmospheric pollutants. The chapter emphasizes the cause and amelioration of ordinary air pollutant emissions as well as the reduction of CO_2 emissions by improvements in vehicle fuel efficiency. We summarize the current state of development of alternative vehicle systems that show promise of significant emission reductions, including electric drive vehicles (battery-powered or fuel-cell-powered) and hybrid electric/ICE drive vehicles.

8.2 INTERNAL COMBUSTION ENGINES FOR HIGHWAY VEHICLES

The most common engine in road vehicles is the gasoline-fueled spark ignition (SI) reciprocating engine. It is economical to manufacture and maintain, provides ample power per unit weight, and has a useful life that equals that of the typical passenger vehicle. The less common alternative, the diesel-fueled compression ignition (CI) reciprocating engine, is more durable and more fuel efficient, but is more expensive to manufacture and provides less power per unit weight (unless turbocharged).[5] Each engine type has different air pollutant emissions and requires different technology to reduce them.

The SI and CI engines, developed at a time when the dominant steam engine was a reciprocating device, utilize the same mechanism of a movable piston within a closed-end cylinder linked by a connecting rod to a rotating crankshaft that converts the reciprocating motion of the piston to the rotary motion of the crankshaft [see Figure 8.2(a)]. High-pressure gas in the cylinder exerts an outward force on the piston, doing mechanical work on the crankshaft as the piston recedes on the power stroke. On the return inward stroke, where the cylinder is filled initially with low-pressure gas, less work is done by the crankshaft on the piston than was done on the crankshaft during the outward stroke, so the crankshaft delivers net positive work to the mechanism the engine is turning. A flywheel is required to smooth out the rotational speed of the crankshaft while the piston exerts variable amounts of torque during its reciprocating motion.

As previously explained in Section 3.10 on the Otto cycle, a high pressure is generated in the cylinder just at the beginning of the outward expansion stroke by burning a fuel with air in the cylinder. In the SI engine, a fuel–air mixture, prepared outside the cylinder, flows into it prior to the inward compression stroke, eventually being ignited by a spark when the piston reaches its innermost position (called top center, TC). On the other hand, for the CI engine, pure air is ingested prior to the compression stroke and the fuel is sprayed into the air at the end of

[5]The CI engine, in contrast to the SI engine, does not lend itself to light, low-power uses such as motorcycles, chain saws, lawn mowers, and so on, or for use in small aircraft, where engine weight is important. On the other hand, large reciprocating engines for construction equipment, ships, and railroad locomotives are invariably CI engines.

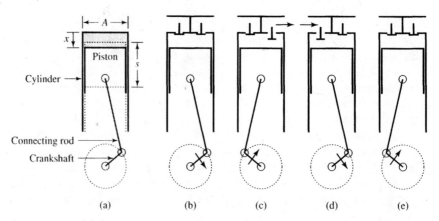

Figure 8.2 (a) A diagram of the mechanical action of a reciprocating internal combustion engine and the four piston strokes of the four-stroke cycle engine: (b) outward power stroke, (c) inward exhaust stroke, (d) outward intake stroke, and (e) inward compression stroke.

compression, whereupon it ignites and burns quickly, without the necessity of a spark, to produce a pressure rise.

To make a repetitive cycle out of this process, it is necessary to replace the burned gases in the cylinder with a fresh charge (fuel–air mixture for the SI engine, pure air for the CI engine). This is accomplished by opening ports in the cylinder connected to ducts (called manifolds) that either conduct the fresh charge into the cylinder or collect the combustion gases leaving the cylinder at the end of the expansion stroke. There are two mechanical schemes for effecting this replacement of burned gas by a fresh charge. In one, called the *two-stroke cycle,* the fresh charge displaces the combustion products during a short time interval when the piston is near its outermost position (called bottom center, BC) and both an inlet port and an exhaust port are open. Only two strokes of the piston, one inward to compress the fresh charge and the second outward to expand the burned gases, are needed to complete this two-stroke cycle, during which the crankshaft turns through one revolution. In the second scheme, called the *four-stroke cycle* [see Figure 8.2(b)–(e)], the outward power stroke is followed by a full inward stroke in which the piston displaces the combustion products, pushing them out through an open exhaust port located in the closed end of the cylinder (called the *cylinder head*). At the end of this inward stroke, called the exhaust stroke, the exhaust port is closed and an inlet port is opened, allowing the piston to suck in a fresh charge as it moves outward during the subsequent intake stroke. By adding these two extra strokes (an extra revolution of the crankshaft), a fresh charge is prepared in the cylinder, permitting the two subsequent strokes to compress, burn, and expand the charge, thereby producing work. These four strokes (two crank revolutions) comprise the four-stroke cycle.

Almost all highway vehicles are powered by four-stroke cycle engines, predominantly gasoline-fueled SI engines for passenger vehicles and diesel-fueled CI engines for trucks. Two-stroke-cycle SI engines are mostly used to power two-wheeled vehicles, such as motorcycles and mopeds. Engines in the latter vehicles have lower fuel efficiency and higher exhaust pollutant emissions, but are lighter and less expensive to manufacture (for a given power). The trend of increasingly stringent government regulation of pollutant emissions and vehicle fuel efficiency has spurred manufacturers to improve both two- and four-stroke-cycle engines, but the predominant use of four-stroke-cycle engines is unlikely to change in the foreseeable future, especially because two-stroke cycle engines have higher exhaust pollutant emissions.[6]

Although SI and CI engines are similar in that they both burn fuel with air in the cylinder when the piston is near TC, so as to create the pressure rise that eventually produces net mechanical work during the four-stroke cycle of events, they differ in three important ways. In the SI engine the fuel and air are mixed together outside the cylinder to form a uniform fuel–air mixture that is ingested during the intake stroke; in the CI engine the fuel is sprayed into air in the cylinder, near TC, forming a nonuniform fuel–air mixture. In the SI engine, the combustible mixture is burned when ignited by a spark at the appropriate time in the cycle near TC, whereas the CI fuel–air mixture ignites spontaneously, shortly after the fuel is sprayed into the engine combustion chamber at the requisite time near TC. A third difference is the method of controlling the power output of the engine, an important consideration because the vehicle engine must operate satisfactorily over a wide range of power in a continuously and frequently variable manner. This is accomplished in the SI engine by adjusting the air pressure in the intake manifold by means of a throttle valve that lowers that pressure below the atmospheric value, the more so as the engine power is reduced. In the CI engine, the power is lowered by reducing the amount of fuel injected into the cylinder at each cycle. These differences have important implications for the formation of pollutants and the fuel efficiency of each type of engine, SI or CI.

The power needed for a passenger vehicle engine is in the range of tens to hundreds of kilowatts. For many practical reasons, the cylinder size in light-duty passenger vehicles and trucks is approximately the same for all engines, with higher-powered engines using more cylinders than small ones. Thus engines of increasing power may contain 2, 3, 4, 5, 6, 8, or 10 cylinders. Geometrically, the piston motions in these cylinders are phased so that the cylinders fire at equally timed intervals during the two revolutions of the crankshaft (720°) that constitute the four-stroke cycle. This also facilitates the mechanical balance of the moving parts, which would otherwise cause excessive engine vibrations. It is general practice to construct 2-, 3-, 4-, 5-, and 6-cylinder engines with all the cylinders aligned one behind the other, while 6-, 8-, and 10-cylinder engines have two banks of cylinders in a V-shaped arrangement. This permits more efficient use of the engine space in the vehicle, especially for front-wheel-drive vehicles.

8.2.1 Combustion in SI and CI Engines

To perform properly, any reciprocating internal combustion engine must burn a mixture of air and fuel in a very short time, completing the combustion soon after the piston begins its outward power stroke. The duration of combustion should not much exceed about 50 degrees of rotation of the crankshaft, which would equal about 2 ms if the crankshaft were rotating at 3600 revolutions per minute (rpm), a typical value for a vehicle traveling at high speed. In order for a fuel molecule to burn, it must be mixed intimately with oxygen at the molecular level. Even then, the conversion of a hydrocarbon molecule to carbon dioxide and water vapor molecules will not occur rapidly enough unless the mixed air and fuel are sufficiently hot. This is assured by the rapid temperature rise of the cylinder charge during the inward compression stroke.

---

[6]For developments in the two-stroke-cycle engine field, see: Heywood, John B., and Eran Shaw, 1999. *The Two-Stroke Cycle Engine: Its Development, Operation, and Design.* Philadelphia: Taylor and Francis.

The combustion process in the SI engine is initiated by a timed electric spark that starts a flame front propagating through the air–fuel mixture. As it sweeps through the combustible mixture, heating the reactants to the point where they burn extremely rapidly, it converts fuel to products of combustion, with each fuel molecule being processed in about 1 μs. When the flame front reaches the walls of the combustion chamber, it is extinguished, but not until all except a tiny proportion of the combustible mixture next to the wall will have been reacted to form combustion products.[7] However, in order to propagate a reliable, rapidly moving flame, the combustible mixture must not be too rich or too lean; that is, it must not have too much or too little fuel, compared to the amount of air, than is needed to completely consume all the fuel and oxygen present, called the stoichiometric mixture. It is for this reason that the fuel and air are mixed in carefully controlled proportions prior to or while entering the engine cylinder. To reduce engine power, the amount of fuel burned per cycle is lowered by reducing the pressure, and hence density, of the incoming charge; its proportions of fuel and air, as well as its temperature, remaining unchanged. Thus the favorable high-speed flame propagation rate and rapid combustion are retained in SI engines, even down to idling conditions.

Fuel combustion in the CI engine proceeds quite differently, without flame propagation, albeit with comparable speed. Fuel injected into the very hot air within the cylinder is quickly evaporated, and the fuel vapor then becomes mixed with the surrounding air and burns spontaneously without the necessity of spark ignition. The surrounding air swirls past the injector nozzle, providing a flow of oxygen needed to oxidize the evaporating fuel droplets as they emerge from the nozzle. When less power is needed, less fuel is injected into the fixed amount of air in the cylinder, consuming less oxygen and reducing the pressure rise in the cylinder. Thus the fuel–air ratio in the CI engine is variable and lean. At maximum power, some excess air is required to burn all the fuel because mixing conditions are not ideal.

At the molecular level, the combustion process is much more complex than might be inferred from the overall stoichiometry of the reaction. For example, the complete oxidation of one octane molecule (C_8H_{18}) requires $8 + 9/2$ oxygen molecules (O_2), forming 8 CO_2 and 9 H_2O molecules. The rearrangement of the carbon, hydrogen, and oxygen atoms among the reacting molecules occurs one step at a time, requiring numerous individual changes as single H and C atoms are stripped away from the fuel molecule and oxidized. These changes are aided by very reactive intermediate species called radicals, such as H, O, OH, C_2, CH, CH_2, and so on, that act to accelerate the molecular rearrangement but that disappear once the reaction is completed. Nevertheless, the combustion process is not perfect, so that small amounts of unreacted or imperfectly oxidized products may remain; molecules that reached dead end paths and were unable to attain the complete thermochemical equilibrium of the bulk of the reactants. These molecules are dispersed among the principal combustion products and, unless removed, enter the atmosphere as air pollutants.

Nitric oxide (NO) is an important air pollutant that is a byproduct of the combustion process. It is formed from nitrogen and oxygen because it is thermochemically favored at the high temperature of the newly formed combustion products. It is produced rapidly by the two following reaction

---

[7]Under certain conditions it is possible for the combustible mixture to ignite spontaneously and burn explosively, in an uncontrolled and destructive manner. This is called combustion knock, which is avoided by controlling the chemical properties of the fuel.

steps, facilitated by the presence of atomic oxygen in the combustion zone:

$$N_2 + O \rightarrow NO + N \tag{8.1}$$

$$N + O_2 \rightarrow NO + O \tag{8.2}$$

There would be no NO in the exhaust gas reaching the atmosphere if these reactions reversed their course as the product gas temperature declined during the expansion stroke and the NO maintained thermochemical equilibrium with the rest of the exhaust products. Unfortunately, this reverse process ceases during the expansion, leaving a residual amount of NO that can be reduced by subsequent treatment in the exhaust system outside the cylinder. This residual NO is sensitively dependent upon the air–fuel ratio and the flame temperature in the combustion process.

8.3   ENGINE POWER AND PERFORMANCE

The burning of fuel in the cylinder of a reciprocating internal combustion engine produces mechanical power by exerting a force on the moving piston. At any instant, if V is the volume of gas in the cylinder and p is the gas pressure, then the increment dW in work done by the gas on the piston when it moves so as to increase the volume by an amount dV is

$$dW = p\,dV \tag{8.3}$$

The total work W done by the gas in one cycle of the engine (two or four strokes) is found by integrating equation (8.3) over a cycle:

$$W = \oint p\,dV \equiv \bar{p}V_c \tag{8.4}$$

where $\bar{p}$ is an effective average gas pressure and V_c is the cylinder displacement—that is, the volume of gas displaced when the piston moves from the innermost to outermost position. If the crankshaft rotates at a frequency N and n is the number of crank revolutions per cycle, then the time for a complete cycle is n/N and the power input to the piston $\mathcal{P}$ is

$$\mathcal{P} = W\frac{N}{n} = \frac{\bar{p}V_c N}{n} \tag{8.5}$$

Figure 8.3 depicts the variation of gas pressure in the cylinder of a four-stroke-cycle SI engine. The uppermost curve traces the pressure during the outward power stroke, which is followed by the inward exhaust stroke, for which the cylinder pressure is about equal to that in the exhaust manifold. The pressure in the following intake stroke is lower, and it is equal to the intake manifold pressure so that the final, compression stroke starts from a low pressure. The work per cycle, $\oint p\,dV$, can be seen to be the difference between the area enclosed by the upper loop of Figure 8.3 minus that of the lower loop.

Only a part of the work done by the gas on the moving pistons of an engine is delivered to the output shaft of the engine, where it can then be connected to the transmission and ultimately to the wheels to propel a vehicle. The friction of the pistons moving along the cylinder surfaces, as well as

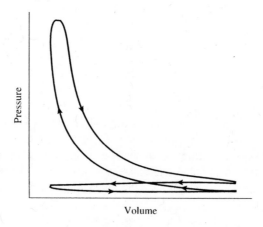

Figure 8.3 Diagram of the pressure, as a function of volume, in a four-stroke-cycle SI engine cylinder.

the friction of the piston, connecting rod, and crankshaft bearings, along with the power consumed in operating the valves, pumping the lubricant, and cooling the water, reduce the power output of the engine below that delivered by the gas in the cylinders [equation (8.5)]. This net output $\mathcal{P}_b$, called the *brake power*, is determined while running the engine in a laboratory at various rotational speeds N and throttle settings. The power is computed as the product of the brake torque T_b (which is measured) and the shaft angular frequency $2\pi N$:

$$\mathcal{P}_b = T_b(2\pi N) \tag{8.6}$$

Sometimes the power and torque are expressed in terms of an average cylinder pressure p_{bmep}, called the *brake mean effective pressure,* using equations (8.5) and (8.6)

$$p_{bmep} \equiv \frac{n\mathcal{P}_b}{V_e N} = \frac{2\pi n T_b}{V_e} \tag{8.7}$$

where V_e is the *engine displacement,* the product of the cylinder displacement and the number of cylinders. The brake mean effective pressure is an indicator of the cycle work per unit displacement volume of the engine, and is little affected by engine size. It directly reflects the amount of fuel burned per unit of cylinder volume.[8]

The size of engine installed in a vehicle is determined by the power required to propel the vehicle at the speed and acceleration needed to perform satisfactorily. An important characteristic of an engine is its power and torque when operated at maximum fuel input per cycle. For an SI engine, this occurs when the throttle is wide open. Figure 8.4 is a sketch of the brake power $\mathcal{P}_b$ and brake mean effective pressure p_{bmep} [which is related to the brake torque T_b by equation (8.7)] as a function of engine speed N. The brake power $\mathcal{P}_b$ rises monotonically with speed, reaching a maximum value *max* $\mathcal{P}_b$ at an engine speed N_m. On the other hand, the brake mean effective

---

[8]Pressure has the dimensions of energy per unit volume.

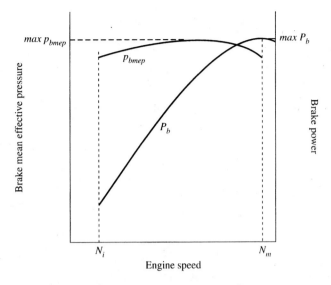

Figure 8.4 A diagram of the brake mean effective pressure p_{bmep} and brake power $\mathcal{P}_p$ of an SI engine at wide open throttle, as a function of engine speed N.

pressure p_{bmep} varies only moderately over the range of speeds between normal idling speed N_i [typically 800 rpm (13.3 Hz)] and the speed N_m at maximum power. It reaches its maximum value $max\ p_{bmep}$ at a speed about two-third of N_m.

At constant throttle setting or inlet pressure, the brake torque does not change much with engine speed. At low speeds, the torque declines somewhat due to heat losses from the cylinder. At high speed, the torque declines more precipitously, as pressure losses in the inlet and exhaust valves and increased piston friction reduce the brake mean effective pressure and cause the brake power to peak. To improve high-speed power, multiple inlet and exhaust valves may be used and the engine may be supercharged.

The characteristics of a sample of model year 2000 passenger cars and light duty trucks equipped with SI engines are listed in Table 8.2, in order of ascending engine displacement. The bigger, more powerful engines employ more cylinders, but the cylinder volume is more nearly the same among these engines than is the size. There are only small differences in the maximum brake mean effective pressure among these engines, and there are comparably small differences in power per unit of engine displacement. As we shall see below, the differences in engine size reflect the differences in vehicle mass, ability to accelerate, and vehicle fuel economy.

8.3.1 Engine Efficiency

The fuel economy of the engine is usually expressed as the *brake specific fuel consumption (bsfc)*, the ratio of the mass of fuel consumed per unit of mechanical work output by the engine shaft. The value of the brake specific fuel consumption depends upon the engine operating conditions. For SI engines, the most economical (i.e., the minimum) bsfc is about 0.27 kg/kWh, while for

TABLE 8.2 2000 Model year Passenger Vehicle Characteristics (SI Engines)

Manufacturer Model Size	Honda Civic Subcompact	Acura Integra Subcompact	Volvo S40[a] Compact	Honda Accord Midsize	Mercedes E320 Midsize
Displacement (L)	1.6	1.8	1.9	2.3	3.2
Number of cylinders	4	4	4	4	6
Maximum power (kW)	94.7	126.8	119.3	111.9	165
Maximum bmep (bar)	11.4	12.1	15.2	11.3	12.4
Vehicle mass (t)	1.14	1.23	1.36	1.37	1.57
Frontal area (m^2)	1.89	1.87	1.94	2.06	2.07
Power/displacement (kW/L)	59.2	70.4	62.8	48.7	51.6
Power/mass (kW/t)	83.0	103.1	88.0	81.7	105.1
Urban/highway fuel efficiency (km/L)	12.8/14.9	10.6/13.2	8.9/11.9	9.8/12.8	8.9/12.8

Manufacturer Model Size	Chrysler LHS Large	Mercedes S340 Large	Honda Odyssey Minivan	Honda Passport SUV	Ford Explorer SUV	Lexus LX470 SUV	Ford F-150 Pickup
Displacement (L)	3.5	4.27	3.5	3.2	4	4.7	5.4
Number of cylinders	6	8	6	6	6	8	8
Maximum power (kW)	188.7	205	156.7	153	119.4	171.6	201.3
Maximum bmep (bar)	12.4	11.8	11.1	11.4	9.6	11.6	
Vehicle mass (t)	1.61	1.88	1.93	1.75	2.04	2.45	2.36
Frontal area (m^2)	2.15	2.68	2.74	2.12	2.2	2.87	2.87
Power/displacement (kW/L)	53.9	48.0	44.8	47.8	29.8	36.5	37.3
Power/mass (kW/t)	117.0	109.0	81.2	87.4	58.5	70.0	85.3
Urban/highway fuel efficiency (km/L)	7.7/11.5	7.2/10.2	7.7/10.6	6.1/8.9	7.0/10.5	5.5/6.8	4.3/9.4

[a]Turbocharged.

CI engines, it is lower, about 0.20 kg/kWh.[9] An alternative measure of engine performance is the thermal efficiency η_e, the ratio of engine mechanical work to the heating value of the fuel mass consumed to produce that work. In terms of the lower heating value (LHV) per unit mass of fuel, these fuel economy measures are related by

$$\eta_e \equiv \frac{3.6}{(bsfc)(\text{kg/kWh}) \times LHV(\text{MJ/kg})} \tag{8.8}$$

The lower heating values of vehicle fuels, per unit mass or volume, are empirically related to the fuel specific gravity SG by

$$LHV = \{28 + 11.2/(SG)\} \text{ MJ/kg}$$
$$LHV = \{11.2 + 28(SG)\} \text{ MJ/L} \tag{8.9}$$

---

[9]The bsfc is sometimes given in units of pounds (mass) per brake horsepower hour, where one pound (mass) per brake horsepower hour equals 1.644 kg/kWh.

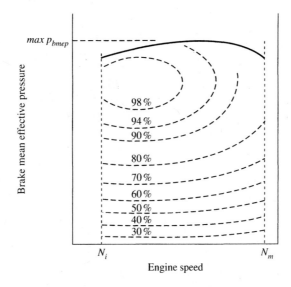

Figure 8.5 A diagram of contours of engine efficiency η_e as a function of engine speed N and brake mean effective pressure for an SI engine. Contour values are expressed as a percent of the maximum value.

For gasoline of $SG = 0.72$, the LHV is 43.55 MJ/kg or 31.6 MJ/L, while for diesel fuel of $SG = 0.85$, the LHV is 41.18 MJ/kg or 35 MJ/L.[10] Converting to thermal efficiency by equation (8.8), the best SI engine efficiency is about 31% while that of a CI engine is 44%.

    The engine efficiency η_e is a function of the engine speed N and brake mean effective pressure, as demonstrated in Figure 8.5. Its peak value occurs at about 35% of the maximum engine speed N_m and 80% of the max p_{bmep}, where the engine power is about 35% of the maximum power. The efficiency declines rapidly with decreasing p_{bmep} but less so with increasing speed. This decrease of efficiency is a consequence of the relative increase of engine friction and flow losses, compared to engine output, as p_{bmep} and engine power are reduced below the optimum value. Because a road vehicle engine must provide the full range of its power over the speed range of the vehicle, it cannot operate at maximum efficiency all the time. Nevertheless, by proper matching of the engine to the vehicle, it is possible to minimize the fuel consumption needed to meet a particular vehicle driving cycle.

8.4 VEHICLE POWER AND PERFORMANCE

Moving a vehicle along a highway requires the expenditure of mechanical power to turn the wheels. Part of this power is needed to overcome the drag force exerted by the air on the moving vehicle. Another part is needed to counter the resistance of the tires moving over the ground, called the rolling resistance. If the vehicle is climbing a hill, additional power is needed to lift it vertically

---

[10]Diesel fuel has more heating value per unit volume than gasoline, by about 11%. Where vehicle fuel efficiency is measured in terms of miles per gallon or kilometers per liter, a diesel-powered vehicle has an inherent 11% advantage, other things being equal.

in the earth's gravitational field. Finally, when the vehicle accelerates to a higher speed, power is needed to increase the kinetic energy of the vehicle.

The aerodynamic drag force acting on a vehicle is conveniently given as the product $\rho C_D A V^2 /2$, where ρ is the mass density of air (kg/m³), A is the frontal area (m²) of the vehicle (about 80% of the height times the width), V is the vehicle speed (m/s), and C_D is the drag coefficient. The latter is empirically determined by testing vehicle models in a wind tunnel, and its value, generally in the range of 0.25–0.5, depends very much on the vehicle shape and external smoothness. Given the fact that a passenger vehicle must enclose the passengers in a safe structure employing good visibility and without excess material, it is difficult to reduce C_D below about 0.25. The corresponding power P_{air} needed to overcome the air resistance is the product of the force times the vehicle velocity:

$$P_{air} = (C_D A)\frac{\rho V^3}{2} \tag{8.10}$$

The power required to overcome air resistance thus increases as the cube of the vehicle speed. At high vehicle speeds, air resistance is the major factor in determining the requirement for engine power.

It might seem that wheels should present little or no resistance to forward motion. While this resistance is small, it cannot be reduced to zero. The source of this resistance lies in the deflection of the tire where it comes into contact with the ground. This deflection is necessary to support the weight of the vehicle and to provide a contact area between the tire and the road that is needed to prevent the tire slipping along the road surface. The deflection is such that, when the wheel rotates, the road surface exerts a retarding torque on the wheel, which must be overcome by the engine drive system, and a corresponding retarding force on the wheel axle. This force is usually specified as $C_R mg$, and the corresponding power P_{roll} becomes

$$P_{roll} = C_R mg V \tag{8.11}$$

where m is the vehicular mass, g is the acceleration of gravity (mg is the total vehicle gravity force supported by the tires), and C_R is a small dimensionless constant whose value depends upon the tire construction and pressure. Stiff, highly pressurized tires will have smaller C_R, but will be more prone to slip and transmit road unevenness to the vehicle. The values of C_R lie in the range 0.01–0.02. Because the rolling power grows only as the first power of the vehicle speed, it is generally smaller than P_{air} at high speeds but can become larger at low speeds.

When climbing a hill of rise angle θ, the force of gravity acting on the vehicle, mg, has a component $mg \sin\theta$ opposing the forward motion. The power required to maintain a steady climb rate is

$$P_{hill} = mg V \sin\theta \tag{8.12}$$

Finally, the instantaneous power required to accelerate the vehicle, P_{acc}, is simply the time rate of increase of vehicular kinetic energy, $m(1+\epsilon)V^2/2$, or

$$P_{acc} = \frac{d}{dt}\left(\frac{m(1+\epsilon)V^2}{2}\right) = m(1+\epsilon)V\frac{dV}{dt} \tag{8.13}$$

where ϵ is a small dimensionless constant that accounts for the rotational inertia of the wheels and drive train, and dV/dt is the acceleration of the vehicle. Of course, when the vehicle decelerates ($dV/dT < 0$) by using the brakes or closing the throttle, the negative power of equation (8.13) does not put power back into the engine, so that P_{acc} is effectively zero during deceleration. In electric drive vehicles, some of the deceleration energy can be recovered, stored in batteries, and used later in the operating cycle, saving fuel.

The total power thus becomes

$$P = P_{acc} + P_{roll} + P_{hill} + P_{air}$$
$$= mV\left((1+\epsilon)\frac{dV}{dt} + C_R g + g\sin\theta\right) + (C_D A)\frac{\rho V^3}{2} \tag{8.14}$$

The first three terms on the right-hand side of equation (8.14), the power required to accelerate the vehicle, overcome the rolling resistance, and climb a hill, are each proportional to the vehicle mass m. In contrast, the last term, the power needed to overcome aerodynamic drag, is independent of the vehicle mass, depending instead on the vehicle frontal area A, which is nearly the same for all light duty passenger vehicles (see Table 8.2) but is noticeably larger for light-duty trucks. For a given driving cycle, heavy cars will require more powerful engines and will consume more fuel, more or less in proportion to vehicle mass (see Table 8.2), than will light vehicles. A very fuel efficient vehicle is necessarily a light one.

For driving at a steady speed, only rolling and aerodynamic resistance must be overcome, the power required being

$$P_{steady} = C_R mgV + (C_D A)\frac{\rho V^3}{2} \tag{8.15}$$

For the typical passenger vehicle, the rolling and aerodynamic resistance are equal at a speed of about 60 km/h. For highway cruising at 120 km/h, the aerodynamic drag would be four times the rolling resistance, the power required being less dependent upon vehicle mass than when driving at low speed.

The parameters of vehicle design that lead to increased fuel economy include low values of vehicle mass m, drag area product $C_D A$, and rolling resistance coefficient C_R. In addition, recovery of vehicle kinetic energy during deceleration for reuse during other portions of the driving cycle will also improve vehicle fuel economy.

8.4.1 Connecting the Engine to the Wheels

Engine power is a maximum at the highest engine speed N_m, but less power is available at lower engine speeds (see Figure 8.4). If we wish to maximize the power available at all vehicle speeds, it is necessary to reduce the ratio of engine speed to wheel speed as the vehicle speed increases. This is accomplished in the transmission, a device attached to the engine that provides stepped speeds to the drive shaft that connects it to the wheels.

There are two forms of transmission: manual and automatic. In a vehicle with a normal transmission, the vehicle operator disengages a clutch and manually shifts to a different gear before engaging the clutch again. In an automatic transmission, a fluid coupling replaces the clutch and gear shifting is accomplished by computer-controlled hydraulic actuators. The more operator-friendly

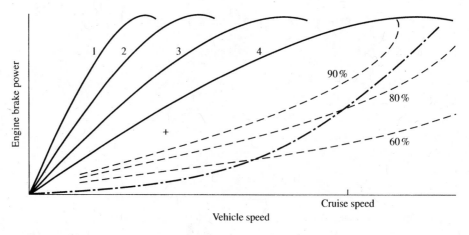

Figure 8.6 A diagram of the maximum engine power provided by a four-speed transmission as a function of vehicle speed (solid lines). The dash–dot line is the steady-speed vehicle power requirement. Contours of relative engine thermal efficiency for lesser power operation in fourth gear are shown as dashed lines, with the 100% peak being marked by +.

automatic transmission is less efficient than the manual one by about 10 percentage points. Most passenger vehicles employ four or five forward speeds and one reverse. The speed ratio between adjacent shifts is about 1.5.

The drive shaft connecting the engine/transmission to the wheels, either front or back (or both in the case of four-wheel drive), utilizes a differential gear to apply equal torque to both drive wheels while allowing different wheel speeds during maneuvering. The drive shaft provides for the vertical motion of wheels with respect to the chassis and for the steering motion when using a front-wheel drive.

The maximum engine power available to a vehicle, as a function of vehicle speed and transmission gear ratio, is shown in Figure 8.6 (solid lines) for a four-speed manual transmission. Each of these four curves is identical to the brake power curve of Figure 8.4, but is plotted here against vehicle speed rather than engine speed. At any vehicle speed, the maximum engine power (which occurs at wide open throttle) is the ordinate of the corresponding gear shift level (1, 2, 3, or 4). At lower vehicle speeds, three or four levels are available, with the lowest yielding the greatest possible power. At the highest speed, only the highest gear can be used to deliver engine power.

For constant-speed travel, the engine power demand, as determined from equation (8.15), is shown as a dash–dot curve in Figure 8.6. It rises rapidly with vehicle speed, as the aerodynamic drag outpaces rolling resistance. At any vehicle speed, the vertical distance between the "demand" (dash–dot) curve and the "maximum supply" (solid) curve is the maximum power available for vehicle acceleration and hill climbing. At low vehicular speeds, this difference is greatest for the lowest shift level, while at the highest speed this difference is least, shrinking to zero at the vehicle's top speed. The lower gear levels are needed to provide high acceleration at low vehicular speed.

It is clear from Figure 8.6 that only a small fraction of the maximum engine power is used at steady vehicle speeds, reaching about 40% at cruising speed. The remainder is available for acceleration and hill climbing. For average vehicle duty cycles, only a fraction of the time is used in acceleration, and the portion of that at which the engine power is a maximum is small. The vehicle capacity factor, the time-averaged fraction of installed power that is utilized by a moving vehicle, is less than 50% for the average duty cycle. Nevertheless, for satisfactory performance reserve power is required over the normal speed range of the vehicle.

Superimposed in Figure 8.6 are the relative efficiency curves of Figure 8.4 (dashed lines), for fourth gear. The 100% peak is marked by +. It is readily apparent that, for a steady vehicle speed, the engine relative efficiency is at best 50% for speeds less than half the cruise speed, rising to about 80% at cruise speed. The normal practice of downshifting at low speeds leads to even lower relative efficiencies. Taking into consideration the short periods of acceleration within a typical driving cycle, the time-averaged relative engine efficiency is certain to be less than 80%, perhaps closer to 60%. This performance could be improved a little by adding a fifth gear level, but that would leave less power reserve for acceleration, necessitating downshifting before accelerating at the higher speeds. An alternative is to employ a continuously variable transmission that can reach higher relative engine efficiencies over a range of steady speeds, but will downshift when acceleration is needed.

8.5   VEHICLE FUEL EFFICIENCY

Traditionally, the efficiency of use of fuel by a vehicle is measured by the distance it moves in a trip divided by the fuel consumed.[11] For road vehicles, this ratio is customarily reported in the units of kilometers per liter (or, in the United States, miles per gallon). The vehicle's fuel efficiency determines both the fuel cost of the trip and the accompanying carbon emissions to the atmosphere, which are almost entirely in the form of carbon dioxide. The vehicle operator is concerned with the fuel cost while national authorities are concerned with the effects of aggregate vehicular fuel consumption on the problems of maintaining a reliable fuel supply and, most recently, on the contribution of vehicles to the national budget of greenhouse gas emissions. In the United States, where fuel retail prices are low compared with most other developed nations, fuel cost is a small fraction of total operating costs of a passenger car, yet it is still a factor in consumer choice of vehicle. Compared with European and Japanese owners, Americans on average drive larger, heavier, less efficient, and more expensive automobiles, but pay less for fuel. National tax policies that greatly affect the price of fuel play a large role in this difference.

The development of vehicle and engine technology that is more fuel efficient and less emitting of pollutants is primarily a response by manufacturers to national policies of regulation and economic tax disincentives that narrow the window of vehicle designs that appeal most to consumers. In this section we consider the technological factors that directly effect vehicle fuel efficiency while meeting the restrictions on vehicle emissions.

8.5.1   U.S. Vehicle Fuel Efficiency Regulations and Test Cycles

In the United States, fuel efficiency of new passenger vehicles and light trucks is regulated by the National Highway Traffic Safety Administration of the U.S. Department of Transportation. This regulation had its origin in the oil shortages of the 1970s caused by an embargo on oil exports

---

[11]We use the technical term *fuel efficiency* interchangeably with the common term *fuel economy,* used in government regulations. A fuel efficient vehicle would be economical to operate, having low fuel cost per distance traveled.

by OPEC nations. To ameliorate the nation's future dependence upon imported oil, Corporate Average Fuel Economy (CAFE) standards were promulgated, beginning with the 1978 model year. These standards required automobile manufacturers to design vehicles so that their sales-averaged vehicle fuel economy[12] did not exceed the level designated for each of two vehicle classes: passenger vehicles and light trucks (pickup trucks, vans, sport utility vehicles).[13] In 2000, the CAFE standards for passenger cars and light trucks were 27.5 miles/gallon (11.7 km/L) and 20.7 miles/gallon (8.8 km/L), respectively.

The measurement of vehicle fuel economy is based upon dynamometer simulations of typical driving cycles for urban and highway travel, originally devised by the U.S. EPA for evaluating vehicle emissions (Urban Dynamometer Driving Schedule, UDDS). The test vehicle is operated in a stationary position, while the drive wheels turn a dynamometer that is adjusted to provide the acceleration/deceleration and air resistance loads as described in equations (8.10) and (8.13). While the dynamometer does not precisely simulate these forces at each point in the test, it suffices to provide a reasonable average for characterizing the emissions and fuel economy of the test cycle.

Two test cycles are used for measurements of vehicle fuel efficiency, one for urban driving and the other for highway travel. Figure 8.7 shows the vehicle speed versus time for each cycle. The urban cycle has many starts and stops (25 in a 17.8-km ride lasting 23 minutes) and a low average speed (while moving) of 31.4 km/h. For an average vehicle, about 34% of the energy needed to propel the vehicle through this cycle is dissipated in braking. In contrast, the highway driving cycle proceeds at a more uniform speed, averaging 78.5 km/h over a 16.5-km run, with only one stop. The two cycles represent two common types of vehicle use: (a) stop-and-go driving in congested urban streets and (b) uncongested freeway travel. For passenger cars, highway travel is between 30% and 50% more fuel efficient than urban travel (see Table 8.2). New vehicle purchasers are notified of the fuel economies for urban and highway travel determined from these tests.[14] For the purpose of enforcing CAFE standards, a weighted average is used (55% urban and 45% highway).[15]

---

[12]While the average fuel economy is reported in units of miles per gallon, the averaging variable is the fuel consumption, measured in units of gallons per mile. Thus the average fuel consumption per mile traveled is regulated.

[13]Passenger cars are called light-duty vehicles (LDV) to distinguish them from heavy-duty passenger vehicles such as buses that carry more than 12 passengers. Light-duty trucks (LDT) are utility vehicles that can carry freight or passengers, whose gross vehicle weight rating (GVWR) is less than 8500 pounds (3.856 t) and frontal area is less than 45 square feet, or can operate off highways (e.g., sport utility vehicle). There are four categories of LDT (LDT1, LDT2, LDT3, and LDT4) depending upon the vehicle GVWR (<6000 lb, >6000 lb) and the loaded vehicle weight (LVW), which is the sum of the vehicle curb weight plus 300 pounds (<3750 lb, <5750 lb). The vehicle curb weight (VCW) is the manufacturer's estimated weight of a fueled vehicle in operating condition.

[14]The consumer guide to vehicle fuel economy discounts the highway and urban test fuel economies by 10% and 22%, respectively, to account partially for disparities in the test procedure compared to average driving conditions. Studies of fuel consumption of vehicles in use indicate higher consumption than the consumer guide values.

[15]The weighting factors are applied to the fuel consumption, which is the reciprocal of the vehicle fuel economy.

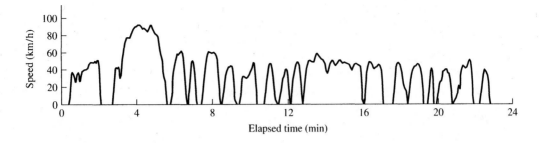

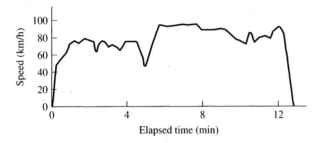

Figure 8.7 U.S. EPA 1972 driving cycles for evaluating emissions and fuel economy. Upper curve: Urban driving cycle (17.77-km length, 22.87-min duration, 31.4-km/h average speed, 91-km/h maximum speed, 1.48-m/s² maximum acceleration, cruise time 7.9%, acceleration time 39.6%, deceleration time 34.6%, idle time 17.8%, 1.43 stops per kilometer). Lower curve: Highway driving cycle (16.5-km length, 12.7-min duration, 78.5-km/h average speed, 96.4-km/h maximum speed, 1.43-m/s² maximum acceleration, cruise time 16.5%, acceleration time 44.4%, deceleration time 38.7%, idle time 0.4%, 0.06 stops per kilometer). (Data from Blackmore, D. R., and A. Thomas, 1977. *Fuel Economy of the Gasoline Engine*. London: Macmillan.)

8.5.2   Improving Vehicle Fuel Economy

In the United States in 1993, a consortium of U.S. vehicle manufacturers and the U.S. federal government undertook to develop a mid-size passenger vehicle that would attain a fuel economy about three times the current value but would have performance properties equal to those of current vehicles, yet would be affordably priced. Known as the Partnership for a New Generation of Vehicles (PNVG), this ambitious program planned for introducing such a vehicle prototype by the model year 2004. The fuel economy goal was judged to be attainable using current or developing technologies. Even if the PNGV program fails to reach its ultimate goal of a highly fuel-efficient vehicle, the technological improvements investigated will make it possible to greatly improve vehicle fuel efficiency of current vehicles.

There are two major thrusts of the PNGV program: improvements to the vehicle design and to the power source. Indeed, both vehicle and power source must be matched to obtain the maximum benefit. Nevertheless, in what follows, we consider these separately, with the understanding that they must not be considered in isolation.

8.5.2.1   Improving Vehicle Performance

In Section 8.4 we described the three components of power consumption in a moving vehicle: vehicle acceleration, rolling resistance, and aerodynamic drag. The first two are proportional to vehicle mass, while the latter depends upon vehicle size and shape, but not mass. The three major parameters determining these components are vehicle mass m, rolling resistance coefficient C_R, and the vehicle drag area $C_D A$. All of these are important.

Vehicle Mass. In current conventional vehicles, mass is the parameter that best correlates fuel economy (see Table 8.2). Large, heavy vehicles require big engines to perform well; they consequently consume more fuel. For a given vehicle size, reducing vehicle mass will permit reductions in engine and transmission mass, tire and wheel mass, braking system mass, fuel storage mass, steering system mass, engine radiator mass, and so on, compounding the gains in direct mass reduction of the vehicle frame. The principal means for reducing mass is the substitution of lighter materials of equal strength and stiffness, such as aluminum alloys or fiber-reinforced plastic for steel and plastic for window glass, as well as the redesign of the vehicle structure to minimize structural mass. Reductions of up to 40% of vehicle mass seem possible.[16]

Reducing vehicle mass by material substitution may have implications for vehicle safety. Vehicle frames are designed to absorb the vehicle's kinetic energy in a crash while protecting the occupants from harm. Substitution of a lighter material of equal strength and energy absorbing capacity in the body structure can maintain the same level of kinetic energy absorbtion and passenger protection, while reducing overall vehicle mass. But in a two-vehicle collision of vehicles of unequal mass, the lighter vehicle absorbs more than its own kinetic energy and thereby suffers a safety disadvantage. In the current U.S. fleet of light-duty passenger vehicles and trucks, the mass ratio of the heaviest to lightest vehicle is about 3. As long as vehicles of different size persist in future fleets, this mass disparity will continue even if all vehicles are made lighter than current ones.

Vehicles with propulsion systems that utilize heavy energy storage systems (electric storage batteries, pressurized gaseous fuel such as natural gas or hydrogen stored in steel cylinders) incur vehicle mass penalties. For lead-acid storage batteries, the battery mass per unit of energy storage is 5000 times that of gasoline, so that vehicle mass for electric battery vehicles is dominated by the battery mass needed to give the vehicle an adequate range. For hydrogen-fueled or natural-gas-fueled vehicles, the storage of the energy equivalent of 60 L of gasoline would add about 300 kg and 130 kg, respectively, to vehicle mass (see Section 4.4.5).

Aerodynamic Resistance. Reduction of aerodynamic resistance is limited to lowering the drag coefficient C_D by careful streamlining of the entire body, because the frontal area A is essentially fixed by the requirements of providing an enclosure for the passengers (see Table 8.2). The bulky form of the automobile limits what can be done to reduce aerodynamic drag, but attention to details can bring the drag coefficient into the range of 0.20–0.25 for passenger vehicles. There does not appear to be a weight penalty attached to low drag coefficients.

---

[16]The cost of direct weight reduction by material substitution is in the range of $1 to $3 per kilogram (Mark, Jason, 1999. *Greener SUVs. A Blueprint for Cleaner, More Efficient Light Trucks.* Cambridge: Union of Concerned Scientists.). But each kilogram of direct weight reduction by material substitution provides an opportunity for additional weight reduction if the engine and drive train are reduced in size in proportion to the vehicle mass reduction, thereby providing a cost saving. This would lower the net cost of the material substitution weight reduction.

Rolling Resistance. The rolling resistance coefficient C_R can be lowered to 0.005 from current values of 0.010–0.014 by improvements in tire design, but it is difficult to maintain durability, performance, and safety (traction) while reducing rolling resistance. Light alloy wheels reduce sprung mass, which is desirable, but more complex suspension systems will be needed to recover normal performance with low-resistance tires.

8.5.2.2 Improving Engine Performance

There are several paths to improving the efficiency of engines while supplying the requisite power needed by the vehicle. One is to improve the fuel (or thermal) efficiency of the engine, especially at off-optimum conditions where the engine is usually operated. A second is to utilize transmissions that permit the engine to maximize its efficiency for a required power output. A third is to reduce engine mass, for a given power, so as to improve vehicle efficiency. However, there are serious constraints imposed on engine efficiency improvements by the requirements for limiting exhaust pollutant emissions. These may limit the possible efficiency improvements.

Reducing Intake Stroke Losses in SI Engines. As explained in Section 8.3, at partial load the cylinder pressure during the intake stroke is lowered to reduce the fuel amount in the cylinder, resulting in a loss of efficiency. This loss may be offset by varying the timing of the inlet and exhaust valves with the engine load. Variable valve timing (VVT) systems are currently used in four-valve engines, adding to the peak power output and engine mass reduction benefit. Another alternative is direct fuel injection (DI) into the cylinder during the intake stroke, forming a nonuniform fuel–air mixture at higher overall pressure and lower intake stroke power loss, albeit at some penalty in NO_x emissions. The recirculation of exhaust gas into the cylinder during the intake stroke so as to reduce exhaust pollutant emissions can be arranged to reduce intake losses and improve engine efficiency at part load.

Replacing SI Engines by CI Engines. The indirect injection CI engine enjoys about a 25% advantage in fuel economy over the SI engine and does not suffer as much efficiency loss at part load as does the SI engine. Direct injection CI engines have even higher fuel economy advantage, about 30–40%.[17] On the other hand, the CI engine is heavier, for a given power, thereby incurring a vehicle mass efficiency penalty, and is more expensive. Also, it is more difficult to reduce NO_x and particulate emissions in the CI engine. If the latter can be overcome, the CI engine provides significant better fuel economy.

Supercharging. Both SI and CI engines can be supercharged (or turbocharged) to increase maximum engine power for a given displacement and engine mass, with some improvement in engine efficiency at higher loads. This provides a vehicle mass efficiency benefit.

Continuously Variable Transmission. As explained in Section 8.4.1, the traditional multistep transmission does not permit the engine to operate at maximum thermal efficiency over the entire

---

[17]In terms of thermal efficiency, the advantage is lower because of the difference in fuel heating value per unit volume of fuel between gasoline and diesel fuel.

vehicle duty cycle. Indeed, the engine seldom operates at best efficiency. A continuously variable transmission (CVT) can be controlled to maximize the engine efficiency at any power level needed by the vehicle. Current CVT transmissions have shown that overall vehicle efficiency can be improved, provided that the transmission efficiency is close to that of conventional transmissions.

Engine Idle-Off. In congested urban driving, considerable time is spent with the engine idling while the vehicle is stationary. If the engine is stopped during these intervals, fuel will be saved and exhaust emissions will be reduced. However, stopping the engine and restarting also consumes fuel, so that the length of the idle period has to be sufficient for there to be a fuel reduction in idle-off control.

8.6   ELECTRIC DRIVE VEHICLES

In the earliest years of the development of the automobile, some were powered by electric drive motors energized by lead-acid storage batteries. Despite their advantages of no noise and exhaust emissions, their limited range and difficulty of recharging confined their use to urban fleet delivery vehicles.

In the last few years, electric drive vehicles have been reconsidered. Advances in electric and mechanical energy storage systems and the energy savings to be made by storing and reusing vehicle braking power make such vehicles attractive from both the emissions and vehicle efficiency point of view. Providing electric drive power from an onboard fuel cell or ICE motor–generator set can provide more acceptable vehicle range. Various combinations of these drive systems might prove even more advantageous. In this section we review the progress made in incorporating technological improvements in vehicles of this type.

8.6.1   Vehicles Powered by Storage Batteries

The development of battery-powered electric drive vehicles was spurred in the United States when the state of California established a future requirement for some zero emission vehicles in its new car fleet mix, beginning with the 2003 model year. Several manufacturers currently produce small passenger cars and pickup trucks that are battery-powered. Their specifications are listed in Table 8.3. Like other vehicles of this type, they have limited passenger- or freight-carrying capability compared to conventional vehicles and have much smaller travel range between recharges of the onboard energy supply.

All these vehicles are powered by cooled AC induction motors drawing their power from battery banks via power conditioning equipment. The traction motor is geared to the drive wheels at a fixed speed ratio, and it can regenerate a partial battery charge during periods of vehicle deceleration.

The EV1 is equipped with either a lead-acid or nickel metal hydride battery bank, with the latter storing 40% more energy with 12% less weight than the former and thereby achieving a 36% greater range. The time needed to charge the battery bank completely is 6–8 hours. After about a thousand charge–discharge cycles the batteries must be replaced by new ones, but of course the

TABLE 8.3 2000 Model Year Electric Vehicle Characteristics

Manufacturer Model Size Battery Type	GM EV1 Two-passenger Lead Acid	GM EV1 Two-passenger NiMH	Toyota RAV4 Five-passenger NiMH	Chevrolet S-10 Pickup Lead Acid	Ford Ranger-EV Pickup Lead Acid	Ford Ranger-EV Pickup NiMH
Maximum electric power (kW)	102	102	50	85	67.2	67.2
Vehicle mass (t)	1.4	1.32	1.56	2.24	2.13	1.86
Frontal area (m^2)	1.8	1.8		1.95		
Battery mass (kg)	595	521		621		
Battery energy storage (kWh)	18.7	26.4		16	23	
Vehicle range (km)	89–153	121–209	201	64–89	123	145
Charge time (h)	5.5–6	6–8	6–8	6–7		
Urban/highway energy efficiency (km/MJ)	1.72/1.60	1.32/1.49			1.18/1.02	1.12/0.99
Urban/highway equivalent fuel efficiency (km/L)[a]	16.3/15.1	12.5/14.1			11.2/9.7	10.6/9.4
Battery/vehicle mass	0.425	0.395		0.318		
Power/mass (kW/t)	56.7	56.7	32.1	43.6	31.5	36.1
Battery energy/mass (Wh/kg)	31	51		26		
Battery energy/maximum power (h)	0.18	0.26		0.19		

[a] Assumes 30% equivalent engine thermal efficiency, or 9.48 MJ/L.

depleted batteries can be recycled to recover battery materials.[18] Battery replacement will cost several thousand dollars every 50,000 miles.

The battery mass of these vehicles constitutes about 40% of the vehicle mass, implying a very large vehicle mass penalty to the vehicle energy efficiency. Indeed, if we convert the vehicle electric energy efficiency of 1.0–1.7 km/MJ to an equivalent fuel efficiency of 10–16 km/L (see Table 8.3), then these vehicles are only slightly more efficient than current vehicles of comparable mass listed in Table 8.2. Despite the recovery of braking energy, the electric vehicles are unable to achieve higher fuel efficiency than comparable conventional ones.

The power/mass ratios of these vehicles, which lie in the range of 30–55, are only half those of conventional vehicles (Table 8.2), limiting their acceleration and grade climbing abilities.

The limited range and fuel efficiency of current battery-powered electric vehicles is tied to the low energy storage densities of currently available batteries, about 25–50 Wh/kg (90–180 kJ/kg). But other electrical storage systems being developed, such as electric capacitors and flywheels, have no higher energy storage densities (see Table 4.1), and other types of high-density storage

---

[18] In a lead-acid battery, the electrolyte sulfate ions are deposited on the surfaces of the electrodes during battery discharge. This process is not entirely reversed upon recharge, gradually leading to the loss of electrode active surface area and the energy storage capacity of the battery with each recharge cycle.

batteries are unsuitable for automobiles (see Table 4.2). Adding battery mass to the vehicle will not bring about a commensurate increase in range, which is better improved by increasing the efficiency with which stored energy is utilized in propelling the vehicle.

8.6.2 Hybrid Vehicles

Hybrid vehicles are those that combine conventional power sources (SI or CI engines) with electric motors to power the vehicle. In one extreme, the SI or CI engine can power a generator that delivers electric power to an electric traction motor; this system is used in railway engines. The more attractive system for automobiles is to attach an electric motor/generator to the combustion engine shaft, between the engine and the transmission. The motor/generator can store energy in a battery bank when excess power is available, during deceleration or when the power need is less than what the combustion engine can deliver, and can deliver extra power to the wheels when it is temporarily needed for acceleration or hill climbing. Such a system can level the peaks and valleys of power demand in the driving cycle, enabling the combustion engine to run closer to average driving cycle power and potentially better average engine efficiency.

The characteristics of two hybrid vehicles of this type are listed in Table 8.4. In the Honda vehicle, the electric motor/generator maximum power is only 10 kW compared to the SI engine maximum power of 47 kW. Nevertheless, this is sufficient to provide extra power for acceleration and absorb braking power in the urban driving cycle. The ratio of electrical energy storage to maximum electrical power is greater than that for the battery-powered vehicles (see Table 8.3). The vehicle fuel economy is impressive, a weighted average of 27.5 km/L (64.7 miles/gallon). While this is still short of the PNGV goal of 35.1 km/L (82.5 miles/gallon), and the vehicle is not mid-size, it is a promising development.

TABLE 8.4 2000 Model Year Hybrid Electric Vehicle Characteristics

Manufacturer Model Size	Honda Insight Two-passenger	Toyota Prius Five-passenger
Displacement (L)	1	1.5
Number of cylinders	3	4
Maximum engine power (kW)	47	43.3
Maximum bmep (bar)	13.3	
Maximum electric power (kW)	10	33
Vehicle mass (t)	0.84	1.24
Frontal area (m^2)	1.83	2.04
Engine power/displacement (kW/L)	47.0	
Engine power/vehicle mass (kW/t)	56.1	
Battery mass (kg)	31.3	62.6
Battery energy storage (kWh)	9.4	18.8
Urban/highway fuel efficiency (km/L)	25.9/29.8	18.3/22.5
Battery/vehicle mass	0.038	0.050
Battery energy/maximum electric power (h)	0.94	0.57

The motor/generator of the Honda hybrid serves two other purposes. The engine is stopped when the vehicle is stationary (idle-off), and it is restarted using the motor while the transmission is in neutral. In addition, the motor/generator armature acts as the motor flywheel. Because the motor/generator is attached to the engine shaft, it operates through the variable speed transmission when adding or subtracting power to the wheel drive system, enabling the motor/generator to operate efficiently over a wide vehicle speed range.

Comparing the Honda and Toyota hybrids of Table 8.4 with the conventional Honda and Acura vehicles of Table 8.2 that are of comparable vehicle mass, we note that hybrid vehicles are about 80% more fuel efficient (for equal mass) and that for either hybrid or conventional vehicle, a small fractional reduction in mass approximately begets a comparable fractional increase in vehicle fuel efficiency. This substantial advantage of the hybrid over conventional design of equal mass is a composite of braking energy recovery, lower rolling and aerodynamic resistance, more efficient engine operating conditions (despite the same driving cycle), and possibly a higher peak engine efficiency.

Evolution of standard vehicles in future years will move in the direction of hybrid drives. It is expected that the electric auxiliary system will change from 14 to 42 volts and triple in power. The current starter motor and belt-driven alternator will be replaced by a motor/generator directly connected to the engine drive shaft, as in the Honda vehicle of Table 8.4, permitting idle-off operation when the vehicle is stationary. Recovery of braking energy would be possible, depending upon the motor/generator power and electric storage capacity. Although the purpose of this development is to utilize electric drive for auxiliary power and thereby improve engine efficiency, it clearly can be extended to become a hybrid system.

8.6.3 Fuel Cell Vehicles

Prototypes of electric drive vehicles whose electric power is supplied by fuel cells have been under development for several decades. Potentially, such vehicles could provide higher vehicle fuel efficiencies than conventional vehicles with little or no air pollutant emissions. Increasingly stringent exhaust pollutant emission standards, especially in California, and national policies in developed nations to secure both economic and environmental benefits of improving fuel economy have increased the incentives for manufacturers to develop vehicle fuel cell technologies.

As explained in Section 3.12, the oxidation of a fuel in a fuel cell has the potential to convert a higher percentage of the fuel's heating value to electrical work than does the typical combustion engine. The upper limit to this proportion is the ratio of the free energy change in the fuel oxidation reaction, Δf, to the enthalpy change, or fuel heating value FHV. For hydrogen, this ratio is 0.83, while for methane and methanol it is 0.92 (see Table 3.1); these upper limits are at least double what could be obtained by burning these fuels in a steam or gas turbine power plant. However, as Figure 3.10 illustrates, this high conversion efficiency is only reached at zero power output; at higher power the cell voltage declines nearly linearly with increasing cell current, resulting in only 50% of the upper value being recoverable at maximum cell power (41.5% for hydrogen and 46% for methane and methanol). Still, these are higher fuel efficiencies than are obtainable in vehicle SI engines at full power; the comparison is even more favorable to the fuel cell at part load because the fuel cell efficiency increases at reduced load.

But there are countervailing factors in vehicle fuel cell systems that lower this fuel efficiency. The only practical fuel cell for vehicle use requires gaseous hydrogen fuel. The most economical and energy efficient source of hydrogen, a synthetic fuel, is by reforming from a fossil fuel such as natural gas, oil, or coal, or from another synthetic fuel like methanol or ethanol. Whether this reforming is done at the fuel depot, where the hydrogen is then loaded and stored on the vehicle, or is accomplished on board the vehicle in a portable reformer, the reforming operation preserves at best only 80% of the parent fuel's heating value (see Table 3.3). This additional loss lowers the maximum power fuel efficiency of hydrogen to 33.2%. Furthermore, if hydrogen is liquified for storage on the vehicle, rather than being stored as a compressed gas in tanks, an additional energy penalty of about 30% is incurred because energy is needed to liquify hydrogen at the very low temperature of $-252.8\,°C$. Altogether, these synthetic fuel transformation penalties diminish the fuel efficiency advantage of fuel cells compared to conventional internal combustion engines in vehicles fueled by conventional hydrocarbon fuels.[19]

The hydrogen fuel cell used in vehicles utilizes a solid polymer electrolyte, called a proton exchange membrane (PEM). Only a fraction of a millimeter in thickness, the membrane is coated on both sides with a very thin layer of platinum catalyst material that is required to promote the electrode reactions producing the flow of electric current through the cell and its external circuit, as described in Section 3.12. Carbon electrodes, provided with grooves that ensure the gaseous reactants [hydrogen at the anode and oxygen (in air) at the cathode], are distributed uniformly across the electrode surfaces and are sandwiched on either side of the PEM, forming a single cell. As many as a hundred or more cells are stacked in series, mechanically and electrically. The fuel and oxidant are supplied under a pressure of several bar, so as to increase the maximum power output per unit of electrode area, which is usually of the order of 1 W/cm^2. Water, formed at the cathode, must be removed, but the PEM must remain moist to function properly. In addition, heat is released in the cell reaction, so that the cell must be cooled, maintaining a temperature less than $100\,°C$.

Table 8.5 lists the salient features of several fuel cell vehicles developed during the 1990s in Europe and the United States: two versions of a compact car (NECAR 4 and NECAR 3), a U.S. compact car (P2000), a van (NECAR 2), and a municipal bus (NEBUS). They all are based upon a conventional vehicle with the replacement of the engine and fuel tank by a fuel cell, electric motor, and fuel storage system, while maintaining the original vehicle's passenger capacity and range. The arrangement of the replacement fuel cell system in the vehicle is shown in Figure 8.8 for the NECAR 4 and NEBUS vehicles.

NECAR 4 is the most recent version of the Daimler compact car. Its hydrogen fuel is stored in liquid form in a cryogenic insulated fuel tank. The previous version of this car, NECAR 3, utilizes liquid methanol as the fuel, which is converted to hydrogen in an onboard reformer. NECAR 2 and NEBUS store hydrogen gas in tanks pressurized to 300 bar and weighing about 75 times more than the fuel they contain, adding to the vehicle mass. In general, these vehicles weigh more than their conventional counterparts.

The values for fuel economy of these vehicles, expressed as kilometers per megajoule of fuel heating value (km/MJ) or kilometers per liter (km/L) of gasoline equivalent (in terms of fuel heating value), are listed in the table as well. For comparison, Table 8.5 also lists the fuel economies of the conventional vehicles of the comparable class, both CI- and SI-powered. The fuel cell vehicles have comparable fuel economies, but are clearly not distinctly superior to their conventional cousins.

---

[19]Fuel cells used in electric power systems do not suffer such disadvantages. They do not require synthetic fuel and operate at high temperatures where coolant streams can be used to generate steam for extra electric power output.

TABLE 8.5 Characteristics of Prototype Fuel Cell Vehicles

Manufacturer	Daimler	Daimler	Ford	Daimler	Daimler
Model	NECAR 4	NECAR 3	P2000	NECAR 2	NEBUS
Vehicle Class	A (compact)	A (compact)	(midsize)	V (van)	O405N(bus)
Passengers	5	5		6	34
Vehicle mass (t)	1.48	1.48	1.52	2.7	14
Frontal area (m^2)	2.2	2.2		3.6	7
Electric power (kW)	55	50	90	50	250
Electric power/vehicular mass (kW/t)	37.2	33.8	59.2	18.5	17.9
Number of cells	320	300	400	300	1500
Cell mass (kg)	275	300	173	300	1400
Fuel cell specific power (kW/kg)	0.20	0.17	43	0.17	0.18
Fuel type	Hydrogen (l)	Methanol (l)	Hydrogen (g)	Hydrogen (g)	Hydrogen (g)
Fuel mass (kg)	5	30.1	1.41	5	21
Fuel heating value (MJ)	709	717.2	200	709	2977
Fuel volume (L)	70.6		38	280	1050
Fuel tank mass (kg)				80	1900
Range (km)	450	400	161	250	250
Top speed (km/h)	145		145	110	80
Vehicle energy economy (km/MJ)[a]	0.635	0.558	0.5	0.353	0.084
Equivalent fuel economy (km/L)[a,b]	20.0	17.6	15.8	11.1	2.65
Equivalent fuel economy (km/L)[b,c]	11.2	14.1	12.6	8.9	2.12
CI class fuel economy[d] (km/L)	20.0	20.0		11.3	
SI class fuel economy (km/L)	14.1	14.1		8.7	

[a] Does not include synthetic fuel energy penalty.
[b] Converted to energy density of gasoline.
[c] Includes penalties of 20% for methane conversion to hydrogen or methanol and 30% for hydrogen gas to liquid.
[d] Adjusted for volume equivalent of gasoline.

Figure 8.8 Left: Artist's sketch of the interior layout of the fuel cell (center) and liquid hydrogen fuel tank (rear) in the NECAR 4 vehicle. Right: The NEBUS municipal bus, showing the fuel cell in the lower rear and the hydrogen gas storage tanks on the roof.

But if one considers the synthetic fuel penalties involved in preparing hydrogen, both gaseous and liquid, or methanol, as done in Table 8.5, then the fuel cell vehicles, at least at this stage of their development, have little or no fuel economy advantage.

There is no question that fuel cell vehicles suffer from the limited availability and high cost of hydrogen fuel. Commercial-scale hydrogen reformers fueled by natural gas are complex systems, with the product stored at high pressure in tanks for transfer to vehicles. Hydrogen is especially prone to burn and explode, raising safety problems even for fleet vehicles that can be more carefully supervised than individually owned vehicles. On the other hand, on-board reformers eliminate the hydrogen supply problem, yet they work satisfactorily only with pure fuel input, such as methanol or ethanol, but not so far with diverse hydrocarbon mixtures like gasoline or diesel fuel. Development of on-board reformers (such as used in NECAR 3) and a common fuel would make fuel cell vehicles suitable substitutes for conventional ones.

8.7 VEHICLE EMISSIONS

By the middle of the twentieth century, vehicle exhaust emissions were recognized to be an important contributor to urban photochemical air pollution, especially in locations like southern California where high insolation and temperature and poor atmospheric ventilation combined with a rapidly growing automobile population to produce record levels of ground level ozone concentrations. Not long thereafter, similar pollution problems appeared in other major cities around the world as urban vehicle populations blossomed. These problems are more acute in lower latitude locations, especially in developing countries where vehicle emission controls are not yet stringent.

Of course, vehicles provide only part of ozone precursor emissions. But they are mobile and more numerous than stationary sources, and they present different problems for abatement. Early in the pollutant regulatory history of the United States, it became obvious that it was more effective to require a few vehicle manufacturers to install control equipment on millions of new vehicles rather than to require millions of vehicle owners to try to reduce their own vehicle's emissions. This scheme, adopted by all developed nations, replaces all vehicles with new and cleaner ones every 12–15 years, providing the opportunity to capitalize on the improvements in emission control technology. In the United States, it has resulted in significant reductions in air pollutant emissions from vehicles.

In this section we discuss the vehicle technologies that are used to reduce the emissions from ICE engines in vehicles in response to national regulations.

8.7.1 U.S. Vehicle Emission Standards

Vehicle emissions to the atmosphere are of two kinds: exhaust emissions and evaporative emissions. The first are the combustion gases emitted while the engine is running, whether or not the vehicle is moving. The second are emissions of fuel vapors from the fuel supply system and the engine, when the vehicle is stationary with the engine not operating.[20] The federal government regulates both of these emissions by requiring the manufacturers of new vehicles sold in the United States

---

[20]Emission of fuel vapor while refueling may be separately regulated by state agencies under state implementation plans for conforming with federal air-quality regulations.

to provide the technology needed to limit these emissions for the useful life of the vehicle and to warrant the performance of these control systems.

To certify a vehicle class for exhaust emissions, the manufacturer must test a prototype vehicle on a dynamometer following the Federal Test Procedure (FTP) (see Section 8.5.1), during which exhaust gases are collected and later analyzed for pollutant content. Regulated pollutants include nonmethane hydrocarbons (NMHC) or organic gases (NMOG), carbon monoxide (CO), nitrogen oxides (NO_x), particulate matter (PM), and formaldehyde (HCHO). The mass of each pollutant collected from the exhaust during the test is divided by the test mileage and is reported as grams per mile. If the prototype vehicle's exhaust emissions do not exceed the standards set for its vehicle type, vehicles of its class and model year may then be sold by the manufacturer. The manufacturer is further responsible for ensuring that their vehicles' control systems continue to function properly during the life of the vehicles, currently set at 100,000 miles. Vehicles must also conform to the exhaust emission limitations of the Supplemental Federal Test Procedure (SFTP), designed to evaluate the effects of air conditioning load, high ambient temperature, and high vehicle speeds (not included in the FTP) on emissions.

Evaporative emissions are tested for two conditions: one where the vehicle is at rest after sufficient use to have brought it to operating temperature, the other for a prolonged period of nonuse. In these tests the vehicle is enclosed in an impermeable bag of known volume, and the organic vapor mass is subsequently determined.

In the United States, vehicle emission standards are set by the U.S. Environmental Protection Agency in accordance with the provisions of federal air-quality legislation. The regulation is based upon the recognition of the ubiquity and mobility of the automobile, its concentration in urban areas, its contribution to urban and regional air-quality problems, and the ability of the manufacturer, and not the owner, to ameliorate its emissions. In the years since the early 1970s, when regulation was first introduced, emission standards have become more stringent as the manufacturers devised better technologies and the difficulty of achieving desirable air quality throughout the United States became more apparent. Given the lead time required by the manufacturers to develop new control technologies and incorporate them in a reliable consumer product, emission standards must be set years in advance of their attainment in new vehicles sold to the consumer.

U.S. exhaust emission standards for vehicles of model year 1996 and beyond are listed in Table 8.6. (Emission standards for earlier model year vehicles are shown in Table 9.2.) The standards apply for two time periods; 1996 to 2007 (Tier 1) and 2004 and beyond (Tier 2). The Tier 2 standards are phased in over the period 2004–2010, during which the Tier 1 standards are simultaneously being phased out.

Tier 1 standards limit four pollutants for five vehicle classes: light-duty vehicles (LDV, which are passenger vehicles for 12 passengers or less) and four types of light-duty trucks (LDT1, LDT2, LDT3, and LDT4, distinguished by the gross vehicle weight rating and the loaded vehicle weight [see Section 8.5.1]). The larger light-duty trucks are permitted greater emissions in recognition of their greater weight-carrying capability. Like earlier standards, the Tier 1 standards apply to several vehicle classes, but within each class each vehicle model, small or large, must meet the same standard.

Tier 2 standards introduce a new method of limiting emissions. Like the fuel economy standard, each manufacturer must achieve a sales-averaged NO_x emissions for all its vehicles of 0.07 g/mile, although individual vehicle models may emit more if they are offset by others that emit less. Each vehicle model is certified in one of seven emission categories (denoted by Bin 1, . . . , Bin 7 in Table 8.6), which ensures that the sales-averaged emission limits for pollutants other than NO_x will

TABLE 8.6 U.S. Vehicle Exhaust Emission Standards

Vehicle Type	Tier 1[a]			
	NMHC[b] (g/mile)	CO (g/mile)	NO$_x$ (g/mile)	PM (g/mile)
LDV	0.25	3.4	0.4	0.08
LDT1	0.25	3.4	0.4	0.08
LDT2	0.32	4.4	0.7	0.08
LDT3	0.32	4.4	0.7	
LDT4	0.39	5.0	1.1	

Vehicle Type	Tier 2[c]				
	NMOG[d] (g/mile)	CO (g/mile)	NO$_x$ (g/mile)	PM (g/mile)	HCHO[e] (g/mile)
All	0.09	4.2	0.07	0.01	0.018
Bin 7	0.125	4.2	0.20	0.02	0.018
Bin 6	0.090	4.2	0.15	0.02	0.018
Bin 5	0.090	4.2	0.07	0.01	0.018
Bin 4	0.055	2.1	0.07	0.01	0.011
Bin 3	0.070	2.1	0.04	0.01	0.011
Bin 2	0.010	2.1	0.02	0.01	0.004
Bin 1	0.000	0.00	0.00	0.00	0.000

[a] Model years 1996–2007. Five years, 50,000 miles.
[b] Nonmethane hydrocarbons.
[c] Model years 2004–2010, except 2008–2010 for LDT3 and LDT4. Full useful life (120,000 miles).
[d] Nonmethane organic gases.
[e] Formaldehyde.

also not be exceeded. This new method of limiting emissions will allow manufacturers to achieve the necessary overall reduction in their fleet's emissions as economically as possible by providing incentives to reduce emissions below the standard in light, low-powered vehicles that may then be credited to heavy, high-powered ones.

The Tier 2 standards, which begin with the 2004 model year, are quite stringent compared with those of the early 1970s, when standards were first applied (see Table 9.2). For the ozone precursor pollutants, NO$_x$ and NMOG, Tier 2 levels are about 2% of those for 1971 model year vehicles and are approximately 0.2% of unregulated 1960's vehicles. This considerable reduction will be required to make it possible to achieve ambient ozone standards in U.S. metropolitan areas in the early part of the twenty-first century, despite increasing vehicle population and increased annual travel per vehicle.

8.7.2 Reducing Vehicle Emissions

Vehicle exhaust pollutants are the remnants of an incomplete and nonequilibrium combustion process in the engine cylinder. Of the mixture of fuel and air introduced into the cylinder, all

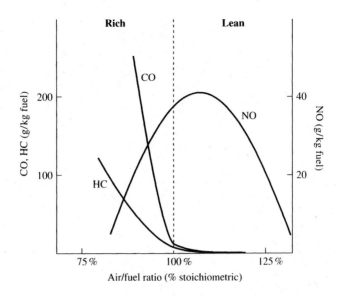

Figure 8.9 A diagram of the mass of exhaust gas pollutants—carbon monoxide (CO), nitrogen oxide (NO), and hydrocarbons (HC)—in a typical SI engine, as a function of the air/fuel ratio.

but a tiny fraction of the fuel is oxidized to CO_2 and H_2O, reaching a state of thermochemical equilibrium after releasing the chemical energy bound up in the fuel. However, a small amount of the reactants do not reach this equilibrium state, but instead remain frozen into a metastable form. The principal molecules of this type are NO, CO, and various kinds of hydrocarbon (HC) molecules, all of which can contaminate the atmosphere into which the vehicle exhaust gas stream is introduced. The purpose of vehicle emission control technology is to reduce the amounts of these pollutants to such low values that the cumulative effects of many vehicles and other sources will not be great enough to cause any damage to living systems, including humans.

In SI engines, the amount of each of these principal pollutants is sensitive to the air/fuel ratio of the mixture inducted into the cylinder prior to ignition by the spark plug. The proportion of air to fuel must not be too far from the stoichiometric value for the engine to function properly and efficiently. If the mixture is fuel-rich (more fuel than can be completely oxidized by the available oxygen), some CO will be formed and not all of the fuel's heating value will be released. If the mixture is fuel-lean (excess, unused oxygen), the combustion product temperature and pressure will be less, resulting in less engine work per cycle. These differences in the chemical and thermodynamic state of the combustion gases influence the amounts of pollutants that leave the engine through the exhaust port.

Figure 8.9 shows how the mass of exhaust gas pollutants varies with the air/fuel ratio in a SI engine. In rich mixtures, there is insufficient oxygen to oxidize completely the fuel molecules, leaving some unburned HC or incompletely oxidized CO and H_2, the more so as the oxygen deficiency becomes larger.[21] When there is surplus oxygen (lean mixture), CO and HC diminish

---

[21]The hydrogen, not shown in Figure 8.9, is formed through the reduction of H_2O by CO at the end of the combustion process. This hydrogen aids the catalytic reduction of NO in the exhaust converter.

to low values as the extra oxygen finds and oxidizes them. On the other hand, NO is formed by the reaction of N_2 with O_2 at the high temperature behind the flame front, but only in very small quantities, and doesn't revert completely to N_2 and O_2 at the much lower exhaust temperature, as it should if thermochemical equilibrium prevailed. It is highest at or close to the stoichiometric mixture where the flame temperature is highest. The mass of HC and NO can be of the order of 1 % of the fuel mass (about 0.1 % of the exhaust gas mass), but the CO mass is 10 times larger. Such values are 10 or more times higher than allowed by future U.S. exhaust emission standards.

For pollutants to reach the very low levels now being required of road vehicle exhaust streams, two steps must be undertaken simultaneously. The first is to reduce as much as possible the pollutant concentrations in the exhaust gas as it leaves the engine (engine-out emissions); the second is to reduce these emissions even further by exhaust gas treatment systems located between the engine and the tailpipe. Neither of these systems by itself can suffice to clean up the emissions to the required low levels.

8.7.2.1 Reducing Engine-Out Emissions

There are several features of modern SI engines that are nearly universally used to improve engine-out emissions.

Precise Control of Air/Fuel Ratio. Low values of the three principal pollutants—HC, CO, and NO—can be maintained if the air/fuel ratio is kept close to its stoichiometric value under all operating conditions. Fuel injection permits close control over fuel flow to each cylinder, and it can be computer-controlled to be proportionate to the intake air flow. An oxygen detector placed downstream of the exhaust ports provides a sensitive signal used to correct the fuel flow so as to home in on the desired air/fuel ratio. A further benefit of this control system is that it can provide optimum conditions for subsequent exhaust gas processing.

Exhaust Gas Recirculation. At the end of the exhaust stroke, when the exhaust valve has closed and the intake valve opens to admit a fresh charge of air–fuel mixture, the residual volume of the cylinder is filled with exhaust gas. This mixes with the incoming fresh charge, diluting it and reducing the temperature and pressure that is reached when that charge is fully burned at the beginning of the power stroke. Because the amount of NO formed is very sensitive to the peak temperature reached during combustion, we can reduce engine-out NO by diluting the fresh charge with even more exhaust gas than is normally encountered. This can be done by varying the exhaust and inlet valve timing or pumping exhaust gas from the exhaust system into the intake system. This is done at part load so the maximum engine torque and power are not compromised, but is acceptable because these maximum values are seldom utilized in standard driving cycles.

8.7.2.2 Catalytic Converters for Exhaust Gas Treatment

The exhaust gas pollutants—HC, CO, and NO—are not in thermochemical equilibrium with the rest of the exhaust gas. It should be possible to oxidize both HC and CO to CO_2 and H_2O if enough oxygen is present, and to reduce NO to N_2 and O_2, because these are thermodynamically favored. To make this happen quickly enough, these molecules must attach themselves to a solid

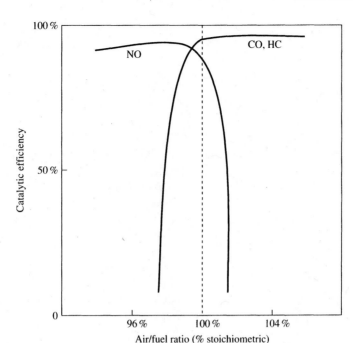

Figure 8.10 A diagram of the catalytic efficiency for CO and HC oxidation and NO reduction in a three-way catalytic converter, as a function of the air/fuel ratio.

surface coated with a catalyst, where they can react and their products evolve into the gas stream. Furthermore, this surface reaction will only occur quickly if the surface is hot enough and the proper catalyst is used. Current three-way oxidation–reduction catalysts utilize such catalysts as platinum and rhodium, and they must be heated to 250 °C or more to be effective.

Simultaneous oxidation of CO and HC and reduction of NO in a catalytic converter requires very close control of the air/fuel ratio in the engine. Insufficient air will inhibit oxidation, whereas too much will prevent reduction. The window of air/fuel ratio that will remove equally all the pollutants is quite small, only a few percent change being allowed. This is illustrated in Figure 8.10 showing how the catalytic efficiency (percent of pollutant removed in the converter) of the oxidation and reduction reactions depend critically upon the air/fuel ratio. Oxygen accumulation on the catalyst surface is built into these catalysts to widen the operating window.

For a catalytic converter to work properly, every pollutant molecule must have the chance to stick to the catalyst surface before it flows through the reactor. This requires that there be a large surface area coated with catalyst and that the flow passages surrounding this surface be finely divided. Either a honeycomb structure or a packed bed of catalyst-coated pebbles satisfies this requirement. Typically, the gas passage dimension is of the order of several millimeters, and the converter volume is about half the engine displacement. This allows the exhaust gas only about one engine cycle period to pass through the converter and be cleansed of most of the pollutants.

When an engine is first started at ambient temperature (called a cold start), the converter does not work until it has been warmed by the hot engine exhaust gas to its "light-off" temperature of about 250 °C. During the minute or two that is required to reach converter light-off conditions, high engine-out pollutant levels are emitted from the tailpipe. Indeed, more than half of the emissions in the Federal Test Procedure may be emitted during this warm-up period. When engines are started in cold winter weather, excess fuel must be injected to achieve sufficient fuel evaporation to start the engine, giving greatly increased CO and HC emissions. Preconverters of low heat capacity, located close to the engine or electrically heated, are required to achieve low emission levels during cold starts.

Catalyst surfaces may be damaged by overheating if the exhaust gas contains excessive unburned fuel, which might occur if the air/fuel control system fails. Also, fuel impurities that leave surface deposits may destroy the catalytic function. Lead additives to gasoline have been phased out to protect converters, and current U.S. fuel regulations will require removal of nearly all sulfur in the near future. Ultraclean fuels will ensure that converters will not deteriorate over the useful life of the vehicle.

8.7.2.3 Evaporative Emissions

Exhaust emissions of nitrogen oxides and incompletely burned fuel contribute to the formation of ground-level ozone in the atmosphere. But the inadvertent escape to the atmosphere of vehicle fuel vapor also can be an important contributor to the reactive organic compounds that participate in the formation of ozone. As a consequence, evaporative emission control systems are required on new U.S. vehicles.

There are several sources of vehicle evaporative emissions. Fuel stored in the fuel tank emits vapor into the air space above the fuel surface within the tank. This vapor can leak to the atmosphere during fuel refilling operations and during diurnal atmospheric temperature and pressure changes. At engine shutdown, unburned fuel remains in the engine and can subsequently leak to the atmosphere from the air intake or exhaust.

Fuel vapor has a different chemical composition than does the fuel because the vapor constituents are much richer in higher-vapor-pressure, lower-molecular-weight components than is the liquid fuel. Some of these components are also more chemically reactive in ozone formation, making it all the more important to prevent their release to the atmosphere. Fuel vapor pressure increases rapidly with temperature, so that uncontrolled evaporative emissions are higher in summer than in winter. Because ozone formation is inherently higher in summer than in winter, escape of fuel vapors exacerbates the summertime smog problem.

Fuel tank vapor emissions are controlled by placing a vapor adsorbing filter in the vent line between the fuel tank air space and the atmosphere. If the fuel and tank air space warm up from solar heating, expelling some vapor–air mixture through the vent line, the filter will retain the vapor molecules on its adsorbing surface. To prevent the adsorbing surface from becoming saturated with fuel molecules, and thereby ineffective for further filtering, air is drawn inward through the filter when the engine is running, cleaning it of adsorbed vapor molecules. This inflow is ducted into the engine intake system so as to incinerate the desorbed vapor in the running engine.

When a fuel tank is filled, the vapor–air mixture in the tank air space is displaced by the incoming fuel. This mixture preferentially escapes through the fuel fill opening; and if it is not collected during the filling process by a vapor control system at the filling station, it will be emitted into the atmosphere. U.S. states with ozone exceedance problems usually require the installation of such equipment by service stations.

8.7.2.4 Reducing CI Engine Emissions

Diesel (CI) engines generally emit lesser amounts of CO and HC than SI engines, but more of NO and particulate matter. The overall air/fuel ratio in a CI engine is always lean, the more so at partial power levels, providing excess oxygen to oxidize HC and CO to CO_2 and H_2O, and N_2 to NO. But the evaporation and mixing of the fuel droplets in a CI engine with surrounding air is uneven, so that some of the fuel burns in an oxygen-deficient atmosphere, giving rise to tiny solid elemental carbon (soot) particles. Most of this particulate matter eventually is oxidized, but some does not and remains unburned in the engine exhaust stream. Soot particles may be coated with low-volatility hydrocarbon molecules (polycyclic aromatic compounds, PAH) which are toxic to humans. Both NO and PM emissions are more difficult to control in CI engines than in SI engines.

Various modifications to direct injection diesel engine combustion are being tried to reduce both NO and PM engine-out emissions. One approach is to increase the fuel injection pressure and control its timing so as to provide more uniform fuel–air mixing and thereby better combustion conditions. Four-valve cylinders and exhaust gas recycling also help to control the combustion process so as to reduce emissions. If the mixture of fuel and air within the cylinder can be made nearly uniform (called homogeneous charge compression ignition), emissions are significantly reduced, at least at part load.

The use of catalytic converters to reduce NO molecules in the engine exhaust is less successful for CI engines than for SI engines because there are fewer hydrogen-containing molecules needed for catalytic reduction of NO. Injecting small amounts of fuel into the exhaust improves NO catalytic reduction. Nevertheless, substantial catalytic conversion will be necessary to meet Tier 2 NO_x emissions standards for CI-powered vehicles.

Catalytic converters oxidize some of the engine-out PM, but not enough to meet future PM emissions standards. Particle filters can further reduce tailpipe PM emissions, but periodic cleaning of the filters by catalytic combustion or other means is necessary to ensure reliable PM reduction. Particle filters have not yet reached the level of development of catalytic converters, but are likely to be necessary for future diesel-powered light duty vehicles in the United States.

CI engine emissions are affected somewhat by the composition of diesel fuel. The greatest effect is due to fuel sulfur, which burns to SO_2 and hampers NO reduction in the catalytic converter. The sulfur content of vehicle fuels is currently regulated by the U.S. EPA to maintain good lifetime performance of catalytic converters.

The superior fuel efficiency of direct injection CI engines is a strong incentive for their use where consumer fuel cost is high. In the United States, where fuel prices are low, the fuel efficiency incentive resides with the manufacturer, who must meet CAFE standards, especially difficult for the current light-duty truck market. Should reduction of vehicle carbon dioxide emissions become a public policy goal in the United States, thereby encouraging greater use of CI engines in the passenger vehicle market, more intense development of CI engine emission control technology will be necessary to meet expected emission standards.

8.7.2.5 Fuel Quality and Its Regulation

We have already noted that fuel anti-knock lead additives and sulfur have been restricted to ensure the successful operation of exhaust gas catalytic converters. Other regulation of fuel properties have been directed at both exhaust and evaporative emissions.

To achieve desirable anti-knock properties, fuel refiners change the composition of the fuel, utilizing more volatile components that increase vapor pressure and thereby evaporative emissions and are more prone to generate ozone. It has been found that the addition of oxygenated fuel components, such as methanol or ethanol, improves fuel performance and reduces exhaust emissions, especially in older vehicles, so incentives to employ these additives have been utilized. Another fuel additive, MTBE (methyl tertiary butyl ether), has been required by some states with ozone problems, but has been found to be environmentally harmful in fuel leaks to ground water, in which it is very soluble.

Natural gas is a clean vehicle fuel, yielding reduced exhaust emissions and no fuel vapor problem because it is very unreactive in photochemical ozone production. But storing natural gas in a vehicle, either as a compressed gas in high-pressure tanks or as a refrigerated liquid at $-253\,°C$, is difficult and expensive, and it limits the vehicle range between fuel refills. At the present time, natural gas vehicles are restricted to fleet vehicles with limited daily range operating out of central fuel depots.

8.8   CONCLUSION

Among all transportation vehicles in the United States, light-duty passenger vehicles and trucks, in aggregate, are the predominant users of fuel and emitters of air pollutants. Transportation accounts for about a quarter of U.S. energy use, so substantial improvements in fuel efficiency and emissions of light-duty vehicles could contribute proportionally to reductions in national fuel consumption and pollutant emissions.

A substantial gain in vehicle fuel efficiency is primarily a matter of improved vehicle design and only secondarily does it turn on the improvement of engine efficiency. Current U.S. vehicles differ much more in vehicle than in engine fuel efficiency, with the larger, more massive vehicles having poorer vehicle fuel efficiency than the smaller, lighter ones.

The vehicle design parameters affecting vehicle fuel efficiency are vehicle mass, aerodynamic drag, and rolling friction (in order of decreasing importance). For a given vehicle size, vehicle mass can be reduced below current designs by substitution of lighter materials of equal strength, particularly in the vehicle frame, without impairing vehicle safety in collisions. As vehicle mass is reduced, less power and mass is needed for the engine, transmission, wheels, tires, fuel tank, and so on, compounding the gain in frame mass reduction. By careful attention to vehicle shape, aerodynamic resistance can be reduced. Efficient tires, in addition to mass reduction, lower the rolling resistance. Altogether, these technologies can improve vehicle fuel efficiencies independent of improvements to engine efficiency.

Improvements in engine fuel efficiency are closely constrained by the requirement to limit exhaust pollutant emissions. In the past, engine fuel efficiency has gradually improved while exhaust emissions were greatly reduced. There is still room for continued improvement in both respects for both SI and CI reciprocating engines.

The most fuel-efficient current vehicle, the hybrid electric vehicle, can achieve two times the vehicle fuel efficiency of current reciprocating engine vehicles of similar size and performance. Utilizing the same principles of vehicle design, vehicles powered by CIDI engine could achieve nearly comparable vehicle fuel efficiencies as current hybrids. Additional improvements seem likely in the future, given the long history of experience with these conventional technologies.

Although electric drive vehicles powered by batteries or fuel cells are still under development, their vehicle fuel efficiencies are not as promising, being scarcely better than existing vehicles. The battery-powered vehicle suffers from an inherent weight problem that limits its equivalent vehicle fuel efficiency and vehicle range. The fuel cell vehicle has a lesser weight problem, but its dependence upon hydrogen fuel, whether generated on board or at fuel suppliers, complicates the vehicle technology. The thermodynamics of synthetic hydrogen production and utilization in the fuel cell does not yet provide a significant fuel efficiency advantage over the conventional utilization of fuel in vehicles to offset the economic and vehicle design advantages that improved conventional vehicles promise, especially considering the infrastructure needs of a hydrogen fuel economy. Furthermore, the very low emissions of these electric drive systems become a less valuable offsetting benefit as the competing conventional vehicles become cleaner.

The technology for reducing exhaust emissions is well-developed, especially for the SI engine. Improvements to the engine and catalytic converter could reduce emissions further, should it prove necessary to go beyond the U.S. national Tier 2 standards. Improvements to CI engine emissions are more difficult to achieve, and they may always have higher nitrogen oxide and particulate matter emissions than their SI counterparts.

PROBLEMS

Problem 8.1

Table 8.2 lists the characteristics of a selection of 2000 model year conventional vehicles. The fuel efficiency (km/L) in highway mode and vehicle mass (t) are listed for 12 vehicles. (a) Plot the fuel consumption FC (L/km), the inverse of the fuel efficiency, as a function of vehicle mass (t). Estimate or calculate by linear regression the value of the slope m, where $FC = m(\text{mass})$ is the best fit for a straight line through these points that passes through the origin. (b) Calculate the average value of FC times the vehicle mass (t km/L) for these vehicles, together with its standard deviation. (c) Discuss whether or not, and why, these figures support the analysis of Section 8.3.

Problem 8.2

Tables 8.3–8.5 list characteristics of battery-powered electric, hybrid electric, and fuel cell passenger vehicles. (a) For each of these three vehicle types, calculate the average ratio of vehicle mass (t) to highway fuel economy (km/L). (b) Rank order these vehicle types according to the value of this ratio.

Problem 8.3

Table 8.2 lists the characteristics of a selection of 2000 model year conventional vehicles. For these vehicles, plot the urban fuel efficiency as a function of the highway fuel efficiency. Calculate the average value of the ratio urban/highway fuel efficiency, along with its standard deviation. Discuss the significance of utilizing both urban and highway fuel efficiencies in comparing the overall efficiencies of different vehicles.

Problem 8.4

A new sports utility vehicle averages 22 miles per gallon. It is expected to travel an average of 12,000 miles per year during a lifetime of 14 years. If fuel sells for $1.50 per gallon, calculate the lifetime expenditure on fuel.

Problem 8.5

Using the data of Table 8.1 for the year 1995, calculate for each vehicle class its fraction of the annual fuel consumed by transportation vehicles.

Problem 8.6

The engine power and size of a light duty vehicle is related to vehicle mass. For the vehicles of Table 8.2, calculate the average values of power/displacement, power/mass, and displacement per cylinder. Using these values, calculate the typical engine power, displacement, and number of cylinders for a 2-ton vehicle.

Problem 8.7

A passenger vehicle diesel engine has a minimum brake specific fuel consumption of 0.22 kg/kWh. Calculate its maximum thermal efficiency.

Problem 8.8

A 1.5-ton SI vehicle accelerates from rest to 100 km/h, then decelerates to a stop by braking. The average vehicle speed during this cycle is 50 km/h, and the cycle lasts 30 seconds. (a) Calculate the kinetic energy of the vehicle at its peak speed. (b) Calculate the time-average (W) and distance-average (J/km) of the energy dissipated in braking. (c) If 25% of the fuel heating value (31.6 MJ/L) is delivered to the wheels, calculate the average vehicle fuel economy in km/L for this start/stop mode, neglecting everything but the dissipation of braking. Compare this with the highway fuel economy of Table 8.2.

Problem 8.9

Motor vehicle manufacturers list the maximum torque as well as the maximum power of the vehicle engine. Explain why the maximum torque has no direct influence on the vehicle performance.

Problem 8.10

A 1.5-ton vehicle has a frontal area of 2 m^2, a rolling resistance coefficient of 0.1, and a drag coefficient of 0.3. Calculate the mechanical power delivered to the wheels at steady vehicle speeds of 50 and 100 km/h, if the atmospheric density is 1.2 kg/m^3.

BIBLIOGRAPHY

Appleby, A. J., and F. R. Foulkes, 1993. *Fuel Cell Handbook.* Malabar: Krieger.

Barnard, R. H., 1996. *Road Vehicle Aerodynamic Design. An Introduction.* Essex: Addison-Wesley Longman.

Blackmore, D. R., and A. Thomas, 1977. *Fuel Economy of the Gasoline Engine.* London: Macmillan.

Heywood, John B., 1988. *Internal Combustion Engine Fundamentals.* New York: McGraw-Hill.

Heywood, John B., and Eran Shaw, 1999. *The Two-Stroke Cycle Engine: Its Development. Operation, and Design.* Philadelphia: Taylor and Francis.

Kordesch, Karl, and Gunter Simader, 1996. *Fuel Cells and Their Applications.* New York: VCH Publishers.

Mark, Jason, and Candace Morey, 1999. *Diesel Passenger Vehicles and the Environment.* Cambridge: Union of Concerned Scientists.

National Research Council. 1997. *Toward a Sustainable Future. Addressing the Long-Term Effects of Motor Vehicle Transportation and a Sustainable Environment.* Washington, DC: National Academy Press.

Poulton, M. L., 1997. *Fuel Efficient Car Technology.* Southampton: Computational Mechanics Publications.

Stone, Richard, 1985. *Introduction to Internal Combustion Engines.* London: Macmillan.

Weiss, Malcomb A., John B. Heywood, Elizabeth Drake, Andreas Schafer, and Felix F. Au Yeung, 2000. *On the Road in 2020.* Cambridge: Energy Laboratory, Massachusetts Institute of Technology.

Environmental Effects
of Fossil Fuel Use

9.1 INTRODUCTION

The use of fossil fuels (coal, oil, or natural gas) almost always entails some environmental degradation and risk to human health. The negative impacts start at the mining phase, continue through
transport and refining, and conclude with the fuel combustion and waste disposal process.

Underground (shaft) mining of coal has claimed thousands of lives throughout the centuries,
because of explosions of methane gas in the mine shafts and also because of the inhalation of coal
dust by the miners. At the mine mouth, mineral and crustal matter, called slag, is separated from the
coal. The slag is deposited in heaps near the mine, thus despoiling the landscape. In recent times,
most of the mined coal is crushed at the mine. The crushed coal is "washed" in a stream of water
in order to separate by gravitational settling the adherent mineral matter, thereby beneficiating the
coal prior to shipment. The "wash" usually contains heavy metals and acidic compounds, which,
if not treated, contaminates streams and groundwater. Surface (strip) mining, which is much more
economical than shaft mining, causes scarring of the landscape. Only recently were there introduced
some regulations in the United States and other countries to ensure the restoration of the wounds
caused by the removal of overburden of the coal seams, and recovery of the pits and trenches after
the coal has been exhausted.

On- and off-shore oil and gas drilling produce piles of drilling mud, along with an unsightly
vista of oil and gas derricks. Also, there is the risk of crude oil spills, and explosions or fire at oil
and natural gas wells.

Transport of coal, oil, and gas by railroad, pipelines, barges, and tankers carry the risk of spills,
explosions, and collision accidents. In the refining process, especially of crude oil, toxic gases are
emitted into the air or flared. Usually, some liquid and solid byproducts are produced that may be
toxic. Strict regulations must be enforced to prevent the toxic wastes from entering the environment
and thereby threatening humans, animals, and vegetation.

The combustion of fossil fuels—whether coal, oil, or gas—inevitably produces a host of
undesirable and often toxic byproducts: (a) gaseous and particulate emissions into the atmosphere, (b) liquid effluents, and (c) solid waste. In many countries, strict regulations were en-
acted as to the maximum level of pollutants that can be emitted into the air or discharged into
surface waters or the ground from large combustion sources, such as power plants, industrial
boilers, kilns, and furnaces. However, the smaller and dispersed combustion devices, such as
residential and commercial furnaces and boilers, are not regulated, and they do emit pollutants
into the air. While great strides have been taken in many countries to control emissions from

automobiles, trucks, and other vehicles, these mobile sources still contribute significantly to air pollution.

Perhaps the greatest long-term threat to the environment is the steadily increasing concentration of carbon dioxide in the atmosphere, for the most part a consequence of the combustion of fossil fuel. CO_2 and some other so-called greenhouse gases may trap the outgoing thermal radiation from the earth, thereby causing global warming and other climate changes.

In this chapter we deal with the environmental problems caused by the use of fossil fuels. The chapter is divided into sections on air pollution (with separate subsections on photo-oxidants and acid deposition), water pollution, and land pollution. The large looming problem of global climate change associated with CO_2 and other greenhouse gas emissions will be addressed in the following chapter.

9.2   AIR POLLUTION

Among the environmental effects of fossil fuel use, those that impair air quality are arguably the most problematic. Most emissions into the atmosphere are a consequence of fossil fuel *combustion*. We are all familiar with the visible smoke that emanates from smoke stacks, fireplaces, and diesel truck exhaust pipes. But in addition to the visible smoke, a plethora of pollutants are emitted from combustion "sources" in an invisible form. Emissions may occur also during the extraction, transport, refining, and storage phases of fossil fuel usage. Examples are fugitive coal dust emissions from coal piles at the mine mouth or storage areas at power plants; evaporative emissions from crude and refined oil storage tanks, as well as from oil and gasoline spills; evaporative emissions from gasoline tanks on board vehicles and during refueling; natural gas leaks from storage tanks and pipelines; fugitive dust from ash piles; and so on.

Air pollution is not a recent phenomenon. Air pollution episodes due to open-fire coal burning were observed in medieval and renaissance England. In 1272, King Edward I issued a decree banning the use of "sea coal," coal that was mined from shallow sea beds and was burned wet in open kilns and iron baskets. In 1661, John Evelyn, a founding member of the Royal Society, wrote ". . . as I was walking in your Majesties Palace at Whitehall . . . a presumptuous Smoake . . . did so invade the Court . . . [that] men could hardly discern one another from the Clowd . . . . And what is all this, but that Hellish and dismall Clowd of Sea-Coal . . . [an] impure and thick Mist, accompanied with a fuliginous and filthy vapour . . . ."[1]

The association of air pollution episodes with human mortality and morbidity was recognized in the late nineteenth and early twentieth century. In 1873, in London, during a typical "fog" episode, 268 deaths occurred in excess of what would be normally expected in that period. In 1930, in the heavily industrialized Meuse Valley, Belgium, during a three-day pollution episode, 60 people died and hundreds were hospitalized. In 1948, during a four-day episode, in Donora, Pennsylvania, where several steel mills and chemical factories are located, 20 persons died and about one-half of the 14,000 inhabitants got sick. A terrible fog episode occurred again in London from 5 to 8 December 1952. The excess death numbered 4000! Most of the dead people had a

---

[1]Quoted from Cooper, C. D., and F. C. Alley, 1994. *Air Pollution Control: a Design Approach,* 2nd edition. Prospect Heights: Waveland Press.

history of bronchitis, emphysema, or heart disease. Apparently, individuals with a previous history of respiratory and cardiac diseases are predisposed to the impact of air pollution.

The 1952 London pollution episode induced the British Parliament to pass a Clean Air Act in 1956. This act focused on the manner and quality of coal burned in Great Britain. This act, and the fact that Great Britain shifted much of her fuel use from coal to oil, "cleaned" the air considerably over the British Isles. The ubiquitous London fog became a much rarer event; and when it occurs, it may be truly a natural phenomenon, rather than of anthropogenic cause as it was often in the past.

In the United States the first Clean Air Act was passed by Congress in 1963. Subsequently, the Clean Air Act was amended in 1970, 1977, and 1990. Other countries followed suit by enacting their own clean air acts and various legislation and regulations pertaining to reducing air pollution. As a consequence, the air quality in most developed countries is improving steadily, although what is gained in reducing emissions from individual sources is often negated by the ever-increasing number of sources, especially automobiles.

We classify air pollutants in two categories: primary and secondary. Primary pollutants are those that are emitted directly from the sources; secondary ones are those that are transformed by chemical reactions in the atmosphere from primary pollutants. Examples of primary pollutants are sulfur dioxide, nitric oxide, carbon monoxide, organic vapors, and particles (inorganic, organic, and elemental carbon). Examples of secondary pollutants are higher oxides of sulfur and nitrogen, ozone, and particles that are formed in the atmosphere by condensation of vapors or coalescence of primary particles. We shall later explain some of the processes that lead to the transformation of primary to secondary pollutants.

Most developed countries prescribe the maximum amount of pollutants that can be emitted from the sources. These are called *emission standards*. For large sources, the emission standards are usually set at a level that, after dispersion in the air within a reasonable distance, the pollutants will not cause significant human health or environmental effects. For small sources, such as automobiles, the emission standard may be set so as to prevent health effects from the cumulative emissions of all sources.

In order to protect human health and biota, most countries also prescribe maximum tolerable concentrations in the air. These are called *ambient standards*. The emission and ambient standards are legal parameters, published in laws and decrees. If these standards are exceeded, the causative sources can be punished or penalized, or their licences can be revoked.

9.2.1 U.S. Emission Standards

In the United States, emission standards have been promulgated for stationary and mobile sources. The Clean Air Act Amendments of 1970, 1977, and 1990 require the Environmental Protection Agency (EPA) to promulgate emission standards, called New Source Performance Standards (NSPS) and National Emission Standards for Hazardous Air Pollutants (NESHAP). Emission standards are specific to certain industrial categories, such as power plants, steel plants, smelters, refineries, pulp and paper mills, chemical manufacturing, and so on, as well to mobile sources—that is, automobiles, trucks, aircraft, and ships. The maximum allowable emission rates are prescribed for a variety of pollutants including SO_2, NO_x (the sum of NO and NO_2), CO, particulate matter (PM), lead, mercury, arsenic, copper, manganese, nickel, vanadium, zinc, barium, boron, chromium, selenium, chlorine, HCl, benzene, asbestos, vinyl chloride, pesticides, radioactive substances, and many other inorganic and organic pollutants.

TABLE 9.1 U.S. NSPS Emission Standards for Fossil Fuel Steam Generators with Heat Input > 73 MW (250 MBtu/h)[a]

Pollutant	Fuel	lb/MBtu Heat Input	g/GJ Heat Input
SO_2^b	Coal	1.2	516
	Oil	0.2	86
	Gas	0.2	86
NO_x	Coal (bituminous)	0.6	260
NO_x	Coal (subbituminous)	0.5	210
NO_x	Oil	0.3	130
NO_x	Gas	0.2	86
PM^c	All	0.03	13

[a]Data from EPA, Standards of Performance for New Stationary Sources, Electric Steam Generating Units, *Federal Register,* **45,** February 1980, pp. 8210–8213.

[b]For power plants built after 1978, Best Available Control Technology (BACT) applies. This means that if the uncontrolled SO_2 emissions were to exceed 1.2 lb/MBtu, 90% emission reduction is required by appropriate control technology (e.g., wet limestone scrubber). For uncontrolled emissions less than 0.6 lb/MBtu, 70% emission reduction is required (e.g., dry scrubber). For uncontrolled emissions between 0.6 and 1.2 lb/MBtu, a specified control technology applies that reduces emissions to a maximum of 0.6 lb/MBtu.

[c]PM = particulate matter. For PM emissions an opacity standard also applies, which allows a maximum obscuration of the background sky by 20% for a 6-minute period.

As an example, Table 9.1 lists the NSPS for fossil-fuel-fired steam generators, including electric power stations, with a thermal power input of more than 73 MW (250 million Btu/hr). The thermal power input equals the fuel heating value times its mass rate of consumption. Thus, a power plant with a 500-MW electrical *output* rating, operating at a thermal efficiency of 33.3%, will have a 1500-MW thermal *input*. Likewise, the NSPS are given in units of pollutant mass per fuel energy (heating value) *input,* grams/Joule or lb/Btu.

The estimation of the mass emission rate of SO_2 from a fossil fuel fired steam generators is quite simple. Because practically all sulfur atoms in the fuel burn up to form a SO_2 molecule, all we need to know is the weight percent of sulfur in the fuel and its heating value. The mass emission rate E_{SO_2} of SO_2 is

$$E_{SO_2} = 2 \times (\% \text{ by wt. S}) \, E(-2) \times FR \tag{9.1}$$

where FR is the firing rate of fuel in lb/s or g/s. The specific emission rate per unit of fuel energy input, e_{SO_2}, is

$$e_{SO_2} = 2 \times (\% \text{ by wt. S}) \, E(-2)/HV \tag{9.2}$$

where HV is the heating value of fuel in Btu/lb or J/g. The factor of two arises because SO_2 has double the molecular weight of S.

The emission rate of NO_x or CO cannot be computed in that manner, because the formation of these pollutants is dependent on the combustion process, and not on the weight percent of the atoms in the fuel. The rate of emission of particulate matter is dependent on the content of incombustible

TABLE 9.2   U.S. Federal Vehicle Emission Standards

Model Year	Light-Duty Vehicles (Auto)				Light-Duty Trucks (Gasoline)			
	HC (g/mi)	CO (g/mi)	NO_x (g/mi)	PM (g/mi)	HC (g/mi)	CO (g/mi)	NO_x (g/mi)	PM (g/mi)
1968	3.2	33						
1971[a]	4.6	47	4.0					
1974	3.4	39	3.0					
1977	1.5	15	2.0					
1978	1.5	15	2.0		2.0	20	3.1	
1979	1.5	15	2.0		1.7	18	2.3	
1980	0.41	7.0	2.0		1.7	18	2.3	
1981	0.41	3.4	1.0		1.7	18	2.3	
1982	0.41	3.4	1.0	0.6	1.7	18	2.3	
1985	0.41	3.4	1.0	0.6	0.8	10	2.3	1.6
1987	0.41	3.4	1.0	0.2	0.8	10	2.3	2.6
1988	0.41	3.4	1.0	0.6	0.8	10	1.2	2.6
1994	0.41	3.4	0.4	0.08	0.8	5.5	0.97	0.1

[a] Test method changed in 1971.

mineral matter in the fuel and on the combustion process. The emission rates of NO_x, CO, and PM would have to be measured at the stack exit.

For mobile sources the U.S. emission standards are given for four major pollutants that are emitted from these sources: carbon monoxide, hydrocarbons (HC), oxides of nitrogen, and particulate matter (PM). Table 9.2 lists the emission standards for mobile sources. The units are grams/mile. For NO_x the units are reckoned in mass of NO_2. Hydrocarbons include all carbonaceous emissions, except carbon monoxide and dioxide, coming from the tailpipe and evaporative emissions from the fuel tank, lines, pump, and injection devices. The hydrocarbon mass unit of measurement is the molecular fragment HC.

Beginning with the Clean Air Act Amendments of 1977, and continuing with CAAA 1990, the U.S. Congress required EPA to adopt a completely new way of regulating emission rates from stationary sources. Instead of defining and measuring emission rates for each industrial category, what is specified is the *control technology* that is presumed to achieve the desired emission standard. Thus, a multitude of acronyms came into being, like BACT (Best Available Control Technology), MACT (Maximum Achievable Control Technology), and RACT (Reasonable Available Control Technology). Depending on the industrial category, the pollutant, and the quality of fuel, as well as on the severity of pollution in an "airshed," the EPA determines which control technology needs to be installed on *new* sources. The CAAA 1990 went even further. For the control of emissions of toxic pollutants and the precursors of photo-oxidants, the act requires that even *existing* sources need the installation of emission control technology. This means *retrofitting* control devices on older sources that in previous acts were "grandfathered." For example, in power plants that use coal in excess of 0.6% by weight of sulfur the BACT for SO_2 control is a wet limestone scrubber; for coal with a lower sulfur content, it is a dry sorbent (usually limestone or lime) injection (see Section 5.2.9.3). For NO_x control the present federal BACT is a low-NO_x burner (LNB), but some states require for NO_x control selective catalytic reduction or nonselective catalytic reduction (see Section 5.2.9.4). For particle control on power plants, BACT is an electrostatic precipitator. For other industrial categories the requirement may be a fabric filter ("baghouse") or a spray chamber scrubber (see Section 5.2.9.1).

Title IV of CAAA 1990 specifically addresses the acid deposition problem. By the year 1995, the average emission level of all major power plants and industrial boilers was to be limited to 2.5 lb SO_2 per million Btu heat input; and by the year 2000, 1.2 lb SO_2/MBtu. It is estimated that in the year 2000 the U.S. emissions of SO_2 were roughly one-half of what they were in 1990, or about a 10 million ton reduction. The utilities and industry could choose any method they wished to achieve that emission reduction, including installing flue gas desulfurization technology, fuel switching, seasonal fuel switching (e.g., using coal in the winter and natural gas in the summer), and marketable permits. The latter means that if one source reduces SO_2 emissions by more than the required quota, the excess can be sold to a source that does not wish to reduce that pollutant.

For toxic pollutants, such as may be emitted from petroleum refineries, chemical manufacturing, aluminum smelters, paper and pulp mills, automobile paint shops, dry cleaners, and so on, the requirement is Maximum Achievable Control Technology (MACT). Depending on the pollutant and the facility, MACT usually consists of adsorption, absorption, or incineration.

The BACTs and MACTs are not supposed to be cast in concrete for all times. As new technologies come into the market that prove to be more efficient and/or more economical, the EPA may adopt, after due process, new technologies in lieu of the old ones. In any case, the introduction of "technology forcing" as manifested by BACT and MACT is a revolutionary concept. It places the onus on government to develop and define appropriate emission control technologies, rather than merely setting an emission standard, and let the "sources" find the devices that meet the standard. Also, the federally required control technologies are uniform across the nation. Individual states may impose even stricter control technologies, but never less efficient ones. This new way of controlling emissions places emphasis not only on measurement of emission rates, but also on monitoring and supervising the installation and proper functioning of the control devices.

9.2.2 U.S. Ambient Standards

The setting of emission standards has as its purpose ensuring that concentrations of air pollutants in the ambient air remain at a sufficient low level so that the population at large—and, especially, sensitive individuals, such as children and the elderly—will not suffer adverse health effects. The appropriate indices for exposure to harmful air pollutants are the ambient concentrations and the averaging time periods for which these concentrations prevail. Therefore, the U.S. Congress mandated the EPA to promulgate ambient concentration standards, called the National Ambient Air Quality Standards (NAAQS). The NAAQS stipulate the concentrations and the averaging time periods for various air pollutants that should not be exceeded. In the United States, current ambient standards exist for five pollutants: particulate matter (PM), sulfur dioxide (SO_2), nitric oxides (NO_x, measured as NO_2), carbon monoxide (CO), and ozone (O_3). These are the so-called *criteria pollutants*. This is not to say that concentrations of other pollutants need not be curtailed. Indeed, some toxic pollutants may be far more injurious to human health and biota than the aforementioned five pollutants. The fact is that for the criteria pollutants a fairly well known dose–response relationship has been established over years of clinical and epidemiological research, while such a relationship may not be known for many other pollutants. For the other pollutants, classified as *toxic pollutants,* there are no specified ambient standards, but their emission into the

TABLE 9.3 U.S. 2000 National Ambient Air Quality Standards (NAAQS)

Pollutant	Primary		Secondary	
	ppm	$\mu g/m^3$	ppm	$\mu g/m^3$
Carbon monoxide (CO)				
8-h average	9	10 mg/m^3		
1-h average	35	40 mg/m^3		
Nitrogen dioxide (NO$_2$)				
Annual arithmetic mean	0.053	100	Same	Same
Ozone (O$_3$)^a				
3-y average of annual fourth highest daily maximum 1-h concentration	0.12	236	Same	Same
Particulate matter, diameter < 10 μm (PM-10)^b				
Annual arithmetic mean		50		Same
Arithmetic mean of 24-h 99th percentile, averaged over 3 y		150		Same
Sulfur dioxide (SO$_2$)				
Annual arithmetic mean	0.03	80		
24-h average	0.14	365		
3-h			0.5	1300

^{a}Proposed new standard: 3-y average of annual 4th highest daily max. 8-h concentration, 0.08 ppm = 157 $\mu g/m^3$.
^{b}Proposed additional new standard for particulate matter, diameter < 2.5 μm (PM-2.5): 3-y annual arithmetic mean, 15 $\mu g/m^3$; arithmetic mean of 24-h 98th percentile averaged over 3 y, 65 $\mu g/m^3$.

atmosphere is to be prevented as much as possible by applying at the sources Maximum Achievable Control Technology (MACT).

The NAAQS for the five criteria pollutants are listed in Table 9.3. Primary standards relate to human health effects, and secondary standards relate to "welfare" effects. The NAAQS definitely have an anthropocentric aspect—that is, they pertain to human health and welfare—but it is understood that the welfare standards are for protecting the environment in general. The averaging times can be as short as one hour (for CO) to one year (for PM, SO$_2$, and NO$_2$). The reason is that the exposure of some pollutants at a high concentration for a short period may cause acute effects, whereas the exposure of others at a relatively low level for longer periods may cause chronic effects (see below). The number of allowed "exceedances" is also listed. For example, the 1-hour standard for CO, 40 mg/m^3 (9 parts per million by volume, ppmV), must not be exceeded more than once per year. The O$_3$ standard is 0.12 ppmV, 1-hour average, not to be exceeded more than once per year. Recent health and ecological studies indicated that ozone may be harmful at a lower level when exposed for a longer period, so a new standard has been proposed of 0.08 ppmV, 8-hour average. Likewise, a new PM standard has been proposed. Currently, only particles with an aerodynamic diameter of less than 10 μm are regulated (PM-10). However, recent epidemiological studies indicated that particles smaller than 2.5 μm are most detrimental to health, because they lodge deeply in the lung's alveoli. Thus, in addition to PM-10 a PM-2.5 standard may be implemented.

The EPA is mandated to revise the NAAQS from time to time, as more results from health and environmental effects studies become available. For example, EPA has been urged to promulgate a short time standard for NO$_2$ instead of the annual standard.

If the NAAQS are exceeded within an Air Quality Control Region (AQCR), the state in which the region is located must develop a plan, called the State Implementation Plan (SIP), which lays out a strategy of how the region will attain compliance with the NAAQS in a reasonable time period. The SIP may include emission curtailments from emitting sources, traffic regulations, tightened inspection schedules and procedures, and other measures. However, research in the past few decades brought out clearly that air pollutants do not respect political and natural geographic boundaries. They travel over control regions, state lines, river valleys, mountains, and even over oceans. Thus, no state can control its air pollution solely by its own means. A regional, national, and even international approach is necessary to control air pollution over a region, over a continent, and, for some air pollutants, over the globe. In part, this is the reason that in the United States, emissions of most air pollutants are regulated on the federal rather than on the state level.

9.2.3 Health and Environmental Effects of Fossil-Fuel-Related Air Pollutants

Air pollutants, when they exceed certain concentrations, can cause acute or chronic diseases in humans, animals, and plants. They can impair visibility, cause climatic changes, and damage materials and structures. Traditionally, the major concern was the impact on human health. That is why in the United States the standards that are supposed to protect humans are called primary standards, and those that protect "welfare" are called secondary standards.

Table 9.4 lists some of the health and environmental effects of air pollutants that are related to fossil fuel use. The listed air pollutants are classified as criteria pollutants by EPA: SO_2, NO_x, O_3, CO, and particulate matter (PM). For these pollutants, EPA promulgated National Ambient Air Quality Standards (NAAQS; see Section 9.2.2). The order of listing does not follow any particular ranking: Some individuals or plants are more sensitive to one kind of pollutant than to another. The pollutants can cause respiratory diseases, and some are suspected toxigens, mutagens, teratogens, carcinogens, and possible animal- and plant-disease-causing agents.[2]

The definition of the deleterious effects of PM is complicated and contentious. The ambient standards are given in units of mass per volume ($\mu g/m^3$). Surely, the effects on health and biota are not dependent as much on mass concentrations of the inhaled particles, but on their quality—that is, their composition. While ordinary soil and road dust may not cause significant health effects, particles that contain acidic species, heavy metals, soot, and polycyclic aromatic hydrocarbons (PAH) may cause respiratory, neurological, and cancerous diseases. It is unlikely that agricultural workers who inhale a great amount of soil dust suffer as much respiratory and neurological diseases as urban dwellers who inhale much smaller amounts but potentially more lethal photochemical smog particles. One reason that EPA and other environmental protection agencies are using mass concentrations as a standard for PM, rather than chemical composition, is that the determination of chemical composition requires complicated and expensive analytical instrumentation. Furthermore, EPA maintains that there is evidence from epidemiological studies that excessive mortality and morbidity is correlated with mass and size of the particles, regardless of chemical composition.

---

[2]Toxigen: a chemical agent that may cause an increase of mortality or of serious illness, or that may pose a present or potential hazard to human health. Mutagen: any agent, including radioactive elements, that may cause biological mutation—that is, alteration of the genes or chromosomes. Teratogen: an agent that may cause defects or diseases of the embryo. Carcinogen: an agent that may cause cancer.

TABLE 9.4 Effects of Criteria Air Pollutants on Human Health, Fauna and Flora, and Structures and Materials

Pollutant	Health Effect	Fauna and Flora Effect	Structure and Material
SO_2	Bronchoconstriction, cough.	Cellular injury, chlorosis, withering of leaves and abscission. Precursor to acid rain: acidification of surface waters with community shifts and mortality of some aquatic organisms. Possible effect on uptake of Al and other toxic metals by plant roots.	Weathering and corrosion. Defacing of monuments.
NO_x	Pulmonary congestion and edema, emphysema, nasal and eye irritation.	Chlorosis and necrosis of leaves. Precursor to acid rain and photo-oxidants.	Weathering and corrosion.
O_3 and photo-oxidants	Pulmonary edema, emphysema, asthma, eye, nose, and throat irritation, reduced lung capacity.	Vegetation damage, necrosis of leaves and pines, stunting of growth, photosynthesis inhibitor, probable cause of forest die-back, suspected cause of crop loss.	Attack and destruction of natural rubber and polymers, textiles and materials.
CO	Neurological symptoms, impairment of reflexes and visual acuity, headache, dizziness, nausea, confusion. Fatal in high concentrations because of irreversible binding to hemoglobin.	NA[a]	NA
Particulate matter	Nonspecific composition: bronchitis, asthma, emphysema. Composition dependent: brain and neurological effects (e.g., lead, mercury), toxigens (e.g., arsenic, selenium, cadmium), throat and lung cancer (e.g., coal dust, coke oven emissions, polycyclic aromatic hydrocarbons, chromium, nickel, arsenic).	NA	Soiling of materials and cloth. Visibility impairment due to light scattering of small particles.

[a]NA, not available.
Adapted from Wark, K., C. F. Warner, and W. T. Davis, 1998. *Air Pollution: Its Origin and Control.* Reading: Addison-Wesley.

9.2.4 Air-Quality Modeling

After leaving the smoke stack or exhaust pipe, the primary air pollutants disperse into the atmosphere by turbulent diffusion, advect by winds, and transform into secondary pollutants by chemical reactions among themselves and with atmospheric species. The estimation of the concentration of pollutants in space and time is called air-quality modeling. It is also called source–receptor modeling or dispersion modeling, where the sources are emissions from point (e.g., a smoke stack), line (e.g., a highway), or area (e.g., an industrial complex) sources, and the receptors are designated human habitats or ecologically sensitive areas.[3] In this section we shall deal only with nonreactive pollutants—that is, those that are not transformed in the atmosphere. In subsequent sections we shall incorporate transformation processes into air-quality models.

9.2.4.1 Air Pollution Meteorology

A basic information necessary for air-quality modeling is the wind statistics for the modeling domain and the dispersion characteristics of the atmosphere. Winds blow from high- to low-pressure regions on earth. Because the earth is a rotating body revolving around the sun, any spot on earth receives constantly changing insolation over day and night and over the seasons. In addition, orographic effects—that is, mountains and valleys—alter the course of winds, as does surface friction, sea–land interfaces, street canyons, and so on. Thus, for air-quality modeling, multiyear wind statistics are necessary for predicting the advection by winds of pollutants from the sources to the receptor.

Meteorological data are available from numerous weather stations operating around the world, especially in the more developed countries. These weather stations measure and record surface and upper air winds, atmospheric pressure, humidity, precipitation, insolation, temperature at the ground, and the temperature gradient in the atmosphere—that is, the temperature variation with altitude. The measurements are usually rendered twice daily at 0000 and 1200 Greenwich Mean Time (GMT), so that measurements are synchronized all over the world. Past and present weather data are available from national repositories—for example, in the United States, the National Weather Service in Asheville, North Carolina.

In the atmosphere, dispersion occurs mostly by turbulent or eddy diffusion. Such diffusion is orders of magnitude faster than molecular or laminar diffusion. The cause for turbulent diffusion is either mechanical or thermal. Mechanical turbulence is due to wind shears in the free atmosphere (adjacent layers of the atmosphere move in different directions or speeds), or friction experienced by winds blowing over the ground surface and obstacles, such as tree canopies, mountains, and buildings. The other cause of turbulence is the thermal gradient in the atmosphere. In the lower troposphere, the temperature is usually higher near the ground and declines with altitude. In a dry atmosphere, the gradient amounts to approximately $-10\,°C/km$. This is the dry adiabatic lapse rate. Occasionally, the gradient is steeper, meaning more negative than $-10\,°C/km$. In a moist atmosphere the gradient is less steep, due to the addition of the latent heat of condensation of water vapor. At night, due to radiative cooling of the surface, the gradient may become positive, with temperature increasing with altitude. This is called an inversion. Inversions can also occur aloft, when a negative gradient is interrupted by a positive one. The bottom layer up to the inversion is called the mixing layer, and the height to the inversion is called the mixing depth. An inversion layer acts like a lid on the mixing layer. With an inversion, atmospheric conditions are especially prone to air pollution episodes, because pollutants emitted at the ground are concentrated in the shallow mixing layer. Later in the day, as the sun rises, the inversion layer may break up, allowing pollutants to escape aloft and thus alleviating the pollution episode. Valleys and urban areas surrounded by mountain chains experience frequently inversion layers, and therefore they are plagued with

---

[3]Another type of modeling is called receptor modeling. These models attempt to identify and quantify the contribution of various sources to the amount and composition of the pollutant concentration at the receptor by using some characteristics of the sources. For example, receptor modeling may compare the distribution of trace elements at the receptor to the distribution of trace elements in the emissions of various sources.

pollution episodes. Los Angeles, Denver, Salt Lake City, Mexico City, and industrial-urbanized river valleys across the continents are examples with frequent inversions, where pollutant concentrations often exceed air-quality standards, and the population suffers from pollutant-caused respiratory and other diseases.

When the temperature in the upper layers is much colder than that in the lower layers, upper air parcels fall downward due to their larger density, and lower air parcels move upward. This movement generates turbulent or eddy diffusion. The steeper the temperature gradient, the more the turbulent intensity. This is called an unstable condition. A temperature gradient that is equal to the dry adiabatic lapse rate is called a neutral condition, and it leads to moderate turbulence. A temperature gradient that is less steep than the dry adiabatic lapse rate, or even a positive gradient, is called a stable condition, in which there is minimal or no turbulence at all.

It is a convention to classify the turbulent conditions of the atmosphere into six stability categories, called Pasquill–Gifford stability categories. They range from A, very unstable (very turbulent), to F, very stable (little turbulence). Category D is called neutral, with moderate turbulence. The negative temperature gradient (lapse rate) of category D coincides with the dry adiabatic lapse, about $-10\,°C/km$. For categories A, B, and C the magnitude of the negative gradient is greater than that for D; for E the gradient is smaller than that for D; and for F the gradient is positive.

While the lapse rate of the atmosphere is measured twice daily at the weather stations, the stability categories can be approximately determined by knowing the surface wind speed, insolation, and cloud cover. Table 9.5 lists the stability categories. It is seen that in daytime, low wind speeds and strong insolation lead to unstable categories A or B; high wind speeds and moderate to slight

TABLE 9.5 Pasquill–Gifford Stability Categories

	Day			Night	
	Incoming Solar Radiation			Cloud Cover	
Surface Wind (m/s)	Strong[a]	Moderate[a]	Slight[a]	Thinly overcast or $\geq$4/8 cloud	$\leq$3/8 cloud
<2	A	A–B	B	b	b
2–3	A–B	B	C	E	F
3–5	B	B–C	C	D	E
5–6	C	C–D	D	D	D
>6	C	D	D	D	D

Stability Category	Stability	dT/dz (°C/100 m)
A	Extremely unstable	≤-1.9
B	Moderately unstable	>-1.9 but ≤-1.7
C	Slightly unstable	>-1.7 but ≤-1.5
D[c]	Neutral	>-1.5 but ≤-0.5
E	Slightly stable	>-0.5 but ≤1.5
F	Very stable	>1.5 but ≤4.0

[a]Zenith angle under clear skies > 60°; moderate 35–60°; slight <15°.
[b]Rural areas = F; large urban areas = D; small urban areas = E.
[c]Category D, neutral, applies to heavy overcast, day or night, all wind speeds.

insolation lead to neutral categories C or D. At night, the stability categories are almost always neutral or stable: D, E, or F. Table 9.5 also lists the temperature gradients which correspond to the stability categories.

9.2.4.2 Modeling of Steady-State Point Source

When an effluent from a power plant stack leaves the top exit of the stack, it immediately begins to mix with the surrounding atmosphere, because the effluent jet velocity is different from that of the surrounding atmosphere. This mixing process begins to dilute the concentration of the stack gas, the more so the more atmospheric air is entrained in the stack plume. Most of the time a wind is present so that the plume, as it mixes with the atmosphere, soon attains the horizontal speed u of the wind, with the plume axis (or centerline) bending sharply and approaching the horizontal direction. At this point in the plume trajectory, mixing of the plume gas with the atmosphere continues at a rate determined by the turbulent motion of the atmosphere. As a consequence of this mixing process, the concentration of the plume gas, a mixture of the principal products of combustion and the air pollutants of interest, steadily declines with downwind distance from the stack exit. The physical evidence of this mixing is visible to the eye for smoky plumes, where it can be seen that the plume width in both vertical and horizontal directions increases with downwind distance, marking the portion of the surrounding atmosphere into which the effluent gas has been mixed and thereby diluted.

Measurements of the pollutant gas concentrations downwind of stacks show that, at any distance x downwind, they are maximum at the plume centerline and decline with vertical or lateral distance from the plume centerline. The centerline concentration and the effective vertical and horizontal widths of plume change gradually with the distance x, in a way that is dependent upon the atmospheric conditions. The concentration profile in both the y and z direction can be approximated by a Gaussian (bell-shaped) curve. To generalize the results of such plume measurements, a mathematical model, called the Gaussian plume model, has been developed.

A smoke stack plume is depicted in Figure 9.1. The stack is placed at the origin of the coordinate system, with the x axis downwind, the y axis crosswind, and the z axis is in the vertical. The stack

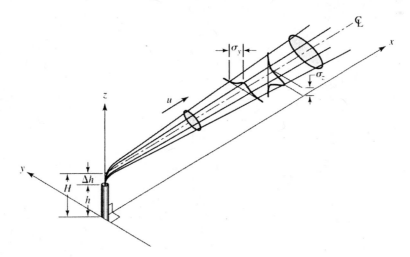

Figure 9.1 The Gaussian plume.

has a height h. Because the exhaust gas is usually warmer than the ambient air, it rises due to buoyancy by an incremental height Δh. This is the "plume rise," which will be evaluated later. The final plume centerline is at height $H = h + \Delta h$. The pollutant time-averaged mass concentration $c(x, y, z)$ at a downwind distance x from a source of pollutant emitted at a mass rate Q_p g s^{-1}, at a height H m, in the presence of wind of speed u m s^{-1}, is given by the following equation, called the Gaussian plume equation (GPE):

$$c(x, y, z) = \frac{Q_p}{2\pi \sigma_y \sigma_z u} \exp\left[-\frac{1}{2}\left(\frac{y^2}{\sigma_y^2} + \frac{(z - H)^2}{\sigma_z^2}\right)\right] \tag{9.3}$$

The σ's are the standard deviations of the Gaussian profile. Their dimension is given in meters. The σ's are called dispersion coefficients, although they should not be confused with diffusion coefficients, which have a dimension m^2 s^{-1}. The σ's are a function of the turbulent condition of the atmosphere and the distance from the source. The dispersion coefficients σ_y and σ_z as a function of downwind distance x are presented in Figure 9.2. It is seen that the σ's are not necessarily equal in the y and z directions. The six curves correspond to the atmospheric stability categories A through F. Because the σ's are largest for unstable conditions and are smallest for stable conditions, stability category A leads to a rapid dispersion of the emitted pollutants and leads to highest ground concentration near the source. Category F leads to a slow dispersion and leads to highest ground concentration far from the source. Mostly, we are interested in concentration of pollutants on the ground; that is, $z = 0$. For ground-level concentrations it is assumed that the pollutant molecules are not absorbed at the ground, but reflected as if they were an optical ray that grazes a mirror. With this assumption, the factor 2 disappears in the denominator of the GPE [equation (9.3)], and ground-level concentrations are twice that without reflection.

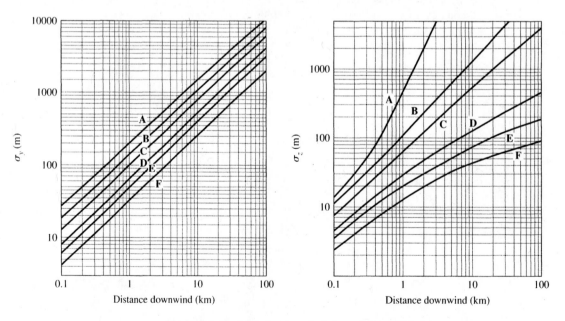

Figure 9.2 Horizontal and vertical dispersion coefficients.

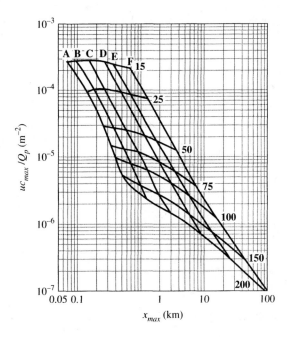

Figure 9.3 (uc_{max}/Q_p) versus x_{max} for six stability classes and a range of plume centerline heights ($H = 15$ to 200 m).

For quick estimation of maximum ground-level pollutant concentrations (with ground reflection) it is convenient to use Figure 9.3. Here c_{max} is the maximum concentration in $\mu g\ m^{-3}$, u is wind speed in $m\ s^{-1}$, and Q_p must be reckoned in $\mu g\ s^{-1}$. The lettered parameters on top of the curves are the stability categories; those on the side are the plume centerline heights H after the plume has risen to its equilibrium value. It is seen that for high stacks ($H \geq 50$ m), the unstable categories cause maximum concentrations on the ground relatively near to the stack, and stable categories cause weak maximum concentration far from the stack.

It must be emphasized that the Gaussian plume model is only an approximation. It works best on level ground. Because the wind speed u appears in the denominator, the GPE cannot be used for calms, when the wind speed is less than approximately 2 m/s. Also, the modeling distance should not be extended further than 20–30 km, because wind direction and speed, as well as the dispersion characteristics (atmospheric stability category), may change over longer distances. Because the GPE is a steady-state model, the emission rate Q_p and plume rise Δh must also remain constant. Experience shows that within a limited distance and on level terrain, the GPE gives concentration contours on the ground that are within a factor of two of measurements. In valleys, hills, and urban areas, aerodynamic obstacle effects need to be considered. Numerous equations exist that work reasonably well when corrections for terrain complexities are incorporated into the Gaussian plume model.

9.2.4.3 Plume Rise

There are several empirical equations that allow the estimation of plume rise Δh from known stack exit conditions. Here we give the Briggs plume rise equation, which is mostly used in

EPA-recommended dispersion models. For *unstable* and *neutral* conditions we have

$$\Delta h = 38.71 F^{3/5} u^{-1} \tag{9.4}$$

where the numerical factor 38.71 has the dimensions $s^{4/5} m^{-2/5}$, where the buoyancy flux parameter F is

$$F = g v_s D_s^2 (T_s - T_a)/4 T_s \tag{9.5}$$

and where

g = acceleration of gravity (9.8 m s^{-2})
v_s = flue gas stack exit velocity, m s^{-1}
D_s = stack diameter, m
T_s = flue gas exit temperature, K
T_a = ambient temperature at stack height, K
u = wind speed at stack height, m s^{-1}

For *stable* conditions we have

$$\Delta h = 2.6[F(uS)^{-1}]^{1/3} \tag{9.6}$$

where the numerical factor 2.6 has the dimension m^2, where S is the atmospheric stability parameter

$$S = g(\delta\theta/\delta z) T_a^{-1} \tag{9.7}$$

and where $\delta\theta/\delta z$ is the potential temperature gradient, whose value is 0.02 K/m for category E and is 0.035 K/m for category F.[4]

9.2.4.4 Steady-State Line Source

A line source pertains to a situation where point sources are aligned in a row, such as several smoke stacks along a river bank, or many automobiles and trucks traveling in both directions along a straight highway. The geometry of a line source is depicted in Figure 9.4. The ground level mass concentration $c(x)$ is estimated by the following variant of the GPE

$$c(x) = \frac{2Q_l}{(2\pi)^{1/2} \sigma_z u \sin\phi} \exp -\frac{1}{2}\left(\frac{H^2}{\sigma_z^2}\right) \tag{9.8}$$

where

Q_1 = average line source mass emission rate per meter, g m^{-1} s^{-1}
H = average release height from sources, m

---

[4]Equation (9.4) is only applicable for a buoyancy parameter $F > 55$ m^4 s^{-3}. For $F \leq 55$ m^4 s^{-3}, and unstable or neutral conditions $\Delta h = 21.425 \, F^{3/4} \, u^{-1}$.

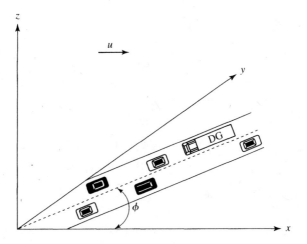

Figure 9.4 Line source.

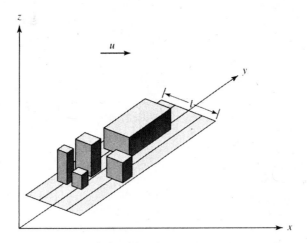

Figure 9.5 Area source.

ϕ = angle between the source axis and the prevailing wind direction (which by convention always blows in the x direction), degrees

x = distance of the receptor from the source axis in the direction of the wind, m

Note that the distance parallel to the source axis does not appear in equation (9.8) because the concentration does not change in that direction. Equation (9.8) should not be used when ϕ is less than 45°.

9.2.4.5 Steady-State Area Source

Emissions from an urban area can be considered as emanating from a bunch of parallel line sources, as depicted in Figure 9.5. The ground-level mass concentration $c(x)$ of pollutant at a downwind distance x is given by the following equation:

$$c(x) = \frac{2Q_a l}{(2\pi)^{1/2}\sigma_z u} \exp\left[-\frac{1}{2}\left(\frac{H^2}{\sigma_z^2}\right)\right] \tag{9.9}$$

Q_a = average mass emission rate per square meter, g m^{-2} s^{-1}

l = width of the urban area along the wind axis, m

Equation (9.9) can only be used at downwind distances $x \gg l$. It is also necessary that the crosswind dimension of the urban area be larger than $\sigma_y(l)$.

9.2.5 Photo-oxidants

Photo-oxidants are a class of secondary air pollutants formed from some of the primary pollutants emitted by fossil fuel combustion. The name arises because these chemicals are formed under the influence of sunlight, and all of them have strong oxidizing capacity. They irritate and destroy (oxidize) the respiratory tract, eyes, skin, animal organs, vegetation tissues, and materials and structures. The major representative of this class of chemicals is ozone, O_3, but other compounds are included: ketones, aldehydes, alkoxy radicals (RO) peroxy radicals, (RO$_2$), peroxyacetyl nitrate (PAN), and peroxybenzoyl nitrate (PBN). The symbol R denotes a hydrocarbon fragment, with one hydrogen missing.

Tropospheric ozone (the "bad" ozone) is to be distinguished from stratospheric ozone (the "good" ozone). Tropospheric ozone is mainly formed as a consequence of fossil fuel combustion, while stratospheric ozone is formed naturally by photochemical reactions under the influence of solar ultraviolet radiation.[5] Some of the tropospheric ozone may be due to intrusion of stratospheric ozone into the troposphere. This constitutes a background level of ozone. However, compared to concentrations of ozone in an urban polluted atmosphere, the background ozone is a small fraction of that concentration.

The current NAAQS (see Section 9.2.2) for O_3 is 120 parts per billion by volume (ppbV), 1-hour maximum concentration. During the 1990s, this standard was exceeded many times per year in most major metropolitan areas of the United States, especially in cities with high insolation and temperature and where topographic conditions are preventing good ventilation, such as Los Angeles, Salt Lake City, Phoenix, Houston, Dallas-Fort Worth, Denver, Atlanta, and other cities. High ozone levels are observed in most metropolitan areas of the world, especially areas with high insolation, such as Mexico City, Sao Paulo, Jakarta, Bombay, Cairo, Istanbul, Rome, Athens, Madrid, and other cities. But rural and remote areas that are downwind from metropolitan areas are also experiencing elevated ozone concentrations. The high concentrations in rural and remote areas may contribute to degradation and die-back of plants and trees.

---

[5]As is well known, stratospheric ozone has been depleted over the past decades due to the penetration into the stratosphere of long-lived industrial chemicals, the chlorofluorocarbons, creating the so-called "ozone hole." Stratospheric ozone shields humans, animals, and vegetation from the penetration to the surface of the earth of harmful ultraviolet radiation. It is an irony that mankind, on one hand, creates tropospheric ozone by emissions of combustion products of fossil fuel and, on the other hand, destroys stratospheric ozone by emissions of some other chemicals.

The only precursor that can initiate ozone formation in the troposphere is nitrogen dioxide, NO_2. As we shall see later, other gases, both man-made and natural ones, can abet ozone formation, but the initiation of the process is solely due to NO_2. Nitrogen dioxide is a brown gas, formed by oxidation of nitric oxide, NO. The sum of NO and NO_2 is termed nitric oxides, NO_x. Nitric oxides are formed primarily in the combustion of fossil fuels. A part of the NO_x is formed on account of the inherent nitrogen content of fossil fuels, especially coal and petroleum. This is called *"fuel NO_x."* A greater part is formed during combustion. At the high flame temperatures some of the air O_2 and N_2 combine to form NO and NO_2. This is called *"thermal NO_x."*

Nitrogen dioxide gas can photo-dissociate in sunlight, at wavelengths shorter than 420 nm. The resulting atomic oxygen combines with molecular oxygen to form ozone:

$$NO_2 + (h\nu)_{\lambda \leq 420 \text{ nm}} \rightarrow NO + O \tag{9.10}$$

$$O + O_2 + M \rightarrow O_3 + M \tag{9.11}$$

where M is an inert molecule which is necessary to bring about the combining of atomic and molecular oxygen. Since the formed O_3 can be destroyed by the same NO which is formed in (9.10):

$$O_3 + NO \rightarrow O_2 + NO_2 \tag{9.12}$$

this cycle of reactions cannot explain the large concentrations of ozone that are observed in various regions of the world. It was assumed in the late 1950s and early 1960s by scientists at the California Institute of Technology that instead of reaction (9.12), NO is reoxidized to NO_2 by some atmospheric oxidant other than O_3, thereby starting reaction (9.10) all over again, allowing the build-up of O_3 from a relatively small concentration of NO_2. That "mystery" oxidant turned out later to be a peroxy radical, RO_2. This radical is formed in the following sequence of reactions:

$$RH + OH \rightarrow R + H_2O \tag{9.13}$$

$$R + O_2 \rightarrow RO_2 \tag{9.14}$$

$$RO_2 + NO \rightarrow NO_2 + RO \tag{9.15}$$

RH designates a hydrocarbon molecule, OH is a hydroxyl radical, RO is a an alcoxy radical, and RO_2 is a peroxy radical. The naturally occurring peroxy radical, HO_2, can also oxidize NO to NO_2. The hydroxyl radical appears to be omnipresent in the atmosphere. It is formed by reaction of water vapor with an excited oxygen atom $O(^1D)$. The latter is formed in the photodissociation of O_3 by sunlight of wavelength less than 319 nm.

Indeed, the peroxy radicals were subsequently found in the lower troposphere and in laboratory experiments, called smog chambers. Now the picture of the formation process of ozone and the other photo-oxidants becomes much more complicated. While NO_2 is the spark that starts the flame, other "fuels" are necessary to sustain the flame and raise its temperature, metaphorically speaking. The other fuels, or precursors as they are called, are hydrocarbon molecules and other organic compounds, collectively known as volatile organic compounds (VOC). Some of the VOC are of anthropogenic origin, such as products of incomplete combustion, evaporation of fuels and solvents, emissions from refineries, and other chemical manufacturing. Others are of biogenic origin, such as evaporation and volatilization of organic molecules from vegetation, wetlands, and surface waters. For example, evergreen trees exude copious quantities of isoprene, terpenes, pinenes, and other compounds, which, by virtue of double bonds in their molecules, are very reactive with a hydroxyl radical.

After the discovery of the presence in the atmosphere of the various oxidizing agents and other compounds that participate in the ozone formation process, researchers developed models that attempted to show how observed concentrations of ozone and other smog ingredients are reached as a function of precursor concentrations, meteorological conditions, and insolation.

9.2.5.1 Photo-oxidant Modeling

Because ozone and the other photo-oxidants are secondary pollutants, the regular dispersion models described in Section 9.2.4, pertaining to primary pollutants that do not transform while dispersing, are not applicable. For photo-oxidant modeling, in addition to meteorological parameters, the reactions of the primary pollutants among themselves, and those with atmospheric species, plus the interactions with sunlight need to be included. A further complication is that some of the chemical kinetic processes are not linear; that is, they are not first-order rate reactions.

The rate of the reaction (9.13) depends on the kind of VOC. Molecules with a double or triple bond are very reactive, followed by aromatics, then by branched- and extended-chain aliphatics. The simple molecule methane, CH_4, reacts very slowly with OH; therefore CH_4 is not included in the rate equations.

The model that has been used most frequently in the past is called the Empirical Kinetic Modeling Approach (EKMA). It is a Lagrangian model; that is, the coordinate system moves with the parcel of air in which chemical changes occur. The architecture of EKMA is as follows. A column of air is transported along a wind trajectory, starting at 0800 local time (LT). The column height reaches to the bottom of the nocturnal inversion layer. The concentration of chemical species in the column is estimated from the 0800 LT emission rates of the following pollutants: NO, NO_2, CO, and eight classes of VOC: olefins, parafins, toluene, xylene, formaldehyde, acetaldehyde, ethene, and nonreactives. Other input parameters are date, longitude, and latitude, which define the insolation rate. As the column moves with the wind, fresh pollutants are emitted into the column. As time progresses, the solar angle increases and the height of the column (mixing height) increases, with a consequent dilution effect. Inside the column the photochemical reactions occur in which ozone is generated. The rate constants for the chemical reactions are empirically determined based on smog-chamber experiments. The model calculations stop when the ozone level reaches a maximum value, which usually occurs between 1500 and 1700 LT.

Figure 9.6 presents an isopleth plot of maximum ozone concentrations versus 0800 LT concentrations of the sum of eight VOC compounds on the horizontal axis, measured in units of parts per million carbon (ppmC), and the NO_x concentration on the vertical axis. The isopleths have the shape of hyperbolas. The diagonals drawn through the isopleths represent the morning ratios of VOC/NO_x. A ratio of 4:1 corresponds to a typical urban ratio; 8:1, suburban; 16:1, rural. Suppose the maximum ozone level reached is 200 ppbV. The "design" (i.e., NAAQS) value is 120 ppbV. To reach the design value, one can go in two directions: reducing either the 0800 LT VOC or the NO_x concentrations, represented by the horizontal and vertical dashed lines in Figure 9.6, respectively. It can be seen that in an urban environment (4:1 diagonal), the 120 ppbV level can be reached with a much smaller reduction of VOC concentrations than NO_x. In a rural environment (16:1 diagonal),

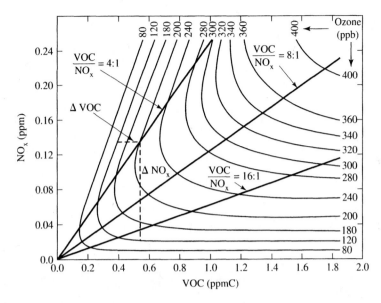

Figure 9.6 EKMA plot of the maximum ozone concentration as a function of the initial NO_x and VOC concentrations.

a smaller reduction of NO_x would be necessary. (A third way would be to go along the diagonal from the 200 ppbV isopleth to the 120 ppbV isopleth. This would require reductions of both VOC and NO_x, albeit in lesser quantities than by reducing only one kind of pollutant.)

The shortcomings of EKMA are that it models the ozone concentrations only within a particular air column, and only for a single day. Because NO_x is not appreciably destroyed in the ozone formation cycle, the NO_x just keeps moving downwind with the column of air. The transported NO_x will participate in ozone formation further downwind on the same day, and even on subsequent days, because its lifetime can be several days, depending on atmospheric conditions and composition. This is the reason that rural and remote areas far downwind from metropolitan areas also experience high concentrations of ozone, even though the emissions of NO_x in those areas may be quite small.

Because of the shortcomings of EKMA, regulatory agencies in the United States and other countries rely now on more sophisticated models that cover much larger areas than a single air column and also cover a longer time period, usually selected to simulate an elevated pollution episode. Such models are of the Eulerian type, in which the coordinate system remains fixed, and the area covered is divided into grid cells. One such model is called the Urban Airshed Model (UAM), and another is called the Regional Oxidant Model (ROM). These models cover an area of several degrees latitude and longitude with variable size grid elements, down to 2 km². Vertically, the models are divided into several layers below and above the mixing height. The model inputs are the wind field in the modeled area, temperature, humidity, terrain roughness and vegetation cover (the latter affect deposition rates), emission inventory, and the background level of ozone. The governing chemical reactions are quite similar to EKMA, with eight VOC categories. These models can predict the time profile of ozone over a selected location, or produce a contour map of maximum concentrations over the whole modeling domain. The time domain of the model is a

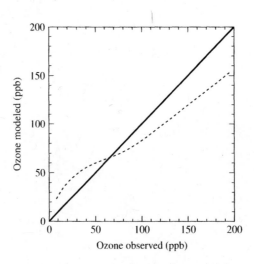

Figure 9.7 Modeled versus observed O_3 concentrations (dotted line) from ROM. (Adapted from National Research Council. 1991. *Rethinking the Ozone Problem in Urban and Regional Air Pollution*. Washington, D.C.: National Academy Press.)

meteorological episode lasting 3–5 days, usually terminating with a precipitation event. Figure 9.7 shows the average comparison between 15,000 measured and modeled data points using ROM. While there is a fairly good correlation between observations and predictions, there is a tendency to overpredict at low concentrations and underpredict at high concentrations.

The grid models bring out clearly that ozone is a regional problem that can only be solved with a regional emission reduction strategy. Because of transport of the precursors NO_x and VOC and of the formed ozone itself, high concentrations in the tens and hundreds ppbV are observed not only in urban areas, but in rural and remote areas, at lower and higher elevations, and even over the sea. Peak concentrations are not only encountered in mid-afternoon hours as the EKMA model would predict, but are also encountered at practically all hours of the day and night.

In areas where the VOC/NO_x ratio is high, NO_x control provides more ozone reduction; in areas with a low VOC/NO_x ratio, VOC control appears more effective. In rural and remote areas where the availability of VOC is plentiful from biogenic sources, NO_x availability is the limiting factor to ozone production. This is so because in chemical kinetics the reactant in short supply determines the rate of the reaction. For a subcontinental region, such as the northeastern or southeastern United States, the diurnal and seasonal average ozone concentrations can only be substantially reduced by region-wide NO_x emission reductions. This is because on a regional basis, biogenic VOC emission can predominate over anthropogenic VOC emissions.

While in the United States and other developed countries great strides have been taken to reduce NO_x emissions from both stationary and mobile sources, ambient NO_x concentrations in the United States and worldwide are either on the increase or at best are level (see Figure 9.10 for NO_x emission trends in the United States). This is because what is gained in NO_x control technology is lost in the ever increasing number of NO_x emitting sources, especially automobiles. In the United States, for example, the highway vehicle population has been increasing at a rate of 3.4 million vehicles per year during the period 1950–1995. In the United States it remains to be seen whether in the future the current ozone standard of 120 ppbV, 1-h average, will not be exceeded, let alone the proposed new standard, 80 ppbV, 8-h average.

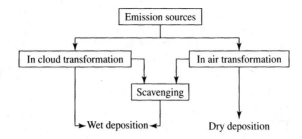

Figure 9.8 Acid deposition schematic, showing transformation and deposition paths.

9.2.6 Acid Deposition

Acid deposition is popularly termed acid rain. Acid deposition is a more appropriate term because acidic matter can be deposited on the ground not only as rain but also in other kinds of precipitation— for example, snow, hail, and fog—and in dry form. The deposition by precipitation is called *wet* deposition; the direct impaction on land and water of acidic gaseous molecules and acidic aerosols (particles) is called *dry* deposition. Acid deposition is a secondary pollutant, because it is a result of transformation of primary emitted pollutants.

In the late seventeenth century, Robert Boyle recognized the presence of "nitrous and salino-sulfurous spirits" in the air and rain around industrial cities of England. In 1853, Robert Angus Smith, an English chemist, published a report on the chemistry of rain in and around the city of Manchester. He later coined the term "acid rain." In the 1960s and 1970s it was first noticed in Scandinavia, then in the northeastern United States and southeastern Canada, that some lakes with a very low buffering capacity (i.e., lakes with low alkalinity), were slowly being acidified with a pH reaching as low as 5.5–6. As most aquatic organisms cannot survive in that kind of acid water, these lakes became devoid of life. Because the acidity of rain precipitation in Europe and in North America was measured to reach a pH value as low as 3, and frequently in the pH 4–5 range, it was soon realized that lake acidification must be a consequence of acid rain. Lakes act like a mildly alkaline solution in a beaker. When the solution is "titrated" over the years with acid deposition, it becomes acidic.[6]

A schematic of acid deposition is presented in Figure 9.8. Power plant, industrial, commercial, residential, and mobile sources emit the precursors of acid deposition, namely, sulfur and nitrogen oxides (SO_x and NO_x). The precursors are advected by winds and are dispersed by turbulent diffusion. During transport in the air, the precursors react with various oxidants present in the air and water molecules to form sulfuric and nitric acid (H_2SO_4 and HNO_3). The acids are deposited on land and water in the dry and wet form. This causes deleterious environmental effects that were listed previously in Table 9.4.

---

[6]pH is the negative logarithm of the hydrogen ion concentration, [H^+] moles per liter, in an aqueous solution; that is, pH = $-\log$ [H^+]. Neutral water has a hydrogen ion concentration of 10^{-7} moles per liter; thus its pH is 7. A one-tenth molar concentration of hydrochloric acid has a hydrogen ion concentration of 10^{-1} moles per liter; thus its pH is 1. Lemon juice has an approximate pH of 3. Carbonic acid has a pH of 5.6. Thus, raindrops in contact with atmospheric carbon dioxide have a slightly acidic pH between 5 and 6 even without addition of other acidic species, such as H_2SO_4 and HNO_3. On the other hand, raindrops in contact with atmospheric alkaline aerosols, such as $CaCO_3$, CaO, $MgCO_3$, and others, may actually have a pH greater than 7.

The exact transformation mechanisms of the precursors to acidic products is still being debated. Evidently, there are two routes of transformation: the "gas-phase" and "aqueous-phase" mechanisms. In the gas phase mechanism the following reactions seem to occur:

$$SO_2 + OH \rightarrow HSO_3 \tag{9.16}$$

$$HSO_3 + OH \rightarrow H_2SO_4 \tag{9.17}$$

and

$$NO_2 + OH \rightarrow HNO_3 \tag{9.18}$$

The acids may directly deposit on land and water as gaseous molecules, or adhere onto ambient aerosols, and then deposit in the dry particulate form. The acid molecules and aerosols may be scavenged by falling hydrometeors and then be deposited in the wet form. This is called "wash-out."

In the aqueous phase mechanism the precursors are first incorporated into cloud drops, a process called "rain-out," followed by reactions with oxidants normally found in cloud drops, namely, hydrogen peroxide (H_2O_2), and ozone (O_3). The distinction between the two mechanisms is of some importance, because the gas-phase mechanism would indicate that the acids are formed in a linear proportion to SO_2 and NO_2 concentrations in the air, whereas in the aqueous phase mechanism the proportion may not be linear. If a nonlinear relationship prevailed, one would not expect that *acid deposition rates* were directly proportional to *acid precursor emission rates*. Trend analyses of acid deposition show that the total amount of deposition integrated over a large region and over a season or year is nearly linearly proportional to the total amount of emissions of the precursors integrated over the same region and time. This does not necessarily mean that all the transformation occurs in the gas phase, but that the aqueous phase transformation is also nearly proportional to the precursors concentration in the cloud drop.

Because the sulfate ion (SO_4^{2-}) is bivalent and the nitrate ion (NO_3^-) is monovalent, if only these two ions were present in equal molar concentrations, two-thirds of the hydrogen ions (H^+) would come from sulfuric acid while one-third would come from nitric acid. However, precipitation also contains other cations and anions; thus, the proper ion balance equation is

$$H^+ = a\,SO_4^{2-} + b\,NO_3^- + c\,Cl^- + d\,HCO_3^- + e\,CO_3^{2-} - f\,NH_4^+ - g\,Ca^{2+} - h\,Mg^{2+} - i\,Na^+ \tag{9.19}$$

where $a, b, \ldots, i$ are factors weighting the respective ion concentrations, and the square brackets have been omitted. Some of the ions are from man-made sources, whereas others are natural (e.g., sea salt, carbonic acid, and ions from earth crustal matter). In Eastern North America (ENA), an approximate empirical ion balance equation appears to hold[7]:

$$H^+ \simeq (1.63 \pm 0.1)SO_4^{2-} + (0.95 \pm 0.1)NO_3^- \tag{9.20}$$

For equimolar concentrations of SO_4^{2-} and NO_3^-, about 63% of the hydrogen ions are due to sulfate ions and 37% are due to nitrate ions.

In ENA, in the 1970s to 1980s, typical SO_4^{2-} and NO_3^- concentrations in precipitation were in the range 15–25 micromoles per liter, each. Using (9.20), the average pH of precipitation can be calculated in the range 4.2–4.4. Figure 9.9(a) shows measurements of the 1986 annual average pH

[7]Golomb, D., 1983. *Atmos. Environ.,* **17**, 1380–1383.

(a)

1986 Annual Precipitation - Weighted pH

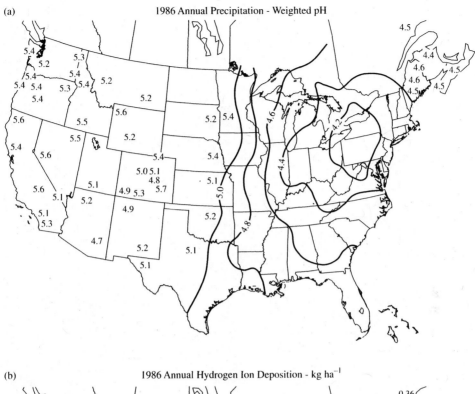

(b)

1986 Annual Hydrogen Ion Deposition - kg ha^{-1}

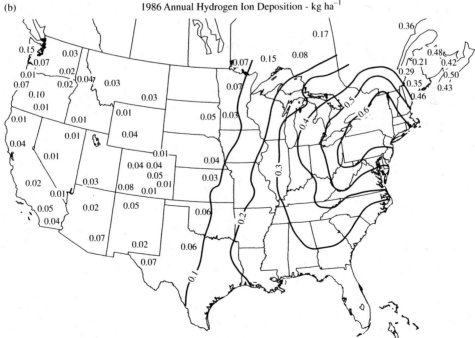

Figure 9.9 (a) 1986 annual precipitation-weighted pH, (b) annual hydrogen ion deposition, kg ha^{-1}. (Adapted from National Acid Precipitation Assessment Program, 1991. Washington, D.C.: Government Printing Office.)

values at hundreds of precipitation monitoring stations in the United States and southern Canada. In the eastern part of the continent the location of the stations was so dense that only the pH isocontours are drawn around the stations. It is seen that in the center of the eastern United States the pH 4.2 isocontour prevailed, whereas the pH 4.4 isocontour encompassed almost all northeastern states and southern Ontario and Quebec. This is in good agreement with the estimates obtained from (9.20). In the western United States the average ion concentrations were lower, and usually the nitrate to sulfate concentration ratios were larger than those in the eastern United States. Figure 9.9(a) shows that in the western United States the measured pH values were in the range pH 5.0–5.5.

The average acidity of precipitation (i.e. the pH) is only a partial indication of the acid deposition problem. A better assessment of the problem is the annual deposition rate of hydrogen ions. Lakes and other surface waters are acidified cumulatively while being exposed ("titrated") to years of acid deposition and exhaustion of their alkalinity content. Episodic deposition rates of hydrogen ions are obtained by multiplying the average hydrogen ion concentrations in each precipitation episode by the amount of precipitation in that episode. Usually, hydrogen ion depositions are measured in a bucket that collects all precipitation that falls over a week. Annual deposition rates are obtained by summing the weekly deposition rates over a year. Figure 9.9(b) shows the 1986 annual hydrogen deposition rates across the United States and southern Canada in units of kilograms per hectare ($10,000 \, m^2$). The isocontours stretch more toward the northeast than the pH isocontours. This is most likely due to the higher precipitation rates in the northeastern United States than in the Midwest.

Figure 9.10 shows the trend of SO_2 and NO_x (reckoned as NO_2) emission rates in the United States for the years 1970–1996. In 1970, SO_2 emission rates reached more than 30 million short tons per year, but fell steadily thereafter as a consequence of (a) installing SO_2 control technology on all new coal-fired power plants and (b) retrofitting existing plants as required by the 1990 Clean Air Act Amendments. Concomitantly, the sulfate and hydrogen ion deposition rates decreased slowly. When the acid deposition section (Title IV) of the 1990 CAAA is fully implemented in the early 2000s, the nationwide SO_2 emission rates are expected to be about one-half those of 1970. NO_x emissions reached a peak in 1980 of about 24 million short tons per year, and they have hovered around that mark ever since. Assuming that sulfate ion depositions will decrease by one-half and that nitrogen ion depositions remain constant, in the early 2000s hydrogen ion deposition rates will

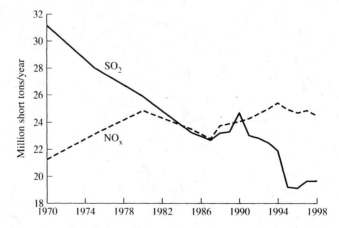

Figure 9.10 Trend of SO_2 and NO_2 annual emissions in the United States, 1970–1998. (Adapted from EPA, 2000. *National Air Pollutant Emission Trends: 1900–1998, EPA 454/R-00-002*. Washington, D.C.: USEPA.)

be about one-third less, and the average pH of precipitation will go up by 0.175 units, compared to 1970s levels.

9.2.6.1 Acid Deposition Modeling

Before enacting costly emission control strategies, government legislators and administrators would like to know the expected environmental benefits of these strategies. For example, if SO_2 emissions were reduced by one-half, will sulfate ion deposition rates decrease proportionally everywhere? This relates to (a) the linearity of transformation of primary emission (SO_2) to secondary pollutant (SO_4^{2-}) and (b) the geographic distribution of the secondary pollutant. To answer these questions, one has to resort to atmospheric transport and transformation models, also called source–receptor models.

In the 1980s in the United States, Canada, and Europe, literally dozens of models were developed, ranging from simple box models (no atmospheric dynamics, just chemical reactions in an enclosed box the size of a subcontinent) to complex Eulerian models stretching over a subcontinent, with individual emission rates in grids of a few km^2 along with simulated wind, diffusion, precipitation, and other meteorologic and topographic factors. These latter models are called supermodels; they require enormous computational capacity to exercise them. However, the supermodels can only simulate a single precipitation episode; they cannot estimate seasonal or annual average deposition rates at widely dispersed receptors from seasonal or annual emission rates at widely dispersed sources. These models have proven, however, that there is a near-linear proportion of SO_4^{2-} deposition rates to SO_2 emission rates. Probably, these models encouraged the U.S. Congress to enact Title IV of the 1990 CAAA, in expectation that reducing about one-half of SO_2 emissions will halve SO_4^{2-} deposition rates and the commensurate hydrogen ion deposition rates.

The authors developed a simple Eulerian model with *time-averaged* emission rates, wind direction and speed, diffusion rate, transformation, and deposition rates.[8] Because all parameters are time-averaged, the model can only be applied to longer periods, such as a season or year. The model quite successfully matched the predicted and observed annual and seasonal deposition rates of sulfate in ENA in the years 1980–1982. Following is a brief description of that model.

The model is a solution to the general atmospheric dispersion equation under steady-state conditions with chemical and physical transformation processes. The latter are assumed to be first order processes. Two solutions are obtained: one for the primary emitted pollutant SO_2, the other for the secondary pollutant SO_4^{2-}:

$$C_p = \frac{Q}{2\pi h D} \exp[ur \cos|\theta - \vartheta| \, (2D)^{-1}] K_0(\gamma r) \tag{9.21}$$

$$C_s = \frac{Q}{2\pi h D^2 \tau_c} \exp[ur \cos|\theta - \vartheta| \, (2D)^{-1}] \frac{K_0(\alpha r) - K_0(\gamma r)}{\gamma^2 - \alpha^2} \tag{9.22}$$

$$\alpha^2 \equiv \left(\frac{1}{\tau_{ws}} + \frac{1}{\tau_{ds}} \right) \frac{1}{D} + \frac{u^2}{4D^2} \tag{9.23}$$

$$\gamma^2 \equiv \left(\frac{1}{\tau_{wp}} + \frac{1}{\tau_{dp}} + \frac{1}{\tau_c} \right) \frac{1}{D} + \frac{u^2}{4D^2} \tag{9.24}$$

[8]See Fay, J. A., D. Golomb, and S. Kumar, 1985. *Atmos. Environ.*, **19**, 1773–1782.

where

C_p = concentration of primary pollutant [SO_2], g m^{-3}

C_s = concentration of secondary pollutant [SO_4^{2-}], g m^{-3}

Q = source emission rate of primary pollutant, g s^{-1}

h = average mixing height of modeling domain, m

D = diffusion coefficient, m^2 s^{-1}

u = average wind speed of modeling domain, m s^{-1}

r = distance between source and receptor, m

θ = azimuthal angle from the source to the receptor, degree

ϑ = azimuthal direction of prevailing wind in modeling domain, degree

K_0 = Bessel function of zeroth order

τ_c = transformation time constant from primary to secondary pollutant, s

τ_{dp} = dry deposition time constant of primary pollutant, s

τ_{ds} = dry deposition time constant of secondary pollutant, s

τ_{wp} = wet deposition time constant of primary pollutant, s

τ_{ws} = wet deposition time constant of secondary pollutant, s

Equations (9.21) and (9.22) give the *concentration* at the receptor of the primary and secondary pollutants. If we are interested in *wet deposition* at the receptor of the secondary pollutant (i.e., the wet deposition of sulfate ions), the following equation is used:

$$D_{ws} = \frac{C_s h R}{\tau_{ws} R_0} \tag{9.25}$$

where D_{ws} is the wet deposition rate in g s^{-1}, R is the annual or seasonal rainfall at the receptor, and R_0 is the annual or seasonal rainfall averaged over the modeling domain. Because h appears in the denominator of equation (9.22) and the numerator of equation (9.25), the wet deposition rate is independent of h. Similarly, the dry deposition rate of sulfate is obtained from

$$D_{ds} = \frac{C_s h}{\tau_{ds}} \tag{9.26}$$

The model parameters were derived by an optimization technique. Several years of observed data of annual wet deposition of sulfate in ENA were compared to model predicted values. Optimized values were obtained from minimizing the root mean square error:

$$E^2 = \frac{\Sigma(\text{observation} - \text{prediction})^2}{\Sigma(\text{observation})^2} \tag{9.27}$$

The optimized model parameters are listed in Table 9.6. The optimized parameters are consistent with independently derived data in the literature. For different averaging periods (e.g., summer or winter seasons) or for different regions, different model parameters must be derived.

TABLE 9.6 Optimized Sulfate Deposition Model Parameters

Parameter	Optimized Value
D_h	$4.3E(6)\ m^2\ s^{-1}$
u	$7.1\ m\,s^{-1}$
δ	$214°$
τ_c	$1.9E(5)\ s$
τ_{wp}	$11.3E(5)\ s$
τ_{ws}	$0.6E(5)\ s$
τ_{dp}	$2.0E(5)\ s$
τ_{ds}	$12.5E(5)\ s$

Data from Fay J. A., D. Golomb, and S. Kumar, 1985. *Atmos. Environ.*, **19**, 1773–1782.

Using optimized model parameters, source receptor distance, and azimuthal angle, wet (and dry) deposition rates at any receptor due to any source in ENA can be estimated. Instead of using individual sources, such as a single power plant, it is convenient to aggregate sources within a whole state by assigning an "emission centroid" to that state. Using state-level emission rates of SO_2 in 31 states east of the Mississippi River, along with observed annual wet sulfate deposition rates at tens of stations in ENA in 1980–1982, the predicted and observed wet sulfate deposition in ENA is presented in Figure 9.11. The observed data points are printed at the approximate location of the stations. The modeled values are represented by the isocontours. The deposition rates are in units of kg SO_4^{2-} per hectare per year. It is seen that a good match is obtained, with a correlation coefficient of 0.88 and an error of 17%, as defined in equation (9.27). Like the hydrogen deposition in Figure 9.9(b), the isocontours are stretched toward the northeast because of the prevailing southwesterly winds.

9.2.6.2 Transfer Coefficients

A transfer coefficient T_{ij} is defined as the ratio of annual wet sulfate deposition at the receptor j to the annual SO_2 emission from source i. It is derived from equations (9.21)–(9.24) by varying the source–receptor distance and the angle subtended between the source receptor line and the prevailing wind. Transfer coefficients for ENA are plotted in Figure 9.12 for three angles: (a) receptor lying directly downwind; (b) receptor lying crosswind; and (c) receptor lying upwind. The units on the y axis are ha^{-1}. If SO_2 emissions are given in kg y^{-1}, depositions of SO_4^{-2} are obtained in units of $1.5\ kg\ ha^{-1}\ y^{-1}$. (The factor 1.5 arises because the ratio of molecular weight SO_4^{2-}/SO_2 is 1.5.) The amount of wet sulfate deposition at receptor j from all sources i is obtained from

$$D_{ws}^j = \sum^i T_{ij}Q_i \tag{9.28}$$

where Q_i is the annual SO_2 emission at source i.

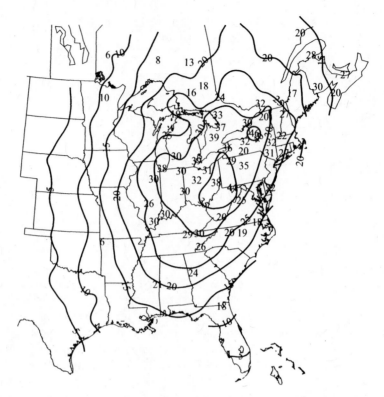

Figure 9.11 Predicted versus observed wet sulfate deposition in ENA (kg SO_4^{2-}/ha y). (Source is same as in Table 9.6.)

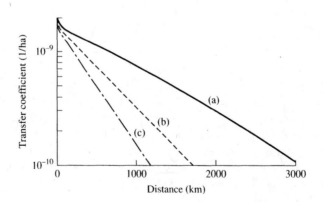

Figure 9.12 Transfer coefficient as a function of source-receptor distance for source: (a) upwind, (b) crosswind, and (c) downwind from receptor. (Source is same as in Table 9.6.)

9.2.6.3 Source Apportionment

The fractional contribution of source i to the wet deposition at receptor j can be determined from

$$S_{ij} = \frac{Q_i T_{ij}}{\sum^i Q_i T_{ij}} \tag{9.29}$$

where the summation is over all contributing sources. This is illustrated in Figure 9.13, which shows the contribution (%) of 31 eastern U.S. states and several subprovinces of Canada to the wet sulfate deposition in the Adirondack region of New York. The percent contribution is dependent on (a) the emission rate of SO_2 in a particular state and (b) the transfer coefficients between the states and the Adirondacks. The high SO_2 emitting states of Pennsylvania, Ohio, West Virginia, Indiana, and New York are the largest contributors to Adirondack sulfate deposition. By coincidence, these states also lie upwind of the Adirondacks. A significant portion of lakes in the Adirondacks have been acidified, probably due to acid deposition.

Thus, acid deposition models, as the one described above, could be used for "targeted" emission reduction strategies, wherein those sources that contribute most to acid deposition to a sensitive area would curtail their emissions to a greater extent than those sources that contribute little or nothing. However, by virtue of Title IV of CAAA 1990, the U.S. Congress decided that SO_2 emissions will be cut by about one-half of 1980 emissions, regardless of the location of a source *vis à vis* a sensitive area. From a political standpoint, a uniform emission reduction strategy is more expedient than a targeted strategy, which would pit some states and regions against others.

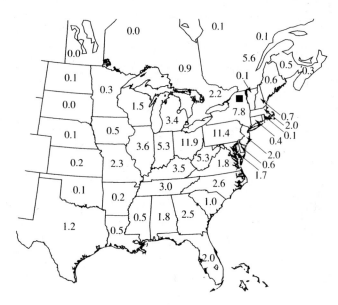

Figure 9.13 Contribution (%) by state and province to sulfate deposition at an Adirondack receptor. (Source is same as in Table 9.6.)

9.2.7   Regional Haze and Visibility Impairment

Small particles (also called fine particles) less than 1–2 μm in diameter settle very slowly on the ground and can travel hundreds to thousands of kilometers from their emitting sources. A part of the fine particles are emitted directly from industrial, commercial–residential, and transportation sources. These are called primary particles. However, the majority of fine particles is a product of gas-to-particle transformation processes, including photochemical processes, where the precursor gases are emitted from the aforementioned sources. These are called secondary particles. The particles can envelope vast areas, such as the northeastern United States and southeastern Canada, California and adjoining states, western and/or central Europe, and southeastern Asia. Satellite photos often show continental areas covered with a blanket of particles, sometimes stretching out hundreds of kilometers over the ocean. This phenomenon is called *regional haze*. The haze is mostly associated with stagnating anticyclones, when a high barometric pressure cell remains stationary over a region, sometimes over a period of a week or more. The high-pressure cell causes air to circulate in a clockwise fashion around the cell, drawing in increasingly large amounts of emissions from the surrounding sources, forming the regional haze. Usually, the regional haze period is terminated by a cold front moving through the region with accompanying convective clouds, thunderstorms, and precipitation that washes out the particles.

The composition of the fine particles varies from region to region, depending on the precursor emissions. In the northeastern United States, central Europe, and southeastern Asia, more than half of the composition is made up of sulfuric acid and its ammonium and sodium salts, largely due to high sulfur coal and oil combustion. The rest is made up of nitric acid and its salts, carbonaceous material (elemental and organic carbon), and crustal matter (fine dust of soil, clay, and rocks). In the western and southwestern United States and in some other urbanized/industrial areas of the world, nitrate and carbonaceous matter makes up the majority of the composition of fine particles. This is due to heavy automobile traffic, chemical industries, refineries, gas-fired power plants, and other urban/industrial sources. A part of the haze may be due to biogenic sources, such as the emissions of isoprene, terpene, pinene, and other isoprene derivatives from coniferous forests, but the predominant cause of regional haze is the emission of gaseous and particulate matter from anthropogenic sources, mostly from fossil fuel combustion. Condensed water is also an ingredient of fine particles, as the precursor gases and the formed particle nuclei attract water molecules from the air, especially during high humidity periods.

The small particles are efficient scatterers of light. The scattering efficiency is dependent upon the wavelength. Maximum scattering efficiency for visible light (400–750 nm) occurs with particles less than 1 μm in diameter, the so-called submicron particles. Light scattering prevents distant objects from being seen. This is called *visibility impairment*. During regional haze periods, one cannot distinguish distant mountains on the horizon, and occasionally one cannot see objects farther than hundreds of meters, such as the other wall of a river valley or buildings a few city blocks away.[9] Also, the increasing concentration of particles in urbanized parts of the continents causes the loss of visibility of the starlit nocturnal sky. These days, small stars, less than fifth order of magnitude, rarely can be seen from populated areas of the world.

---

[9]Visibility impairment became a significant problem in some national parks in the United States. For example, on some hazy days the northern rim of the Grand Canyon in Arizona cannot be seen from the southern rim.

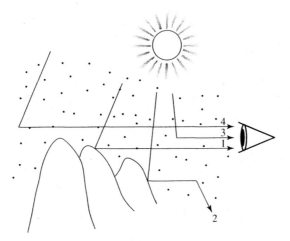

Figure 9.14 Reflection and scattering of light rays. Ray 1, reflected from object; Ray 2, reflected but scattered out of view; Ray 3, scattered by particles between object and observer into view; Ray 4, scattered by background particles. Adapted from Seinfeld, J. and S. N. Pandis, 1998. *Atmospheric Chemistry and Physics.* New York: Wiley Interscience Pub.

The scattering of light causes a loss of contrast. This is depicted in Figure 9.14. A part of the incoming sunlight is reflected from the object and reaches the retina of the eye (ray 1). A part of the reflected light is scattered out by particles in the air and does not reach the retina (ray 2). A part of the incoming sunlight is scattered by particles between the object and the eye, reaching the retina, but this ray causes a loss of contrast (ray 3). Finally, a part of the incoming sunlight is scattered by particles behind the object, reaching the retina, but with a loss of contrast (ray 4). The consequence of rays 2 to 4 is a loss of contrast of the object against the background. Empirically, it has been shown that an object is no longer visible when the contrast against the background is less than 2%. This is called the threshold contrast and can be expressed as

$$C_x = \frac{I_{bkg} - I_{obj}}{I_{obj}} \geq 2\% \tag{9.30}$$

where C_x is the contrast at distance x meters, I_{bkg} is the intensity of light rays from the background, and I_{obj} is the intensity of light rays from the object. Alternatively, the contrast relationship can be written

$$C_x = C_0 \exp(-b_{ext}x) \tag{9.31}$$

where C_0 is the contrast at zero distance, and b_{ext} is the extinction coefficient in m^{-1}. Threshold contrast is when $C_x/C_0 = 0.02$, thus

$$-b_{ext}x = \ln(0.02) = -3.912 \tag{9.32}$$

Equation (9.32) is called the Koschmieder relationship. It gives the extinction coefficient b_{ext} when an object loses its contrast against the background at a distance x meters. In clear air $b_{ext} = E(-5)$

to 5E(-5) m^{-1}, and objects can be discerned as far as 80 to 400 km (of course, the latter distance is only visible from a height or to a height, so as to reach over the earth's curvature). In polluted air, b_{ext} can be as large as E(-3) m^{-1}, when objects disappear at a few kilometers distance.

Visibility impairment is a significant and unpleasant side effect of fossil fuel use. Visibility improvement can be accomplished by reducing fine-particle and gaseous precursor emissions from fossil fuel combustion and other fossil fuel usage. In practice this means improved emission control devices for primary particles, SO_2, NO_x, and VOC (see Chapters 5 and 8). In the United States, visibility improvement is addressed in Section 169A of the Clean Air Act Amendment of 1977, which requires that visibility impairment at U.S. national parks must be lessened by reducing particles and their precursor emissions from nearby sources. As a consequence of better and wider use of emission control devices on stationary and mobile sources, visibility has improved steadily in the past decades, and regional haze occurrences in the United States are now less frequent than they were in the 1960s and 1970s.

9.3 WATER POLLUTION

The consumption of fossil fuel entails a significant impact on water quality and water usage. The contamination of water starts at the mining and extraction stage, through transport and refining, all the way to leaching into the groundwater of ash and scrubber sludge left behind after combustion of fossil fuels. We shall limit this section to the effects of acid mine drainage, coal washing, leaching from coal and ash piles, and water pollution due to atmospheric deposition of toxic byproducts of fossil fuel combustion.

The major environmental disasters due to collisions or grounding of supertankers carrying crude oil have been documented in scientific journals and the press and will not be treated here in detail. Suffice it to say that in 1978, the *Amoco Cadiz* ran aground off the coast of France, spilling 223,000 metric tons of crude oil. In 1979, in the collision of the *Atlantic Empress* with the *Aegean Captain,* 50 miles northwest of Tobago, more than 287,000 t were spilled. In 1983, the *Castillo de Bellver* spilled 252,000 t off the coast of South Africa. In 1989, the *Exxon Valdez* spilled 37,000 t of crude in the environmentally sensitive Prince William Sound in Alaska. In 1991, the *ABT Summer* spilled 260,000 t off the coast of Angola. It is estimated that between 3 to 4 million metric tons of oil is spilled annually into the world rivers, lakes, seas and oceans. Oil spills seem to be a high environmental risk associated with fossil oil usage.

9.3.1 Acid Mine Drainage and Coal Washing

In terms of sheer quantity, the most serious water pollution problem associated with coal use is acid drainage from mines, especially surface mines, coal piles, and coal washing. Precipitation falling on open coal seams and on coal piles will leach out mineral matter. The leachate contains acids, toxic elements, and often radioactive isotopes. Leachates with a pH as low as 2.7 have been measured. The acid is a product of oxidation and hydrolysis of the pyritic sulfur in coal. The following toxic elements are found in coal mine and pile drainage in concentrations that exceed drinking water standards: arsenic, barium, beryllium, boron, chromium, fluorine, lead, mercury, nickel, selenium, vanadium, and zinc. U.S. federal and state regulations now require that the leachate be either treated or disposed of in a secure impoundment. Coal contains the radioactive isotopes ^{14}C and ^{40}K, because these elements were inherent in the original biologic tissue of the antecedents of coal. These radionuclides do not pose an environmental hazard. However, frequently uranium, thorium, and their radioactive daughter elements are also found in the mineral matter adhering to the coal. Thus, leachate from coal mines, piles, and slag needs to be monitored for its radioactivity before entering into the environment and for preventing mine worker exposure.

About 50% of all mined coal in the United States is now "washed" at the mine mouth prior to shipment to the user (see Section 5.2.9.3). Coal washing increases the heating value per unit mass of coal by removing the incombustible mineral matter. More importantly, coal washing removes pyritic sulfur, which can amount up to 50% of the sulfur content of the raw coal. The most widely used technique for coal washing is mass separation. Mineral matter has a higher specific gravity than coal. By flushing crushed coal in a stream of water, the mineral matter settles out, while the lighter coal particles float in the stream. The settled mineral matter contains a high concentration of acidic, toxic, and possibly radioactive compounds and elements. This slag must be analyzed for its content. If the content is toxic, the slag must be treated or safely disposed of in a hazardous waste impoundment. In the United States, discharges from mining activities are regulated under the Clean Water Act, specifically, the National Pollution Discharge Elimination System (NPDES).

9.3.2 Solid Waste from Power Plants

Although much of the mineral matter of coal is removed at the mine mouth, coal delivered to a power plant or other facilities still contains adhering mineral matter, simply called ash. The ash content can amount to anywhere between 1% and 15% of the coal weight. Even oil contains ash, amounting to 0.01–0.5% by weight. After combustion of the coal particle or oil droplet, the mineral matter remains uncombusted, and either falls to the bottom of the boiler or is blown out with the flue gas as *fly ash*. In modern pulverized-coal-fired boilers, about 90% of the mineral matter forms fly ash, and 10% bottom ash. Most of the fly ash is collected in electrostatic precipitators (ESP, see Section 5.2.9.2), except the very small particles (less than 1 μm in diameter), which escape into the air because they are not removed efficiently by the precipitator. The fly ash contains approximately the same acidic, toxic, and radioactive matter as the original mineral matter in coal. The fly ash is required to be chemically analyzed and, if found hazardous to human health and the environment, properly disposed of. Nonhazardous bottom and fly ash is often used as aggregate material in concrete or asphalt.

As a consequence of acid deposition regulations in the United States and many other countries, coal-burning power plants and industrial boilers are required to be equipped with *flue gas desulfurization* devices (see Section 5.2.9.3). For high-sulfur-content coal ($\geq$0.6% by weight), a wet limestone scrubber is necessary. After passing the ESP, the flue gas enters a scrubber tower. A slurry of limestone is injected from sprinklers at the top of the scrubber. Flue gas containing SO_2 and other sulfur compounds flows countercurrent to the limestone spray. A sludge is collected in the bottom of the scrubber, consisting of wet calcium sulfate (gypsum) and calcium sulfite along with unreacted limestone. The sludge is dewatered as much as possible (unfortunately, gypsum is very difficult to dewater), then disposed of. Because the sludge also may contain toxic elements—notably arsenic, cadmium, mercury and selenium—its disposal may require a specially designed secure impoundment.

9.3.3 Water Use and Thermal Pollution from Power Plants

It is a consequence of the second law of thermodynamics that heat engines, such as power plants, must reject some of the fuel energy in the form of heat transfer to a cold reservoir. Roughly about one-third of the inherent heating value of the fuel is rejected via the steam condenser to the cold reservoir. Another third is rejected to the atmosphere via the stack gas, and only one-third is transformed into useful work. The cold reservoir is usually a water body. Some power plants and industrial boilers are located near surface waters, such as a river, lake, or ocean, which can be used for absorbing the heat from the condenser. Other facilities need to use a cooling tower or a cooling pond. The EPA requires that all new power plants use cooling towers for heat rejection, rather than once-through cooling water from adjacent surface waters. In cooling towers, cold municipal or well water percolates around the hot condenser tubes, taking up the rejected heat. The water is chilled in a draft of cold air and then is recycled. Alternatively, spent steam from the steam turbine is directly condensed in cold air that is drawn through the cooling tower by huge fans. In both cases a part of the cooling water is evaporated by taking up the heat from the condensing steam. Evaporated water rises from the cooling tower into the cold ambient air and condenses, thus forming the visible "steam" plume that emanates from the tower. A 1000-MW electric power plant working at a thermodynamic efficiency of 33% and ambient temperature of 15 °C loses about $1.7 E(7)$ m^3/y (about 4.4 billion gallons/y) of cooling water due to evaporation. In the United States, cooling towers alone consume between 2% and 3% of total withdrawals from surface waters. Some critics question the wisdom of using precious water resources for evaporative cooling purposes, rather than surface waters (where available), even though the latter entails (a) some risk of thermal pollution and surface water contamination by leaching of the heat exchanger components and (b) some increased evaporation of the surface waters to the atmosphere. It is seen here that there is no simple or unequivocal solution to environmental and natural resource problems engendered by fossil fuel usage.

9.3.4 Atmospheric Deposition of Toxic Pollutants onto Surface Waters

In Section 9.2.6 we dealt with one form of water pollution due to combustion of fossil fuels, that of acid deposition. But in addition to sulfur and nitrogen oxides, there are other combustion products that escape from smoke stacks and eventually are deposited on land and water, which may cause deleterious health and environmental effects. Two cases in point are the atmospheric deposition of toxic metals and polycyclic aromatic hydrocarbons (PAH).

9.3.4.1 Toxic Metals

We noted previously that fly ash particles may contain toxic metals, such as arsenic, cadmium, mercury, lead, selenium, vanadium, and zinc. These metals are found in small particles, less than 1 μm in diameter. The small particles are not efficiently collected by the ESP and thus escape into the atmosphere. Because of their small size, these particles are little affected by gravity and can be transported over large distances, hundreds to thousands of kilometers. Eventually, they are deposited in dry or wet form on land and water. From the land, toxic metals may leach into groundwater, or run off into streams, lakes, or ocean. Thus, they may enter the food chain, either directly by drinking water or via aquatic organisms that drink or feed in the water. For example, fish from the Great

Lakes and other lakes in the northeastern United States and southeastern Canada often contain a concentration of mercury which is deemed unhealthy, especially for pregnant women. Children born to mothers who consumed mercury-laced fish may show birth defects or mental retardation. In addition to coal-burning power plants, municipal incinerators are sources of mercury, because batteries, switches, and fluorescent bulbs that are tossed into the garbage may contain mercury, which, when incinerated, escapes into the atmosphere. Lead may also cause mental retardation and other brain and central nervous system effects. Part of the lead enters surface and groundwater via atmospheric deposition. The lead used to originate from leaded gasoline. With the phasing out of leaded gasoline, lead deposition and lead-related diseases are on the decline.

9.3.4.2 Polycyclic Aromatic Hydrocarbons

PAH are organic compounds consisting of two or more fused benzene rings. Naphthalene has two rings, anthracene three, pyrene and chrysene four (in different arrangement), benzopyrenes (there are several) five, perylene six, and coronene seven. Some of the PAH are known or suspected carcinogens, notably benzo(a)pyrene. PAH are a product of incomplete combustion. Whenever we see a sooty smoke, we can be sure that it contains PAH, because PAH adhere to soot particles. PAH in the vapor or particulate phase are carried by winds and subsequently deposited on land and water in the dry form, while those scavenged by precipitation are deposited in the wet form. Because PAH are unstable (they decompose in sunlight, or by oxidation with atmospheric oxidants), they do not travel as far as metals, and they are deposited closer to the combustion sources. Because PAH are very hydrophobic, they do not readily incorporate into rain droplets; therefore dry deposition predominates over wet deposition, especially near polluted cities. Rivers, lakes, and coastal waters surrounded by urban-industrial areas are especially affected by PAH deposition from the myriad of combustion sources in those areas. Soot particles with adsorbed PAH settle in the sediments of these waters. Bottom feeding fish and shellfish that live in the sediments may ingest PAH and are often found with cancerous lesions and other diseases. Because PAH dissolve in fatty tissues, these organisms may bioaccumulate PAH and transfer them to the food chain, including humans.

The irony is that PAH are not emitted as much from big centralized combustion sources as from small, dispersed sources. Large power plants, industrial boilers, and municipal incinerators are easily controlled for preventing emissions of products of incomplete combustion, of which PAH are a part. Basically, the control involves "good engineering practice," which is combustion in excess air, thorough mixing of fuel and air, high flame temperature, and sufficient residence time in the combustion chamber. Under those conditions, practically all the carbonaceous matter burns up into CO_2 and H_2O, and no organic molecules or radicals are left which may recombine to form soot and PAH. Most of the soot and PAH emissions come from residential and commercial furnaces, wood stoves and fireplaces, open fires, barbecues, aircraft jet engines, gas turbines, and, most copiously, internal combustion engines, such as diesel and gasoline-fueled trucks and automobiles. We are all familiar with the cloud of soot emanating from a diesel truck under accelerating drive, when the fuel-to-air ratio becomes very rich, or the black trail left behind a jet aircraft during take-off. If we want to minimize PAH emissions, we must find better ways of controlling these dispersed stationary and mobile sources. Cigarette smoke also contains PAH; the latter are probably the cause of lung and throat cancer of smokers.

9.4　LAND POLLUTION

In regard to fossil fuel use, the heaviest toll on land impact is due to coal mining—in particular, surface mining, also called strip mining. In the United States, more than 60% of about 750 million metric tons of coal mined annually is surface mined. Because the coal seam is rarely on the surface itself, this requires removing the "overburden." The latter can amount to up to 100 m of soil, sand, silt, clay, and shale. Some coal seams appear in hill sides and river banks. Here the coal is removed by auger mining, which uses a huge drill to dig the coal out of the seams.

Surface mining leaves behind enormous scars on the landscape, not to mention the disruption of the ecosystem that existed before mining started, or the loss of other possible uses of land instead of mining. For that reason, the U.S. Congress in 1977 passed the Surface Mining Control and Reclamation Act. The law delegates to the states the authority to devise a permitting plan that must include an environmental assessment of the mining operations, and procedures for reclamation of the land after cessation of operations. The permitting plan must describe (a) the condition, productivity, uses, and potential uses of mine lands prior to mining and (b) proposed post-mining reconditioning of the mined lands, including revegetation. Stringent operational standards apply to mining in critical areas, including those with slopes greater than 20 degrees and alluvial valleys. The land must be restored to a condition capable of supporting any prior use and possible uses to which the land might have been put if it had not been mined at all. Needless to say that this law, and the requirements for safe and ecologically sound surface mining practices, imposed a great effort and economic burden on surface mining companies and indirectly raised the price of surface-mined coal.

Deep shaft mining also places a burden on the land. We mentioned earlier that the mined coal is brought to the surface, where it is crushed and washed. The removed mineral matter accumulates in enormous slag piles that mar and despoil the landscape. Nowadays, strict regulations pertain in the United States and other countries as to the disposal of the coal mine slag and restoration of the landscape.

Finally, we should mention that the vast number of derricks associated with oil and gas exploration and exploitation is neither an aesthetically pleasing sight nor helpful to the ecology that existed on the land prior to oil and gas mining.

9.5　CONCLUSION

The consumption of vast quantities of fossil fuel by mankind causes many deleterious environmental and health effects. These effects start from the mining phase of the fossil fuels, through transportation, refining, combustion, and waste disposal. When coal is mined in deep shafts or in strip mines, mineral matter is separated at the mines by milling and washing. The residual slag may contain toxic, acidic, and sometimes radioactive material, which needs to be properly disposed of without endangering the environment and humans. Oil and gas well derricks, on- and off-shore, are aesthetic eyesores and are the cause of oil spillage. Oil and gas pipes may rupture and leak their contents. Oil tankers and barges spill on the average 4 million tons per year of crude and refined petroleum on our waterways and oceans. Oil refineries are sources of (a) toxic emissions through vents, leaks, and flaring and (b) toxic liquid effluents and solid waste.

By far, the greatest environmental and health impact is due to fossil fuel combustion in furnaces, stoves, kilns, boilers, gas turbines, and the internal combustion engine, which powers our automobiles, trucks, tractors, locomotives, ships, and other mobile and stationary machinery. The combustion process emits pollutants through smoke stacks, chimneys, vents, and exhaust pipes, such as particulate matter, oxides of sulfur and nitrogen, carbon monoxide, products of incomplete combustion, and volatile toxic metals. Some of these pollutants are toxic to humans, animals, and vegetation per se, whereas others transform in the atmosphere to toxic pollutants, such as ozone, organic nitrates, and acids. The pollutants are advected by winds and dispersed by atmospheric turbulence over hundreds to thousands of kilometers, affecting sensitive population and biota far removed from the emission sources. Particles, besides containing toxic and potentially carcinogenic agents, often envelope whole subcontinental areas in a haze that reduces visibility and the enjoyment of the landscape and a starry sky.

Great strides have been taken, especially in the more affluent countries, for limiting the emissions of air pollutants from the combustion sources. For example, most particles can be filtered out of the flue gas by electrostatic precipitators or fabric filters. Sulfur oxide emissions can be reduced by wet or dry limestone scrubbers. Nitric oxide emissions can be reduced by catalytic and noncatalytic injections of chemicals into the flue gas. The catalytic converter, which is now applied to gasoline fueled automobiles in many countries, reduced the emissions of automobiles significantly compared to the uncontrolled predecessors. Diesel engines, with proper tuning, are also emitting less pollutants, although a magic box like the catalytic converter has yet to be found for diesel engines. Alas, little or no emission control devices are applied to the myriad of dispersed sources, such as residential furnaces, stoves and fireplaces, and smaller industrial facilities.

The less affluent countries, because of more pressing economic needs, do not yet avail themselves of emission control devices. Consequently, air quality in those countries is much worse than in the more affluent ones. Because air pollutants do not stop at national boundaries, surrounding countries may feel the effect of emissions from neighboring states. It is incumbent on the more affluent societies to help the poorer ones in controlling air pollutant emissions, because the health of the whole human population and ecology of the planet is at stake.

Finally, this chapter has not addressed the great looming risk of global warming due to anthropogenic greenhouse gas emissions from fossil fuel usage. This will be the subject of the next chapter.

PROBLEMS

Problem 9.1

Calculate the emission rate of SO_2 (lb/s) from a large coal-fired power plant that uses 2 million short tons (1 short ton = 2000 lb) per year having a coal sulfur content of 2% by weight.

Problem 9.2

Calculate the emission rate of SO_2 per fuel heat input (g/GJ) of a large coal-fired power plant that uses coal having a heating value of 30 MJ/kg and a sulfur content of 2% by weight.

Problem 9.3

A vehicle traveling along a highway at 90 km/h emits 80 g CO and 10 g NO per liter of fuel burned. The vehicle travels 8 km per liter of fuel burned. Calculate the emission rate of CO and NO in units of g/km and g/mi. Compare the latter with the U.S. federal vehicle emission standards for model year 1994.

Problem 9.4

Table P9.4 shows ozone concentrations measured every hour on the hour on a high pollution day in Los Angeles. Plot the diurnal ozone profile. Determine the arithmetic average and geometric mean ozone concentrations for that day. Determine the maximum 8-h arithmetic average concentration and compare it with the proposed U.S. ambient O_3 standard of 80 ppbV.

t (h)	1	2	3	4	5	6	7	8	9	10	11	12
O_3 (ppbV)	12	8	4	4	6	10	23	42	63	88	112	183

t (h)	13	14	15	16	17	18	19	20	21	22	23	24
O_3 (ppbV)	212	254	302	312	315	312	204	162	88	34	23	18

Problem 9.5

A coal-fired power plant has a smoke stack 30 m high and 3 m in diameter. The flue gas exit velocity is 5 m/s, the exit temperature is 300 °C, the ambient temperature is 30 °C, the wind speed is 3 m/s, and the atmospheric stability category is B. Calculate the effective height above ground level (m) of the plume center line using the Briggs plume rise equation.

Problem 9.6

Using the calculated emission rate of the plant with characteristics given in Problem 9.1, the Gaussian Plume Equation, stability category B, and the effective plume height calculated in Problem 9.5, calculate the SO_2 concentration ($\mu g\ m^{-3}$) at a downwind distance of 1 km, 100 m perpendicular to the wind direction at the top of a 100 m high hill.

Problem 9.7

For the same power plant, using Figure 9.3, estimate the highest concentration of SO_2 ($\mu g\ m^{-3}$) on level ground for stability category B. Estimate the downwind distance (m) at which the highest concentration (x_{max}) would occur.

Problem 9.8

A highway with 8 lanes has traffic of one vehicle per 30 m per lane. Each vehicle emits on the average 10 g/s CO at an average height of 2 m. The wind speed is 6 m/s, wind direction at an angle of 45° toward the highway, stability category C. Calculate the CO concentration (μg m^{-3}) at a receptor 100 m downwind from the highway.

Problem 9.9

Using the EKMA plot of Figure 9.6, assuming the peak afternoon concentration of O_3 is 200 ppbV, estimate the percentage of the necessary reductions of initial concentrations of VOC and NO_x to achieve an afternoon peak concentration of O_3 of 120 ppbV for an urban (VOC:$NO_x = 4{:}1$), sub-urban (VOC:$NO_x = 8{:}1$), and rural (VOC:$NO_x = 16{:}1$) environment. Estimate the concentration reductions in two ways: VOC only and NO_x only. Explain why in one environment it is more effective to reduce VOC concentrations, whereas in another environment it is more effective to reduce NO_x concentrations. Is it possible to achieve in a certain environment the 120 ppbV concentration by actually increasing morning NO_x concentrations? In which environment is that possible, and what would that do to downwind locations?

Problem 9.10

Measurements show that in an environmentally sensitive area the average sulfate ion concentration in precipitation is 25 μmol/L and the nitrate ion concentration is 15 μmol/L. Using the approximation equation (9.20), calculate the pH of precipitation.

Problem 9.11

The state of Ohio emits about 2.5 million tons per year of SO_2. The emission centroid of Ohio is about 1000 km from the center of Massachusetts. Ohio is exactly upwind from Massachusetts. (a) Using the transfer coefficients plotted in Figure 9.12, estimate the annual rate of wet deposition of sulfate ions (kg ha^{-1} y^{-1}) in Massachusetts arising from Ohio SO_2 emission sources. (b) Massachusetts receives on the average 25 kg ha^{-1} y^{-1} wet sulfate deposition. What percentage of that does Ohio contribute? Note that the molecular weight of sulfate is 1.5 times the molecular weight of SO_2.

Problem 9.12

The annual average airborne sulfate particle concentration in Massachusetts is 7 μg m^{-3}, the dry deposition time constant of sulfate is $\tau_{ds} = 2E(5)$ s, and the average mixing height is 1000 m. What is the dry deposition rate (kg ha^{-1} y^{-1}) of sulfate particles in Massachusetts. Compare this with the wet deposition rate of sulfate.

Problem 9.13

(a) Estimate the maximum visibility distance (km) for a polluted atmosphere with $b_{ext} = 6E(-4)$ m^{-1} and for a clean atmosphere with $b_{ext} = 6E(-5)$ m^{-1}. (b) If the maximum visibility is 6 km, what is the extinction coefficient (m^{-1})?

BIBLIOGRAPHY

Heinsohn, R. J., and R. L. Kabel, 1999. *Sources and Control of Air Pollution*. New York: Prentice-Hall.

Kraushaar, J. J., and R. A. Ristinen, 1993. *Energy and Problems of a Technical Society,* 2nd edition. New York: John Wiley & Sons.

National Acid Precipitation Assessment Program, 1990. *Acidic Deposition, State of Science and Technology,* in four volumes. Washington D.C.: U.S. Government Printing Office.

National Research Council, 1991. *Rethinking the Ozone Problem in Urban and Regional Air Pollution.* Washington, D.C.: National Academy Press.

Seinfeld, J., and S. N. Pandis, 1998. *Atmospheric Chemistry and Physics,* New York: Wiley Interscience.

Turner, D. B., 1994. *Workbook of Atmospheric Dispersion Estimates,* 2nd edition. Chelsea, Michigan: Lewis Publishers.

Wark, K., C. F. Warner, and W. T. Davis, 1998. *Air Pollution, Its Origin and Control.* Reading: Addison-Wesley.

Global Warming

10.1 INTRODUCTION

Of all environmental effects of fossil fuel usage, global warming, including its concomitant climate change, is the most perplexing, potentially most threatening, and arguably most intractable. It is caused by the ever-increasing accumulation in the atmosphere of CO_2 and other gases, such as CH_4, N_2O, chlorofluorocarbons (CFCs), and aerosols, largely due to emissions of these gases from anthropogenic activities, and reaches levels that exceed those that existed for centuries before the beginning of the industrial revolution. Called *greenhouse gases*, these substances augment the *greenhouse effect* of the earth's atmosphere, which provides a warmer climate at the earth's surface than would exist in an atmosphere-free earth.

Global warming is an enhancement of the greenhouse effect of the earth's atmosphere, resulting in an increase of the annual average surface temperature of the earth on the order of 0.5–1 °C since the middle of the nineteenth century. While yet small, this temperature rise might reach 2–3 °C by the end of the twenty-first century, an amount believed almost certain to cause global climate changes affecting all biological life on the planet with uncertain consequences.

Most of the emissions of greenhouse gases and aerosols are a consequence of fossil fuel usage to satisfy our growing energy needs. The burning of 1 kg of coal liberates about 3.4 kg of CO_2; that of oil, 3.1 kg; and that of natural gas, 2.75 kg. In 1996, the global anthropogenic emissions of CO_2 amounted to over 25 gigatons per year (Gty^{-1}). Methane emissions are in part due to fossil energy usage, as CH_4 leaks from gas pipes, storage tanks, tankers, and coal mine shafts. In 1996, the global anthropogenic emissions of CH_4 amounted to 0.4 Gty^{-1}. Nitrous oxide (N_2O) is a minor product of combustion of fossil fuels. CFCs are not directly associated with fossil fuel usage; however, they are emitted inadvertently from energy-using devices, such as refrigerators, air conditioners, chillers, and heat pumps.

The prevention of global warming will require a very significant shift from our present energy use pattern toward one of lesser reliance on fossil fuels. Inevitably, such a shift will cause a higher cost for energy commodities, employment disruptions (not necessarily job losses, but different employment patterns), development of alternative technologies, efficiency improvements, and conservation measures. Because of the expected socioeconomic disruptions, there is considerable opposition to implementing preventative policies. Opposition comes from affected interest groups, including coal, oil, and gas suppliers; automobile manufacturers; steel, cement, and other heavy industries; their financiers and shareholders; and their political representatives. Opposition comes also from developing countries, which claim that their peoples' economic standard is so much lower compared to that in more developed countries that increased use of fossil fuel is necessary to elevate their standard. On the other hand, most scientists consider preventative measures a necessity in order to avoid significant changes in global climate and the consequent impact on human habitat and ecological systems.

In 1999 the American Geophysical Union (AGU) issued a position statement on global climate change, part of which we quote[1]:

> ... present understanding of the earth climate system provides a compelling basis for legitimate public concern over future global- and regional-scale changes resulting from increased concentrations of greenhouse gases. These changes are predicted to include increases in global mean surface temperatures, increases in global mean rates of precipitation and evaporation, rising sea levels, and changes in the biosphere. .... The world may already be committed to some degree of human-caused climate change, and further buildup of greenhouse gas concentrations may be expected to cause further change. Some of these changes may be beneficial and others damaging for different parts of the world. However, the rapidity and uneven geographic distribution of these changes could be very disruptive. AGU recommends the development and evaluation of strategies such as emissions reduction, carbon sequestration, and the adaptation to the impacts of climate change. AGU believes that the present level of scientific uncertainty does not justify inaction in the mitigation of human induced climate change and/or the adaptation to it.

Similarly, in 1995, the Intergovernmental Panel on Climate Change (IPCC) concluded[2]:

> Our ability to quantify the human influence on global climate is currently limited because the expected signal is still emerging from the noise of natural climate variability, and because there are uncertainties in key factors. These include the magnitude and patterns of long-term natural variability and the time-evolving pattern of forcing by, and response to, changes in the concentrations of greenhouse gases and aerosols, and land surface changes. Nevertheless, the balance of evidence suggests a discernible influence on global climate.

Based on the IPCC recommendations, the Framework Convention on Climate Change convened in Kyoto in late 1997 to discuss international efforts on curbing greenhouse gas emissions. The developed countries and the so-called "transition countries" (mainly the former USSR and eastern European countries) agreed that by 2010 they would reduce their CO_2 emissions to an average of 6–8% below 1990 emission levels. (Because CO_2 emissions in the developed countries are increasing on average by 1.5–2% per year, the emission reduction target in 2010 would be much greater than 6–8% below unconstrained 2010 levels.) Unfortunately, the lesser developed countries, including China, India, Indonesia, Mexico, and Brazil, did not sign the Kyoto agreement. Also, some developed countries, notably the United States, had not ratified the agreement by 2000.[3] Thus, it is doubtful that significant CO_2 and other greenhouse gas emission reductions will be achieved in the near future, except in countries where there is an economic downturn. (Historically, economic growth was always associated with increased CO_2 emissions, whereas economic recession was also associated with decreased emissions.)

In this chapter we explain the characteristics of the greenhouse effect, the predicted trend of global warming and climate change, and the possible means to ameliorate these effects.

---

[1]*EOS, Trans. Am. Geophys. Union,* **80,** 49, 1999.

[2]Houghton, J. T., G. J. Jenkins, and J. J. Ephraums, Eds., 1995. *Climate Change, the IPCC Scientific Assessment.* Cambridge: Cambridge University Press.

[3]In 2001, President George W. Bush declared that the Kyoto agreement is "fatally flawed," and that the United States will not abide by its recommendations of greenhouse gas emission reductions.

10.2 WHAT IS THE GREENHOUSE EFFECT?

The term greenhouse effect is derived by analogy to a garden greenhouse. There, a glass covered structure lets in the sun's radiation, warming the soil and plants that grow in it, while the glass cover restricts the escape of heat into the ambient surroundings by convection and radiation. Similarly, the earth's atmosphere lets through most of the sun's radiation, which warms the earth's surface, but certain gases, called greenhouse gases (GHG), trap outgoing *radiative* heat near the surface, causing elevated surface temperatures.

The warming effect on the earth's surface by certain gases in the atmosphere was first recognized in 1827 by Jean-Baptiste Fourier, the famous French mathematician. Around 1860, the British scientist John Tyndall measured the absorption of infrared radiation by CO_2 and water vapor, and he suggested that the cause of the ice ages may be due to a decrease of atmospheric concentrations of CO_2. In 1896, the Swedish scientist Svante Arrhenius estimated that doubling the concentration of CO_2 in the atmosphere may lead to an increase of the earth's surface temperature by 5–6 °C.

The major natural GHG are water vapor and carbon dioxide. If not for these gases, the earth would be quite inhospitable, with a temperature well below freezing. So, why are we concerned about adding more man-made GHG to the atmosphere? Will not the earth become even more comfortable to living creatures? The answer is that humans and ecological systems have adapted to present climatic patterns. Perturbing those patterns may result in unpredictable climatic, ecologic, and social consequences.

10.2.1 Solar and Terrestrial Radiation

The sun emits electromagnetic radiation ranging from very short wavelength gamma rays, X rays, ultraviolet through visible, to infrared radiation. A section of the solar spectrum reaching the earth, which contains the near-UV, visible, and near-IR spectrum up to 3.2 μm, is shown in Figure 10.1. Three spectra are presented. The uppermost curve gives the solar irradiance outside the earth's atmosphere; the lowest curve is the spectrum of incident radiation at sea level. The dashed curve, which follows closely the upper curve, is the spectral *radiance* of a black body heated to a temperature of 5900 K, scaled to equal the total solar *irradiance* at the earth.[4] This tells us that the sun's surface temperature is approximately 5900 K (the interior of the sun is much hotter, by many millions K, because of the nuclear fusion reactions occurring there). A black body heated to 5900 K emits its peak radiance in the visible portion of the electromagnetic spectrum, at about 500 nm (0.5 μm). The reason the solar irradiance curve does not follow the black-body radiance curve exactly is that at some wavelengths there is excess radiance, at others a deficit.[5]

The solar irradiance spectrum at sea level (lower curve) is much different from the spectrum at the top of the atmosphere. This is because gases in the earth's atmosphere absorb or scatter

---

[4] A black body is a perfect radiator. It absorbs all the incident radiation of any wavelength, and it emits radiation at any wavelength commensurate with its temperature. A radiation field enclosed by a black body of temperature T has an energy density (and distribution of that energy with wavelength) that depends only upon the temperature and not on any other characteristics of the black body.

[5] Excess radiance results from excitation of atoms and ions in the solar corona (e.g., the sodium-D lines around 589 nm). Deficits result from self-absorption of the incident radiation by other gaseous atoms and ions which are also present in the solar corona. These are called Fraunhofer lines.

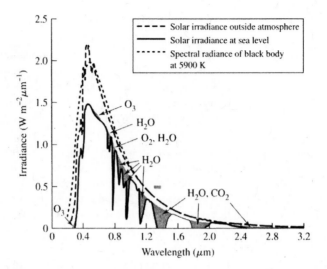

Figure 10.1 Solar spectrum at the top of the atmosphere (upper curve) and at sea level (lower curve). Shaded areas indicate absorption of radiation by atmospheric molecules. The dashed curve is the radiance spectrum of a black body heated to 5900 K, scaled to the solar irradiance curve at the top of atmosphere. (Data from Valley, S. L., Ed., 1965. *Handbook of Geophysics.* Bedford: United States Air Force.)

radiation. On the UV side, oxygen and ozone in the stratosphere are the major absorbing gases. In the visible portion, density fluctuations of atmospheric molecules scatter sunlight.[6] In the infrared, polyatomic molecules present in the lower atmosphere (troposphere), such as H_2O, CO_2, O_3, CH_4, N_2O, and others, absorb solar radiation.

The earth radiates outward to space a spectrum commensurate to her surface temperature. Figure 10.2 presents the earth's radiance ("*earth-shine*") as observed by a spectro-photometer housed on an artificial satellite looking toward the earth's surface. The solid curves in Figure 10.2 represent black-body radiances at various temperatures. The black-body radiance curve that best represents the earth's radiance at the spot where the spectrum was taken is that for a temperature of about 280–285 K.

The radiance curves span the far-infrared region of the electromagnetic spectrum, from about 5 to 50 μm, peaking at about 18 μm. But the earth's radiance curve has several deficits. These deficits are the result of absorption, and emission at reduced levels, of the *outgoing* thermal radiation by greenhouse gases.

10.2.2 Sun–Earth–Space Radiative Equilibrium

The sun–earth–space radiative equilibrium is depicted in Figure 10.3. The numbers on arrows are average annual radiation received from the sun or emitted from the earth, in watts per square meter. (The intensities in Figures 10.1 and 10.2 are instantaneous radiation fluxes from the sun and

---

[6]Scattered sunlight causes the sky to be blue. Blue is the preferred scattered radiation of a pure atmosphere, devoid of particles. With increasing particle concentration the scattered radiation becomes colorless (i.e., white).

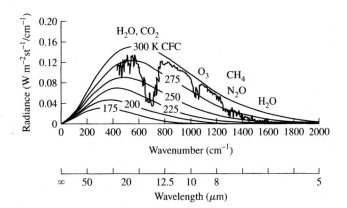

Figure 10.2 Earth spectrum ("earth-shine") as observed from the Nimbus-7 satellite. Major absorption bands by greenhouse gases are indicated. (Adapted from Liou, K. N. 1980. *Introduction to Atmospheric Radiation.* New York: Academic Press.)

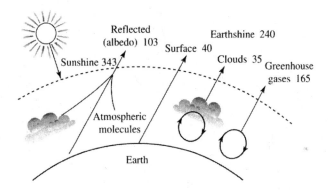

Figure 10.3 Sun–earth–space radiative equilibrium. Of the incoming global annual solar radiation, about 30% is reflected into space (albedo); the rest reaches the earth's surface or is scattered by clouds, aerosols, and atmospheric molecules. Of the outgoing terrestrial thermal radiation, about 17% goes directly into space; the rest emerges after redistribution by clouds, aerosols, and greenhouse gases. Units are W m^{-2}.

earth, respectively.) About 30% of the incoming solar radiation is reflected into space by clouds and the earth's surface. This factor is called the *albedo*. Another part (25%) is absorbed by gases and aerosols, including clouds, in the earth's upper and lower atmosphere, exciting and splitting molecules, and ionizing molecules and atoms (for example, solar UV radiation causes the formation of ozone in the stratosphere and free electrons in the ionosphere). The rest (about 45%) reaches and is absorbed by the earth's surface, thereby elevating land and water temperatures. Much of the absorbed radiation goes into evaporating water from the oceans and other surface waters.

Quantitatively, the solar energy input to the earth on top of the atmosphere is

$$I_S = S_o \pi R_E^2 (1 - \alpha) \qquad (10.1)$$

where the factor πR_E^2 is the disk area presented by the earth toward the sun and where

S_o = annual average solar energy impinging on top of the atmosphere, called the solar constant, 1367 W m^{-2}

R_E = earth's radius, 6371 km

α = the present average albedo of the earth, 0.3 ± 0.03

The earth's radiative output to space is

$$R_E = 4\pi R_E^2 \sigma T_E^4 \qquad (10.2)$$

where $4\pi R_E^2$ is the total area of the earth which radiates into space and where

σ = Stefan–Boltzmann constant, 5.67 E(-8) W m^{-2} K^{-4}

T_E = equivalent black-body radiative temperature of earth, K

From 10.1 and 10.2 we can calculate the earth's radiative temperature T_E:

$$T_E = \left[\frac{S_o(1-\alpha)}{4\sigma} \right]^{1/4} = 255 \text{ K } (-18\,^{\circ}\text{C}) \qquad (10.3)$$

where we used an albedo $\alpha = 0.3$. The temperature $T_E = 255$ K is the average radiative temperature of the earth and her atmosphere as it would appear to a space observer looking toward the earth. Presently, the earth's average surface temperature is $T_S = 288$ K (15 °C). The difference $T_S - T_E = 33$ K is a consequence of the greenhouse effect. This effect is present on other planets. Mars has a very thin atmosphere; therefore the effective radiative temperature and the surface temperature are very close: 217 and 220 K, respectively. On the other hand, Venus, which has a very dense CO_2 atmosphere, the effective radiative temperature is 232 K, and the surface temperature is about 700 K.

The earth's average surface temperature has varied over past geological ages, as evidenced by the glacial and interglacial periods. In part, these temperature variations may have been caused by changing GHG concentrations in the atmosphere. Other factors for global climate change may have been related to the variations of the earth's orbit around the sun and the tilt of her axis. First, the eccentricity of the earth's elliptical orbit varies with a period of about 100,000 years. That changes the average insolation of the earth. Second, the axis of rotation of the earth vis-à-vis the ecliptic varies between 21.6° and 24.5° (currently 23.5°) with a period of about 41,000 years. That changes the amount of insolation of the hemispheres. A third possible factor is a change in the solar constant.

10.2.3 Modeling Global Warming

Modeling of global warming is based on calculations of radiative transfer between the radiating body (in this case the earth's surface) and atmospheric molecules, together with other energy transport processes in the atmosphere. The change caused by extra amounts of GHG is called *radiative forcing*. Modeling requires a detailed knowledge of the spectroscopic characteristics of GHG and the structure of the atmosphere, as well as very high capacity computers on which to exercise the radiative transfer models.

The absorption bands of atmospheric molecules in the far IR, between 5 and 50 μm, where the earth's outgoing thermal radiation is prominent, can be seen as the deficits in Figure 10.2. Water vapor absorbs at 5–7 μm and all wavelengths above 10 μm; CO_2 at 12–18 μm; CH_4 and N_2O at 7–8 μm; O_3 at 9–10 μm; the major CFCs ($CFCl_3$ and $CF_2 Cl_2$) at 10–12 μm. Each absorption band is composed of many individual absorption lines, due to vibrational–rotational transitions of the molecule. At some wavelengths there is overlap of absorption by individual species. Water vapor provides a quasi-continuum absorption (crowding of individual rotational lines) at the longer wavelengths. The CO_2 band at 12–18 μm is so strong that it is almost saturated at the center. Further additions of CO_2 to the atmosphere will result in more absorption at the wings of the bands, but little at the center. Thus, a doubling of the CO_2 concentration in the atmosphere will not necessarily result in doubling of the earth's thermal radiation absorption, but only a fraction thereof. This is analogous to a window that is painted black at the center, but shades of gray at the periphery. Adding another coat of black on the entire window will not double its opaqueness. On the other hand, CH_4 bands are not saturated at all; thus doubling of CH_4 concentrations will nearly double CH_4 absorption. It is estimated that adding one molecule of CH_4 is as effective in absorbing the far IR radiation as adding 20 molecules of CO_2.

The absorption of energy at a particular wavelength by an individual molecule depends on line intensity, line half-width, and lower state energy. The latter is a function of the temperature of the surrounding atmosphere with which the molecule is in equilibrium. The atmosphere is modeled by many layers, in each of which the absorbing gas concentration and temperature determine the radiative transport away from and towards the earth, and from which the model predicts the surface temperature and atmospheric temperature profile.

The earth's atmosphere has a complicated temperature structure. In the troposphere the temperature decreases with altitude; in the stratosphere the temperature increases; in the mesosphere it decreases; in the thermosphere it increases again with altitude. The GHG molecules can reduce the outgoing terrestrial radiation only if they are at a colder temperature than the radiating earth. We noted before that the earth's effective radiating temperature corresponds to a black-body temperature $T_E = 255$ K. With an average surface temperature $T_S = 288$ K and an average temperature gradient (lapse rate) in the troposphere of about -6 K/km, the apparent height (called the *mean radiating height*), at which the temperature level of 255 K originates is 5.5 km. Because the effective radiating temperature of the earth will remain nearly at $T_E = 255$ K, addition of GHG to the atmosphere will change the surface temperature T_S, the temperature profile of the atmosphere, and the mean radiating height.[7] The sun–earth–space system remains in radiative equilibrium; what changes with the addition of GHG to the atmosphere is the *redistribution* of energy, manifested by a redistribution of temperature, between the earth's surface and her atmosphere. Increased GHG concentrations have the effect that in the higher troposphere, where the GHG molecules are colder, more outgoing terrestrial radiation is absorbed. That radiation is re-emitted in all directions. About one-half is radiated downward to the earth's surface, raising T_S; the other half is radiated upward, warming the even colder upper layers of the troposphere. On the other hand, the lower stratosphere will be cooler because those layers receive less outgoing radiation from the surface, due to absorption of the intervening GHG. Taking the whole atmospheric temperature profile into account, the mean radiating height will be relatively higher than at present. These changes are depicted in Figure 10.4.

---

[7]The radiative temperature of the earth can only change if the albedo changes [see equation (10.3)].

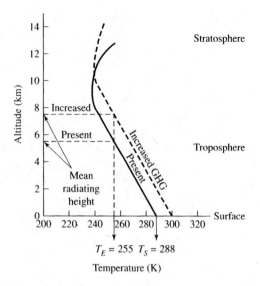

Figure 10.4 Effect of increased GHG concentrations on surface temperature and vertical temperature profile of the atmosphere.

10.2.4 Feedback Effects

In addition to radiative forcing due to enhanced concentration of GHG, the modeling of global warming is further complicated because a rising surface temperature inevitably will cause secondary effects, called *feedback effects*. This can be expressed by the proportionality

$$\Delta T_S \propto \Delta Q / \beta \qquad (10.4)$$

where ΔT_S is the rise of surface temperature, ΔQ is the radiative forcing due to GHG alone, and β is a factor accounting for the feedback effect. If $\beta < 1$, the feedback is positive, which will increase the surface temperature even more than by increasing greenhouse gas concentrations alone. If $\beta > 1$, the feedback is negative, which will cause a smaller surface temperature increase. There are several feedback effects possible: water vapor, cloud, aerosol, ice-albedo, and ocean circulation feedbacks.

10.2.4.1 Water Vapor Feedback

This mechanism may be the most important feedback effect. Water vapor is a strong infrared absorbing gas at wavelengths between 5 and 7 μm and above 10 μm. As the average earth's surface temperature rises because of increasing GHG concentrations, more evaporation will occur from the vast ocean surfaces, hence the atmosphere will become laden with more water vapor; that is, the humidity will increase. This would result in stronger absorption of the outgoing far-IR radiation, thus a positive feedback. Radiative models predict that the water feedback effect may increase the GHG caused global warming by about 60% ($\beta \approx 0.6$).

10.2.4.2 Cloud-Radiation Feedback

The feedback effect of clouds is complicated. Clouds can have both a negative and positive feedback effect. The negative effect is due to reflection by clouds of the incoming solar radiation, contributing to an increase of the earth's albedo and thus reducing the earth's surface temperature. The positive effect is due to the clouds reflecting the outgoing earth's thermal radiation. The balance of the two effects is dependent on cloud characteristics, their altitude, and droplet or crystal size. In general, low, cumulus-type clouds reflect solar radiation: a cooling effect. High-altitude, cirrus-type clouds reflect the earth's radiation: a warming effect. Modeling of cloud feedback effects is highly uncertain. At this stage a best "guestimate" is that in the overall the positive feedback effect wins (more high altitude clouds as a consequence of GHG caused surface water warming), so that the cloud feedback effect will add another 20% ($\beta \approx 0.8$) to the surface temperature increase due to increasing GHG concentrations.

10.2.4.3 Aerosol Feedback

Similarly to clouds, natural and man-made small particles suspended in the air, called aerosols, can interfere with the incoming solar and outgoing terrestrial radiation. The average composition of aerosols is about one-third crustal material (fugitive dust from soil, sand, rocks, and shale), one-third sulfate (mainly from sulfur emissions associated with fossil fuel combustion), and one-third carbonaceous matter and nitrate (also from fossil fuel combustion). The diameter range of the aerosols is much less than a micrometer. This diameter range is more effective in scattering incoming solar radiation than in reflecting outgoing terrestrial radiation. Thus, the aerosol feedback effect is thought to be negative, reducing the GHG effect by 10–15% ($\beta \approx 1.1$).

The aerosols may also have an indirect feedback effect. Aerosols serve as condensation nuclei for clouds. The more aerosols in the air, the higher the probability of cloud formation. Depending on the formed cloud height, and droplet or crystal size, clouds may reflect incoming solar radiation or outgoing terrestrial radiation. This indirect effect is presently under intense study, and the resulting feedback cannot yet be assessed.

10.2.4.4 Ice–Albedo Feedback

As the earth's surface warms due to increased GHG concentrations, the fringes of the Arctic and Antarctic ice caps may melt. Also, glaciers, which already are receding in this interglacial period, may recede even faster. Because ice has a higher albedo (reflects more sunlight) than water and land, the disappearance of ice will lead to a decrease in the earth's albedo [α in equations (10.1) and (10.3)]. This will cause the earth's radiative temperature T_E, and concomitantly the surface temperature T_S, to increase slightly. The ice–albedo effect may add 20% to the GHG-caused surface warming ($\beta \approx 0.8$).

Altogether, the various feedback effects may double the GHG-caused global warming.

10.2.4.5 Ocean Circulation Feedback

Another possible feedback mechanism is the alteration of the ocean circulation and currents. In general, cold and highly saline water sinks to greater depths, and warm, less saline water rises. The cold and larger-than-average saline waters are generated in the Arctics, as ice is formed at the surface. These waters sink to the ocean bottom and move toward the equator. There, warmer and less saline waters rise and move toward the poles, completing a loop. With melting ice caps, along with an increase of precipitation at high latitudes (which sweetens the surface layer), the normal ocean circulation pattern may be altered, with possible changes in the average global surface temperature. This feedback effect on average surface temperature is very difficult to predict; it may be negative, positive, or neutral. The disruption of the ocean circulation may have other consequences, such as enhanced El Niño effect and changing storm system patterns (see Section 10.3).

10.2.5   Results of Global Warming Modeling

Based on radiative forcing models, the Intergovernmental Panel on Climate Change (IPCC) projects the average earth's surface temperature increase in the twenty-first century as shown in Figure 10.5. This projection is based on estimated increase of CO_2 and other GHG emissions, as well on various feedback effects. The "best" estimate predicts a rise of the earth's surface temperature by the end of this century of about 2 °C; the "optimistic" estimate predicts about 1 °C, and the "pessimistic" estimate predicts about 3 °C. The optimistic estimate relies on slowing of CO_2 and other GHG emissions, the pessimistic estimate relies on "business-as-usual" (i.e., on continuing rate of growth of CO_2 and other GHG emissions), and the best estimate is somewhere in between.

10.2.6   Observed Trend of Global Warming

Because atmospheric CO_2 concentrations have risen from about 280 ppmv from the start of the industrial era to about 370 ppmv to date, models show that to date there already should have been a global warming of about 0.5 °C to 1 °C, depending on model and assumed feedback effects. Has such warming actually occurred?

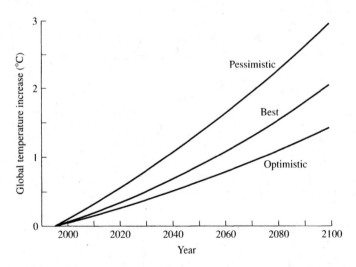

Figure 10.5 Projected trend of the earth's surface temperature increase. Upper curve: pessimistic scenario, based on "business-as-usual" fossil fuel consumption. Lower curve: optimistic scenario, based on slowing down of fossil fuel consumption. Middle curve: in between worst and best scenario. (Data from Houghton, J. T., G. J. Jenkins, and J. J. Ephraums, Eds., 1995. *Climate Change, the IPCC Scientific Assessment.* Cambridge: Cambridge University Press.)

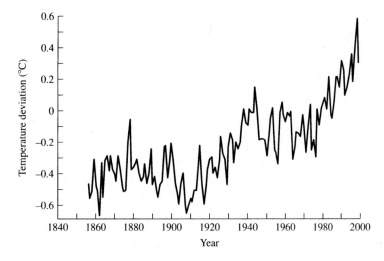

Figure 10.6 Observed average global surface temperature deviation 1855–1999. (Data from Carbon Dioxide Information Analysis Center, 2000. *Trends Online: A Compendium of Data on Global Change.* Oak Ridge: Oak Ridge National Laboratory.)

Figure 10.6 shows the observed temperature trend from 1855 to 1999. Measured temperature data from hundreds of stations have been averaged. As can be seen, there are wide fluctuations from year to year. In spite of the fluctuations, there is an unmistakable upward trend. Averaging the fluctuations, the mean global temperature appears to have risen by about 0.5 °C to 1 °C. Radiative forcing models, considering the effects of greenhouse gases alone, predict a rise of about 1 °C over that period. Assuming that man-made aerosols have a negative feedback effect, other models predict that the average temperature should have risen by only about 0.5 °C. Thus, modeling results are more or less in line with the observed global warming trend over the last century.

10.2.7   Other Effects of Global Warming

As a consequence of increased GHG concentrations in the atmosphere, the earth's surface temperature may rise as discussed in previous sections. The surface temperature rise may cause several ancillary effects on global climate and hydrogeology, which in turn will affect human habitat, welfare, and the ecology.

10.2.7.1   Sea Level Rise

With increasing surface temperatures the average sea level will rise because of three factors: melting of polar ice caps, receding of glaciers, and thermal expansion of the ocean surface waters.

The sea level varied greatly over past geological periods. Before the onset of the last glacial period, about 120,000 years ago, the global average temperature was about 2 °C higher than today. The average sea level was about 5–6 m higher than today. At the end of the last glacial period, about 18,000 years ago, the summer air temperatures were lower by 8 °C to 15 °C over most of North America and Eurasia south of the ice sheets, and sea surface temperatures were about 2 °C to 2.5 °C below present. The average sea level was over 100 m lower than at present. At that time the

British Isles were joined with the European mainland, and the polar ice sheets extended in Europe as far as southern England and Switzerland, and in North America they extended to the Great Lakes and southern New England.

The melting of ice caps may actually contribute the smallest fraction of sea level rise. This is because of two competing effects. With increasing surface temperatures there will be more evaporation and precipitation, including snowfall, which enlarges the ice caps. On the other hand, some melting of the fringes of ice caps will occur. The net result may be a few centimeters of sea level rise by the end of the next century.

In recent times, glaciers over the continents have been receding noticeably. Over the last century this receding has contributed between 2 and 5 cm of sea level rise. If all glaciers outside Antarctica and Greenland were to melt, the sea level rise would be in the range 40–60 cm.

The largest contribution to sea level rise will come from the thermal expansion of the surface layer of the oceans. The estimation of this contribution is quite complicated, because the coefficient of thermal expansion of water is a function of temperature, and the depth of the mixed surface layer which varies over the globe.

Combining all three factors—melting of ice caps, receding of glaciers, and thermal expansion of water—it is estimated that by the end of the twenty-first century the average sea level may be 30–50 cm higher than it is today. This can seriously affect low-lying coastal areas, such as the Netherlands in Europe, Bangladesh in Asia, and low-lying islands in the Pacific and other oceans.

10.2.7.2　Climate Changes

Predicting global and regional climatic changes as a consequence of average surface temperature rise is extremely difficult and fraught with uncertainties. It is expected that regional temperatures, prevailing winds, and storm and precipitation patterns will change, but where and when, and to what extent changes will occur, is a subject of intensive investigation and modeling on the largest available computers, the so-called supercomputers. Climate is influenced not only by surface temperature changes, but also by biological and hydrological processes and by the response of ocean circulation, which are all coupled to temperature changes.

It is expected that temperate climates will extend to higher latitudes, probably enabling the cultivation of grain crops further toward the north than at present. But crops need water. On the average, the global evaporation and precipitation balance will not change much, although at any instant more water vapor (humidity) may be locked up in the atmosphere. However, precipitation patterns may alter, and the amount of rainfall in any episode may be larger than it is now. Consequently, the runoff (and soil erosion) may be enhanced, and areas of flooded watersheds may increase.

Hurricanes and typhoons spawn in waters that are warmer than 27 °C, in a band from 5° to 20° north and south latitude. As the surface waters will become warmer and the latitude band expands, it is very likely that the frequency and intensity of tropical storms will increase.

The well-known ocean currents, such as the Gulf Stream, Equatorial, Labrador, Peru, and Kuroshoi currents, are driven by surface winds and density differences in the water. A typical example is the El Niño event. Along the coast of Peru and Chile, prevailing winds blow offshore, driving the surface waters westward from the South American coast. This sets up a three-dimensional pattern of ocean circulation in which the warm surface waters are replaced by upwelling of colder water from depths reaching to 300 m. These colder waters are rich in nutrients on which pelagic (water column) fish and other aquatic organisms feed—for example, the anchovies. This event occurs cyclically around Christmas, hence the name El Niño. Recently, major El Niños occurred in 1957–1958, 1972–1973, 1976–1977, 1982–1983, the early 1990s, and 1997–1998. The benefits of El Niño are a richer anchovy harvest. The disadvantages are wide-reaching climatic effects, such as cool, wet summers in Europe; increased frequency of hurricanes and typhoons in the western Pacific, South China Sea, southern Atlantic, and Caribbean Sea; and heavy storms and precipitation battering the eastern Pacific coast from Mexico to British Columbia. The 1997–1998 El Niño, which caused winter storms along the Pacific coast of North America with heavy precipitation, mud slides, and resulting property damage, is a recent reminder of what may have been a periodic natural event, but which could be exacerbated and made more frequent in the future with enhanced global warming.

The sea level and climatic changes may cause considerable shifts in population centers and distribution of agricultural and forestry resources, and they may require incalculable investments in habitat and property protection.

10.3 GREENHOUSE GAS EMISSIONS

The increase of GHG concentrations in the atmosphere is a consequence of rising emissions of these gases from anthropogenic sources. The most significant of these gases is CO_2, but CH_4, CFCs, and N_2O emissions also need to be considered, as well as changes in O_3 concentrations.

10.3.1 Carbon Dioxide Emissions and the Carbon Cycle

Because carbon constitutes a major mass fraction of all living matter, on earth there is an enormous reservoir of carbon in the living biosphere and its fossilized remnant. Sedimentary limestone, $CaCO_3$, contains about 12% of its mass as carbon. This limestone originates in part from shells and skeletons of past living creatures, and in part from precipitation from a supersaturated aqueous solution of $CaCO_3$.

There is a continuous exchange of CO_2 between the biosphere and atmosphere. Carbon dioxide is absorbed from the atmosphere during photosynthesis of land vegetation and phytoplankton living in oceans and other surface waters. Carbon dioxide is returned to the atmosphere during respiration of animals and decomposition (slow combustion of carbonaceous matter) of dead plant matter and animals.

Figure 10.7 shows the rates of carbon exchange between the biosphere and atmosphere and between ocean and atmosphere, and it also shows emissions from fossil fuel combustion and forest burning (Gt y^{-1}). The figure also shows the carbon reservoirs residing in biota and soil litter on land (2000 Gt), dissolved in the ocean (40,000 Gt), in fossil fuels (5000–10,000 Gt), and in the atmosphere (750 Gt).[8]

The respiration and decomposition of land organisms emit about 60 Gt y^{-1} of carbon into the atmosphere, while photosynthesis absorbs about 62 Gt y^{-1}. Thus, there is a small net absorption of CO_2 from the atmosphere by land-based biota. The oceans and other surface waters absorb 92 Gt y^{-1} of carbon by dissolution of CO_2 and by photosynthesis of phytoplankton. The oceans

---

[8]Carbon and carbon dioxide will be used interchangeably in subsequent discussions. To convert from carbon to CO_2, multiply by $44/12 = 3.67$.

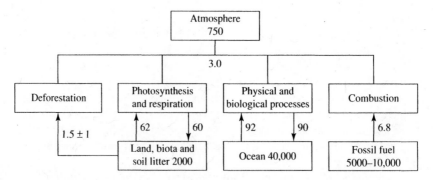

Figure 10.7 Rates of carbon exchange between biosphere and atmosphere and between ocean and atmosphere. Also shown are emissions from fossil fuel combustion and forest burning (Gt y^{-1}). Carbon reservoirs in soil, ocean, fossil fuel, and atmosphere (Gt).

return into the atmosphere about 90 Gt y^{-1} by respiration and outgassing. Thus, the oceans are also net absorbers of CO_2. We note that the net increase of atmospheric carbon, about 3 Gt y^{-1}, is quite small compared to the annual cycling of 150 Gt y^{-1} via land and water. It is the extremely slow secular change in the latter that has accompanied glaciation periods.

Currently, about 6.8 Gt y^{-1} of carbon (25 Gt y^{-1} CO_2) are emitted into the atmosphere by fossil fuel combustion (see Chapter 2). Another 1.5 ± 1 Gt y^{-1} are emitted due to deforestation and land use changes, mainly artificial burning of rain forests in the tropics and logging of mature trees, which disrupts photosynthesis. The atmospheric carbon content is increasing by about 3 Gt y^{-1}. Because the biosphere and oceans absorb about 2 Gt y^{-1} each, about 1.3 ± 1 Gt y^{-1} of carbon are unaccounted for. Most likely, the biosphere and oceans absorb more CO_2 than is indicated in Figure 10.7.

Figure 10.8 shows the growth of CO_2 concentrations in the global atmosphere from 1000 to 2000. The atmospheric concentrations before 1958 were estimated from ice cores. During descent in the atmosphere, snow flakes equilibrate with the atmospheric concentration of CO_2. Subsequently, the snow fall is compressed into ice, with its CO_2 content preserved. Knowing the age of the ice layer, it is possible to reconstruct the prevailing atmospheric concentration of CO_2 when the snow flakes were falling. In 1958, Charles Keating established an accurate atmospheric CO_2 concentration measurement device using infrared absorption on top of Mauna Loa, Hawaii. This instrument is still operating, and the measurements thereof provide an accurate record of global average CO_2 concentrations ever since.

The historic record indicates that before the twentieth century the atmospheric concentration of CO_2 hovered around 280 ppmV, with a dip in the sixteenth and seventeenth centuries, corresponding to the "little ice age." Starting around 1900, when the use of fossil fuels accelerated, the CO_2 concentration began a steady increase of about 0.4%/y, reaching close to 370 ppmV in 2000. If that rate of increase were to continue into the future, a doubling of CO_2 concentrations would occur in about 175 years. However, if no measures are taken to reduce CO_2 emissions, then due to the population increase and the concomitant enhancement of fossil fuel use, the rate of growth of CO_2 concentrations will increase more than 0.4% per year, and the doubling time will be achieved sooner.

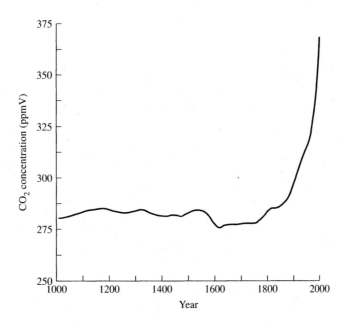

Figure 10.8 Trend of CO_2 concentrations in the atmosphere 1000–2000 A.D. (Data from Carbon Dioxide Information Analysis Center, 2000. *Trends Online: A Compendium of Data on Global Change.* Oak Ridge: Oak Ridge National Laboratory.)

10.3.2 Methane

Methane is emitted by natural processes, such as (a) the anaerobic decomposition of organic matter in swamps and marshes and (b) enteric fermentation of fodder in guts of animals. Anthropogenic emissions of CH_4 come from leakage of oil and gas wells, gas pipe lines, and storage and transportation vessels. Accumulated CH_4 in coal seams is mostly vented into the atmosphere when mine shafts are dug into the seams. Worldwide concentrations are presently about 1.7 ppmv and are growing by about 0.6% per year. Considering that CH_4 is a stronger absorber of IR radiation *per molecule* than CO_2 (see Section 10.2.3), its contribution to the greenhouse effect is significant even though its present concentration in the atmosphere is only about 5E(−3) times that of CO_2. Assuming continued growth of CH_4 and CO_2 in the atmosphere, by 2100 CH_4 may contribute as much as 15% to global warming.[9]

10.3.3 Nitrous Oxide

Nitrous oxide (N_2O) is to be distinguished from the air pollutants NO and NO_2, emitted during fossil fuel combustion.[10] Nitrous oxide is emitted naturally during bacterial nitrogen fixation in soils

---

[9]Houghton, J. T., G. J. Jenkins, and J. J. Ephraums, Eds., 1995. *Climate Change, the IPCC Scientific Assessment.* Cambridge: Cambridge University Press.

[10]N_2O, called laughing gas, is used as an anesthetic in dental surgery.

and by lightning discharge. Minor quantities are emitted from fossil fuel combustion and in some chemical manufacturing processes (e.g., nitric acid, chemical fertilizer, and nylon production). Its current concentration in the atmosphere is about 0.3 ppmv and is growing by about 0.25% per year. While N_2O is a strong far-infrared absorbing gas in the 7- to 8-μm range, currently measured small concentrations of this gas and its slower growth rate compared to CO_2 mean that in about 2100 it may contribute about 10% to global warming.

10.3.4   Chlorofluorocarbons

Chlorofluorocarbons (CFC) are entirely man-made products, because they are produced in chemical factories for use as refrigerants, propellants in spray cans, foam-blowing agents, solvents, and so on. By international conventions, they are gradually being phased out of production. However, because of their slow venting from existing appliances and foam insulation materials, coupled with their long lifetime in the atmosphere (hundreds of years), they will contribute to global warming for a long time after they cease to be produced. In 2100, chlorofluorocarbons may contribute about 5–10% to global warming.

In addition, chlorofluorocarbons reduce the concentration of ozone in the stratosphere (see Section 9.2.5).

10.3.5   Ozone

In Chapter 9 we noted that there are two layers of O_3 in the atmosphere: one in the stratosphere (the "good" ozone), the other in the troposphere (the "bad" ozone). Stratospheric O_3 is produced naturally from molecular oxygen under the influence of solar UV radiation. Some tropospheric O_3 has diffused down from the stratosphere, and the rest is produced from anthropogenic precursors, nitric oxides (NO_x), and volatile organic compounds (VOC), under the influence of solar radiation. About 10% of the total column density of ozone resides in the troposphere. Because the upper troposphere and lower stratosphere is colder than the earth's surface, ozone molecules residing in those layers absorb part of the outgoing far-infrared terrestrial radiation and then re-radiate back to the surface, thus adding to global warming.

Ozone column densities vary greatly as a function of latitude, altitude, and seasons, and they also vary from urbanized-industrialized parts of the continents to remote areas. Thus, it is difficult to assign a global, annual average column density and altitude profile for O_3. Currently, the stratospheric ozone is being depleted by chlorofluorocarbons, which results in a small negative feedback to global warming because less of the outgoing terrestrial infrared radiation is absorbed by stratospheric ozone. With the phasing out of CFC production worldwide, it is estimated that stratospheric ozone concentrations will return slowly (tens to hundreds of years) to pre-CFC times. On the other hand, tropospheric ozone is on the increase because of increased emission of the ozone precursors, NO_x and VOC (see Section 9.2.5). In the upper troposphere, where ambient temperatures are lower than at the earth's surface, the anthropogenic ozone absorbs some of the outgoing terrestrial radiation, causing a positive feedback to global warming. On balance, one may assume that in the foreseeable future the stratospheric ozone deficit and the tropospheric ozone surplus will cancel each other, and no appreciable contribution to global warming is to be expected from ozone.

10.3.6 GHG Control

In conclusion, by the year 2100, CH_4, N_2O, and CFC may contribute about one-third to global warming, with the remaining two-thirds being contributed by CO_2. In regard to methane, efforts can be made to limit its leakage from gas wells, transport pipelines, storage tanks, LNG tankers, coal mines, and other sources of anthropogenic CH_4. The production of anthropogenic N_2O is poorly understood; hence it will be difficult to control its emission. Because CFC production is being phased out worldwide, it will eventually disappear from the atmosphere. The greatest technical and economic problem will be the reduction of CO_2 emissions. This is the subject of the next section.

10.4 CONTROLLING CO₂ EMISSIONS

Global warming can be ameliorated by reducing the emissions of CO_2 and other greenhouse gases into the atmosphere. In this section we shall discuss the possibilities of slowing the build-up of CO_2 concentrations in the atmosphere.

About 86% of the world's present primary energy usage comes from the consumption of fossil fuels. Worldwide the energy consumption in general—and, concomitantly, fossil energy consumption—is growing on the average by 1.5% per year (see Section 2.3). As fossil fuels will become scarcer and more expensive to recover and other energy technologies become available, the fossil fuel portion of the total energy consumption may become smaller. In the next decades, however, we only can hope to ameliorate the rate of increase of global temperatures by slowing down and then reversing the rate of growth of fossil energy consumption. Also, some technologies are becoming available for capturing CO_2 from combustion sources and sequestering it in terrestrial and deep ocean repositories. Emission reductions of CO_2 can be accomplished by a combination of several of the following approaches:

- End-use efficiency improvements and conservation
- Supply side efficiency improvements
- Capture and sequestration of CO_2 in subterranean reservoirs or in the deep ocean
- Utilization of CO_2 for enhanced oil and natural gas recovery and for enhanced biomass production (photosynthesis)
- Shift to nonfossil energy sources.

10.4.1 End-Use Efficiency Improvements and Conservation

The simplest and most cost-effective approach to reducing carbon emissions is by end-use efficiency improvements and conservation. In Chapter 2 we discussed various end-use measures that would result in carbon emission reductions. In the residential–commercial sector, they range from lowering the thermostat in the winter (less heating), raising it in the summer (less air conditioning), better insulation, less hot water use, replacement of incandescent with fluorescent lighting, replacement of electric clothes dryers with gas dryers, and so on. It is estimated that in the United States, in the residential–commercial sector, by the year 2010, carbon emissions could be reduced by 10.5% below 1990 levels with judicious and cost-effective conservation measures.[11]

In the industrial sector the largest savings could come from reductions in direct use of fossil fuels (e.g., coal for process heat or smelting), process modification, energy-efficient motors, better heat exchangers, and so on. A "high-efficiency–low-carbon" scenario predicts that as much as 12.5% energy savings could be realized in the U.S. industrial sector by the year 2010, translating to a 62-Mt carbon emission reduction.[12]

In the transportation sector, fossil fuel energy consumption is growing by leaps and bounds all over the world. Increasing population and living standards, coupled with the movement from agricultural to urban-industrial societies, puts more and more people and cargo in automobiles, trucks, trains, airplanes, and ships. The most convenient fuel for the transportation sector is a fluid fuel (liquid or gaseous)—for example, gasoline, diesel, jet-fuel (kerosene), alcohol, methane, propane, or synthetic fluid fuels, which usually are derived from fossil fuel. Because it is unrealistic to expect that the *number* of transportation vehicles, or the *distances* covered, will diminish, the only chances for reducing carbon emissions in the transportation sector lie in efficiency improvements. Such improvements are discussed in Chapter 8. They range from smaller, lightweight automobiles to hybrid electric-internal combustion engine or fuel-cell-powered vehicles. It is estimated that in the United States, in the year 2010, carbon emissions from the transportation sector will be 26% higher than in 1997 under "business-as-usual" conditions, whereas if possible efficiency improvements were introduced, they would be only 5% higher.[13]

10.4.2 Supply-Side Efficiency Improvements

By supply-side efficiency improvements we mean principally electricity supply. In the United States, 36% of carbon emissions come from electricity generating plants, and a similar percentage is applicable worldwide (see Section 2.4). The electricity industry has many options to reduce carbon emissions while supplying an ever-increasing electricity demand. These options, discussed in Chapter 5, include the following:

- Shift from coal to natural gas. Per unit of energy, gas emits roughly half as much carbon as coal.
- Replacement of single-cycle gas-fired steam power plants with combined cycle gas turbine plants (CCGT). Because single-cycle power plant thermal efficiencies are in the range 35–40%, whereas the combined cycle plants can achieve 50–55%, the carbon emission savings are in the range of 10–20%.
- Replacement of single-cycle coal-fired power plants with gas-fired CCGT. The carbon emission savings are in the range 60–70% (50% on account of shift from coal to gas, and 10–20% on account of higher efficiency).
- Replacement of single-cycle coal-fired power plants with coal-derived synthetic gas-fired combined cycle gas turbine plants. The efficiency of such plants is 40–45% based on the

[11]Brown, M. A., et al., 1998. *Annu. Rev. Energy Environ.*, **23**, 287–385.
[12]Brown, M. A., et al., 1998. *Annu. Rev. Energy Environ.*, **23**, 287–385.
[13]Ibid.

coal input energy. Here, the carbon emission savings would be only on the order of 5–10%. While a CCGT has a higher efficiency than a single-cycle plant, coal gasification requires some of the coal energy to be spent on gasification.

10.4.3 CO$_2$ Capture

In general, the volume fraction of CO$_2$ in the flue gas of fossil fuel electric power plants ranges from 9% to 15%, depending on fuel (e.g., gas versus coal) and excess air used for combustion. Because a relatively small fraction of one gas (CO$_2$) needs to be separated from the majority of other gases (N$_2$, H$_2$O, and excess O$_2$), the capture of CO$_2$ is quite difficult and expensive, and it requires extra energy, which reduces the thermal efficiency of a power plant.

The capture of CO$_2$ is only worthwhile in large power plants, especially those burning coal (because coal emits more CO$_2$ than oil or gas). A 1000-MW coal-fired power plant emits between 6 and 8 Mt y^{-1} of CO$_2$. The capture of CO$_2$ from all the world's large coal-fired power plants would make a significant dent in the global carbon emissions. The following technologies for CO$_2$ capture from power plants are being developed:

- Air separation–CO$_2$ recycling
- Solvent absorption
- Membrane gas separation

10.4.3.1 Air Separation–CO$_2$ Recycling

This method is based on combustion of the fossil fuel in pure oxygen, instead of air. A plant using this method requires an air separation unit (ASU). Here, the gas separation shifts to precombustion, rather than postcombustion. This does not save much energy in gas separation, but there are several other beneficial effects to this method. First, the ASU produces useful byproducts, such as nitrogen and argon. Second, part of the oxygen produced in the ASU can be used for coal gasification, thus enabling the power plant to operate in a combined cycle gas turbine mode with a somewhat higher thermal efficiency than a single-cycle coal combustion–steam turbine mode. (Some efficiency is lost on account of using coal energy for gasification.) Third, the combustion of "syngas" (a mix of CO and H$_2$) with pure oxygen releases no SO$_2$, NO$_x$, or particulate matter. All the contaminants in coal are removed during and after the coal gasification process. The combustion products of syngas with pure oxygen are almost entirely CO$_2$ and H$_2$O. The water vapor is condensed, and CO$_2$ is captured. Thus, there is no exhaust gas. In fact, the power plant itself requires no smoke stack.

A schematic of an integrated air separation coal gasification combined cycle power plant with CO$_2$ capture is shown in Figure 10.9. The ASU is a standard commercial unit, as employed on a large scale in the steel industry and for coal gasification. After nitrogen separation (depending on use, the separated nitrogen can be either liquid or gaseous), part of the oxygen is used for coal gasification (see Chapter 5), and the rest is used for combustion. The fuel (syngas) comes from the gasification unit. Because the combustion of syngas with pure oxygen would yield too high a temperature for conventional materials, the flame needs to be cooled. Therefore, a part of the captured CO$_2$ is recycled into the combustion chamber in order to achieve tolerable combustion temperatures. The combustion gases expand in a gas turbine, producing a part of the power. The residual heat is recovered in a heat recovery steam generator (HRSG). The generated steam runs a steam turbine, producing more power. After heat exchange in the HRSG, water is condensed from

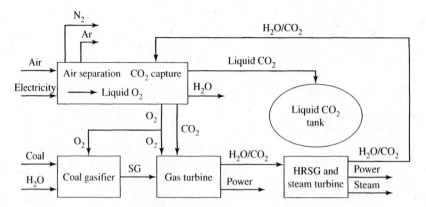

Figure 10.9 Schematic of an integrated coal gasification combined cycle power plant with CO_2 capture.

the combustion gases. The remainder consists almost entirely of CO_2. The CO_2 is condensed in a heat exchanger with liquid oxygen that comes from the ASU. A part of the condensed CO_2 is recycled into the combustion chamber, and the rest is stored as a liquid for subsequent sequestering.

The estimated thermal efficiency of such a plant is about 37%, which includes the energy spent on air separation and coal gasification. This compares to 40–45% estimated for a plant with coal gasification, air combustion, and no CO_2 capture. The plant has a thermal efficiency comparable to modern pulverized coal single-cycle steam plants, which emit not only the full amount of CO_2, but other air pollutants as well.[14]

Obviously, the cost of an integrated coal gasification combined cycle plant with CO_2 capture is higher than one without capture, and much higher than a pulverized-coal single-cycle plant. The estimated levelized cost of electricity would increase by 10–20% compared to a coal gasification plant without CO_2 capture, and 20–30% compared to a pulverized-coal plant with flue gas desulfurization. The incremental cost does not include the cost of transporting and sequestering of CO_2, which may add another 10–20% to the cost of electricity over the no-capture plant.

10.4.3.2 Solvent Absorption

Carbon dioxide is soluble in some solvents, notably ethanolamines; e.g., monoethanolamine (MEA). Absorption–desorption is a reversible process; absorption proceeds at low temperatures, whereas desorption occurs at elevated temperatures:

$$C_2H_4OHNH_2 + H_2O + CO_2 \rightleftharpoons C_2H_4OHNH_3^+ + HCO_3^- \tag{10.5}$$

A flowsheet of the MEA process is shown in Figure 10.10. The key process units are the absorption tower and the regeneration (stripper) tower. The absorption tower operates at 40–65 °C, which means that the incoming flue gas must be cooled before entering the tower. Some pressurization of CO_2 is necessary to overcome the pressure drop in the tower. The regeneration tower operates at 100–120 °C, which is the major source of efficiency loss for the power plant that uses this method. The reboiler provides the heat for the regeneration tower and for vaporizing water.

---

[14]Golomb, D., and Y. Shao, 1996. *Energy Convers. Manage.*, **37**, 903–908.

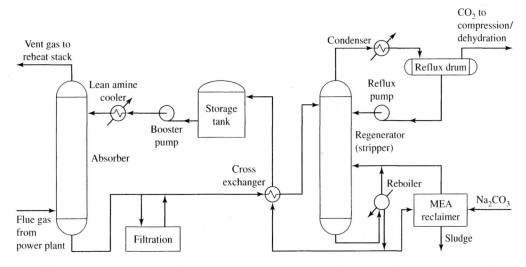

Figure 10.10 Flowsheet of the monoethanolamine (MEA) process for CO_2 capture.

Solvent absorption has been used for decades for producing CO_2 from flue gas. The produced CO_2 is used for dry ice manufacturing and for carbonated drinks, and it is used in large quantities for enhanced oil recovery at semi-depleted oil reservoirs where the injection of high-pressure CO_2 into the reservoir helps to bring forth more oil to the surface. However, for all these uses, CO_2 is captured from the flue gas of *natural gas* combustion. The reason is that coal and oil combustion produce copious quantities of SO_2, NO_x, and other contaminants that "poison" the solvent. If coal or oil were used as the source of CO_2, the flue gas would have to be thoroughly purified before entering the absorption tower, or the coal would have to be gasified before combustion.

The thermal efficiency of a coal gasification combined cycle power plant with CO_2 capture by MEA is 30–35% compared to 40–45% without capture, and the cost increment of electricity production is around 50%. Thus, this method is less efficient and costlier than estimated for air separation–CO_2 recycling, but the technology is well established, whereas the oxygen combustion technology is still in the development stage.

10.4.3.3 Membrane Separation

Gas separation by membranes relies on the different permeation rates of gases through the membrane pores. Hydrogen, in particular, because of its small molecular size, permeates faster than any other gas through small pores. Polymer membranes are being used commercially in recovery of hydrogen from a purge stream in ammonia synthesis, methanol synthesis, and oil refineries. Typical membrane examples are polysulfone/silicon, cellulose acetate, polyphenyleneoxide, polyalkene, polyimide, polydimethylsiloxane, and polyetherimide.

The membrane method could be used for capturing CO_2 from a mixture of CO_2 and H_2, which is the product of coal gasification and the water shift reaction (see Section 5.3.2). This gas mixture is subjected to membrane separation. The hydrogen permeates much faster through a membrane than does CO_2, and only a few stages of membrane separation would be necessary for almost 100% separation of the gases. The separated H_2 can be used as a fuel in a gas turbine or in a fuel cell for power production. The separated CO_2 would be compressed and sequestered.

The membrane separation method for capturing CO_2 is furthest from realization compared to the other two methods. Preliminary estimates predict a thermal efficiency of a power plant with coal gasification and CO_2 capture by membrane separation of 25–30%, along with an increase of electricity production cost over 100% compared to coal gasification combined cycle without CO_2 capture.[15]

10.4.4 CO_2 Sequestration

After capture, the CO_2 needs to be sequestered in a reservoir for an indefinite period, so it will not reemerge into the atmosphere. The following reservoirs are being investigated, and in some instances already employed, for sequestering CO_2:

- Depleted oil and gas reservoirs
- Deep ocean
- Deep aquifers

10.4.4.1 Depleted Oil and Gas Reservoirs

Oil and gas reservoirs are usually covered by an impenetrable layer of rock, so that CO_2 deposited into the reservoirs would not reemerge into the atmosphere. In respect to sequestering CO_2, oil and gas reservoirs behave differently. Whereas CO_2 can be injected into oil reservoirs while the oil is being pumped out of it, it can be injected into gas reservoirs only after depletion of the gas (because of the miscibility of CO_2 and methane). In fact, injecting CO_2 into semi-depleted oil reservoirs is a well-established technology. It is not done for sequestering CO_2, but rather for enhanced oil recovery (EOR).

Worldwide there are some 71 oil fields where carbon-dioxide-enhanced oil recovery (CO_2-EOR) is used. The majority of these fields are in the United States, in Texas and Colorado, whereas others are in the North Sea. In the United States, total production from CO_2-EOR wells is not large compared to total oil production (about 2%). Other oil wells use "water flooding"—that is, pressurized water injection. Water-EOR accounts for roughly 50% of U.S. oil production.

For EOR, carbon dioxide is injected above its critical point (31 °C, 73 atm). Its viscosity is lower than the oil it displaces. This means that CO_2 flows more easily through the reservoir rock, causing "fingers" of the injectant to move through the oil and to occasionally bypass the oil altogether. Therefore, CO_2 must be recovered at the well head and reinjected. On the average, 0.1–0.3 standard cubic meters of CO_2 are injected for the recovery of one barrel of oil.

In the United States, most of the injected CO_2 comes from natural sources, rather than from flue gas capture. The natural sources are subterranean aquifers saturated with CO_2. At two fields in Texas, CO_2 comes from flue gas capture using the solvent absorption method. Supercritical CO_2 is transported from a gas-fired power plant overland a distance of 350 km in a 40-cm-diameter pipe. The exploitation of CO_2 for enhanced oil recovery is largely dependent on the price of crude oil. Carbon dioxide injection costs about $5 to $8 per barrel of oil produced, so it is only worth it if the price of crude oil is above $25/bbl.

The potential storage capacity in depleted oil and gas reservoirs is about 40 Gt carbon. This compares with an annual worldwide emission of 6.8 Gt carbon per year. Thus, depleted oil and gas

---

[15]Riemer, P., 1993. *The Capture of CO₂ from Fossil Fuel Fired Power Stations*. Cheltenham: IEA Publications.

reservoirs have a limited capacity for sequestering CO$_2$. However, the potential storage capacity of the world's proven oil and gas fields is about 140 Gt carbon, assuming they all will be depleted eventually (which will occur in the not-too-distant future, say within 50–100 years). The problem with using oil and gas reservoirs for sequestering CO$_2$ is not only the limited capacity, but also the fact that very few existing large coal-fired power plants are within transport distance to the reservoirs. Furthermore, urban-industrial areas where new power plants are going to be built are usually far away from oil and gas fields. The transport cost of piping supercritical CO$_2$ is significant. Currently it is estimated at \$2 to \$7 per metric ton of CO$_2$ per 250-km distance, depending on pipe diameter (the larger the diameter, the lower the cost). Thus, a 1000-MW coal-fueled power plant may have to spend in the range of \$12 to \$56 million per year for transporting the CO$_2$ to a depleted oil or gas field that is 250 km away. This cost is in addition to the cost of capturing and compressing the CO$_2$ at the power plant.

Sequestering CO$_2$ in depleted and semi-depleted oil and gas reservoirs can play a role in mitigating global warming, albeit on a limited scale and at an economic cost that could not be recovered from the price of electricity presently charged to customers.

10.4.4.2 Deep Ocean

The ocean is a natural repository for CO$_2$. The ocean is vast: It covers about 70% of the earth's surface, and the average depth is 3800 m. There is a continuous exchange of CO$_2$ between the atmosphere and the ocean. We have seen in Section 10.4.1 that the ocean absorbs about 92 Gt y^{-1} of carbon from the atmosphere, while it outgasses into the atmosphere about 90 Gt y^{-1}. Thus, the ocean is a net absorber of carbon, which probably is part of the reason that CO$_2$ concentrations in the atmosphere do not increase as fast as expected from anthropogenic emissions. Most of the ocean–atmosphere carbon exchange occurs within the surface layer of the ocean, which is on the average about 100 m deep. This layer is more or less saturated with CO$_2$. Between 100- and 1000-m depth the temperature of the ocean is steadily declining (the so-called *thermocline*). Beneath about 1000 m depth, the temperature is nearly constant, between 2 °C and 4 °C, and the density increases slightly because of hydrostatic pressure. This makes the deeper layers of the ocean very stable; and the "turnover" time—that is, the time it takes for the deep layers to exchange waters with the surface layer—is long, on the order of hundreds to thousands of years.

The deep layers of the ocean, 1000 m or deeper, are highly unsaturated in regard to CO$_2$. The absorptive capacity of the deep ocean is estimated on the order of E(19) tons of carbon, so conceivably all the carbon residing in fossil fuels on earth could be accommodated there, without reaching even near the saturation limit. Because CO$_2$ emitted into the atmosphere eventually will wind up in the deep ocean, an artificial injection of CO$_2$ at depth would merely short-circuit the natural cycle that takes hundreds to thousands of years.

Carbon balance models predict that the atmospheric concentrations of CO$_2$ will increase more or less in proportion to the amount of fossil fuel combustion. Peak concentrations will occur somewhere in the twenty-third century, after which they will slowly decline because (a) mankind will run out of fossil fuels and (b) the ocean will slowly absorb the excess CO$_2$ that built up in the atmosphere. In about a 1000 years, a new equilibrium will be reached with atmospheric CO$_2$ concentrations slightly higher than today. The purpose of deep ocean injection of CO$_2$ is to shave off that peak that would be building up in the next 200–300 years. Figure 10.11 shows how the peak could be minimized by injecting incremental amounts of CO$_2$ into the deep ocean.

An injection of CO$_2$ at 1000 m or deeper is deemed necessary not only because the deep layers are unsaturated with regard to CO$_2$, but also on account of the physical properties of CO$_2$.

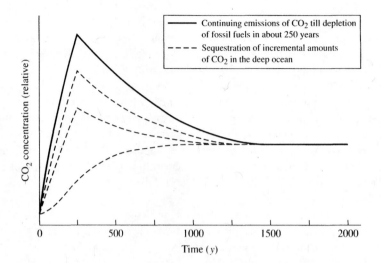

Figure 10.11 Qualitative illustration of the effect of ocean sequestration on atmospheric concentrations of CO_2. Note that in about 1000 years a new level of CO_2 concentration will be reached regardless of ocean disposal. (Adapted from Wilson, T. R. S., 1992. *Energy Convers. Manage.*, **33**, 627–633.)

If injected at 500 m or less, liquid CO_2 would immediately flash into gaseous CO_2 and bubble up to the surface. Between 500 and 3000 m the liquid CO_2 injected from a diffuser at the end of a pipe would disintegrate into droplets of various diameters, depending on hydrostatic pressure of the liquid CO_2 and the release orifices' diameter. Because at these depths the density of liquid CO_2 is less than that of seawater, the droplets would ascend due to buoyancy and may reach the 500-m level, where they would flash into gaseous CO_2. It is estimated that a minimum injection depth of 1000 m and a maximum initial CO_2 droplet diameter of 1 cm would be necessary for complete dissolution in seawater of the ascending CO_2 droplets before they reach a depth of 500 m, where they would flash into gaseous bubbles. Beneath about 3000 m, liquid CO_2 becomes denser than seawater, so the injected droplets would sink to the ocean bottom.

Another feature of CO_2 is that it forms a clathrate, also called hydrate. At a hydrostatic pressure of about 50 atm (corresponding to a depth of about 500 m) and temperatures below 10 °C, a solid CO_2 hydrate is formed, the crystalline structure of which contains one CO_2 molecule surrounded by 5–7 H_2O molecules tied to CO_2 by hydrogen bonds. Laboratory and pilot-scale releases of CO_2 in the deep ocean confirm the formation of CO_2 hydrate. At this time it is not clear whether hydrate formation is good or bad for CO_2 sequestration. If the hydrate crystals were heavier than seawater, they would sink to greater depth from the release point, thereby increasing the sequestration period. If the hydrate crystals were to occlude liquid or gaseous CO_2, they would ascend by buoyancy and would hinder the dissolution of the released CO_2 in seawater. The hydrates may also clog up the release pipe and diffuser. Further pilot-scale releases are planned at various depths to ascertain the optimum depth and release configuration before large-scale injection of CO_2 in the deep ocean is to be implemented.

The laying of large-diameter pipes on the continental shelf reaching to 1000-m depth is a formidable task, but pipes have been laid to off-shore oil wells at such depths in the Gulf of Mexico, the North Sea, and elsewhere. The cost of such pipelines is estimated between $1 and $2 million per kilometer length. Therefore, it is desirable to reach depths of 1000 m or more as close to the shore as possible. Potential sites are as follows:

Americas

- Hudson Canyon, Delaware Canyon, and Hatteras Canyon
- Mississippi Canyon
- Baja California Trench, Monterey Canyon, and Columbia River Canyon
- Several canyons along the coast of Mexico toward the Gulf of Mexico and the Pacific
- Along the west and east coasts of South America

Europe

- Outflow of the Mediterranean Sea at Gibraltar
- Canyons along the coast of Portugal and the Bay of Biscayne
- Coast of Norway
- From western Scotland and Ireland over the Hebridian Shelf into several canyons

Asia, Australia, and Oceania

- Several canyons from the coast of India into the Arabian Sea and Bay of Bengal
- Several canyons from the coast of China into the South China Sea
- Japan is surrounded with the deepest trenches in the world
- Canyons leading to the Japan Abyssal Plain from Korea and eastern Siberia
- Canyons along the southeastern coast of Australia
- Trenches along the Philippine and Indonesian islands

Altogether the access to the deeper layers of the ocean from industrial-urbanized continents is quite limited. For example, the industrial countries of central and eastern Europe are too far from the deep ocean. Because overland pipelines would be even more expensive to lay than underwater pipes, it is unlikely that power plants in those countries would utilize ocean sequestration. The same is true for power plants in the midwestern United States and central Canada. It is conceivable, though, that new power plants will be built along the coasts of continents where access is available to the deep ocean. From these plants, *electricity* can be transmitted deeper into the continents, because it is cheaper to transmit electricity in cables than to transport *carbon dioxide* in pipes overland. Nevertheless, the deep ocean will be accessible to only a limited number of power plants; therefore, ocean sequestration can be only one of several options to be employed in the amelioration of the greenhouse effect.

Apart from the technical and geographical problems, there is a concern about the environmental impact of deep sea sequestration of CO$_2$. Even though CO$_2$ is not toxic, large-scale injection of CO$_2$ into the deep ocean would create localized regions of high carbonic acid concentrations. It is estimated that the discharge of CO$_2$ from ten 500-MW coal-fired power plants will create a volume of 100 km^3 in which the pH is less than 7. In that volume, fish and other creatures must either escape or die. However, the impacted regions would be minuscule in comparison to the world's ocean volume. Nevertheless, there is a general public opposition to disposing of anything in the ocean, so it is not clear whether an international agreement on ocean sequestration of CO$_2$ can be forged.

10.4.4.3   Deep Aquifers

Deep aquifers may underlay vast areas under the continents and oceans. Such aquifers usually contain saline water (brine) and are separated from shallower aquifers—the source of much of the drinking water—by impermeable rock. The deep aquifers themselves consist of permeable, porous rock, such as sedimentary shale-, lime-, or sandstone, the pores of which are saturated with brine. Such aquifers are found at depths of 800 m or deeper. The injected CO_2 (in liquid or supercritical phase) would dissolve in the brine as carbonic acid. In the case of limestone formation, some of the carbonate (CO_3^{2-}) would dissolve into bicarbonate (HCO_3^-), furthering the absorption capacity of the reservoir and reducing the risk of leakage.

There is limited information available on the extent and CO_2 absorption capacity of deep aquifers. In the United States, the storage capacity of subterranean aquifers ranges from 5 to 500 Gt carbon (compared to annual emission rates from power plants of 1.7 Gt y^{-1} of carbon). Other estimates for worldwide capacities range from 100 to 3000 Gt carbon. The problem with deep aquifers is not so much their capacity, but their location vis-à-vis power plants, the difficulty of drilling large diameter pipes into the overlaying strata, the cost of compressing and pumping liquid CO_2 into the pipes, and constructing an appropriate diffuser at the end of the pipe, so that CO_2 disperses throughout the aquifer without leaking through possible permeable overlay formations.[16] Intensive research is ongoing to establishing the location and capacity of the deep aquifers for potential sequestration of CO_2 in them.

One of the subocean floor aquifers has been used for CO_2 disposal since 1996. At the Sleipner gas fields in the North Sea off the coast of Norway, natural gas contains 9.5% by volume CO_2. The CO_2 is separated by MEA absorption, compressed and injected into the Utsira undersea aquifer at a rate of 1 Mt y^{-1}. The cost of the injection is about $15 per ton CO_2, which compares favorably with a tax of $50 per ton of CO_2 that the government of Norway would levy if the CO_2 were emitted into the atmosphere. Another plan for undersea aquifer injection exists for the Natuna gas field off the coast of Borneo in Indonesia, where natural gas contains about 70% by volume of CO_2.

10.4.5 CO_2 Utilization

In Section 10.4.4.1 we discussed the use of flue gas CO_2 for enhanced oil or gas recovery. Other uses would be for dry ice manufacturing, for carbonated drinks, and as a raw material for chemical products, such as urea, methanol, or other oxygenated fuels. The problem with such propositions is twofold: (a) most of the carbon in the product would eventually burn up or decompose back to CO_2 and would wind up in the atmosphere; (b) the reduction of CO_2 into the useful product requires virtually the same amount of energy as was given off when carbon oxidized into CO_2. Also, the present market for chemical products that could be based on CO_2 is quite limited, amounting perhaps to less than 50–70 Mt y^{-1}, whereas the emission of CO_2 from a single 1000-MW coal-fired power plant amounts to 6–8 Mt y^{-1}.

Let us take the example of converting CO_2 to methanol:

$$CO_2(g) + 3H_2(g) \rightarrow CH_3OH(l) + H_2O(l) - 171\,kJ\,mol^{-1} \qquad (10.6)$$

---

[16]Pipe diameters of 50–100 cm are deemed necessary for disposing the CO_2 output of a single 1000-MW power plant.

where (g) means gaseous phase and (l) means liquid phase. The minus sign indicates that energy is liberated; that is, the reaction is exothermic, with 171 kJ evolved per mole of CO$_2$ reacted. However, the reaction as written requires three moles of hydrogen for each mole of CO$_2$ to produce one mole of CH$_3$OH(l). The production of one mole of hydrogen from dissociation of water requires 286 kJ of energy; three moles require 858 kJ. Thus, the energy balance of equation (10.6) is actually negative, requiring 687 kJ mol^{-1} CH$_3$OH produced. The production of methanol from CO$_2$ and H$_2$ would only make sense if that hydrogen were derived from nonfossil energy, such as solar (e.g., photodissociation of water) or nuclear (e.g., electrolysis of water using nuclear electricity). Even if hydrogen is derived from nonfossil sources, the question is whether it should not be utilized directly, such as in fuel cells, rather than producing methanol. Furthermore, when methanol is burned in a heat engine, carbon is reoxidized to CO$_2$, and nothing has been gained in terms of global warming mitigation.

Another example is the production of urea from CO$_2$. Urea is an important industrial chemical, because it is used in chemical fertilizers, in polyurethane foam production, and as an intermediate in a host of other chemicals. The production of urea can be written in the following simplified reaction:

$$CO_2(g) + 3H_2(g) + N_2(g) \rightarrow NH_2CONH_2(s) + H_2O(l) + 632 \text{ kJ mol}^{-1} \qquad (10.7)$$

where (s) means solid phase. This reaction is highly endothermic, with 632 kJ of energy required per mole of solid urea produced, including the 858 kJ of energy necessary for the production of three moles of hydrogen. This example shows again that unless hydrogen is produced from nonfossil energy sources, the utilization of CO$_2$ for converting to other chemicals makes no sense.

Nature uses CO$_2$ as a raw material for producing biomass by means of photosynthesis:

$$nCO_2 + mH_2O \xrightarrow{h\nu} C_n(H_2O)_m + O_2 + 570.5 \text{ kJ mol}^{-1} \qquad (10.8)$$

where $h\nu$ represents a solar photon, and C$_n$(H$_2$O)$_m$ is a basic building block of biomass. This reaction is endothermic; it requires 570.5 kJ of energy per mole of CH$_2$O(s) produced. That energy comes from the sun. All the vegetation and phytoplankton in the world are based on reaction (10.8). Because plants and planktons are at the bottom of the food chain, all animals and mankind are dependent on reaction (10.8) for their sustenance. In fact, all fossil fuels were created over the eons by geochemical conversion of biomass to coal, oil and natural gas.

The utilization of biomass for fuel in boilers or heat engines is CO$_2$ neutral. Every carbon atom burnt or utilized from biomass is reabsorbed in the next generation of plants or planktons. The utilization of biomass as a renewable energy source is discussed in Section 7.3.

Here we should also mention that fostering biomass growth without its utilization would have a positive effect on reducing CO$_2$ concentrations in the atmosphere. It is estimated that typical coniferous and tropical forests absorb between 6 and 10 metric tons of carbon per hectare per year (t C ha^{-1} y^{-1}). Planting new forests, without using the wood until atmospheric CO$_2$ concentrations start to decline because of depletion of fossil fuels or other CO$_2$ emission abatement measures, would absorb part of the anthropogenic emissions of CO$_2$ due to the photosynthesis reaction (10.8). Indeed, the Kyoto Convention of 1997 and the subsequent Buenos Aires Conference in 1998 envisaged that part of the greenhouse effect mitigation efforts will come by afforestation. Large CO$_2$-emitting countries are encouraged to reduce global concentrations of CO$_2$ by investing in other countries (mostly tropical, less-developed countries) to plant new trees. The emitting countries would obtain credit for CO_2 emission reductions equivalent to the amount of CO_2 absorbed by the new trees. This is called emission trading. The exact accounting procedures for such emission trading have not yet been established.

Finally, one must note that deforestation has the opposite effect. Every hectare cleared of forests will reduce the absorption of CO_2 by 6–10 t C ha^{-1} y^{-1}. And if the trees are burned, as it is still widely practiced in tropical countries of the world, there is a double effect: cessation of absorption of CO_2 and emission of CO_2 due to wood burning. It is estimated that forest burning alone adds about 1.5 Gt y^{-1} of carbon to the atmosphere.

10.4.6 Shift to Non-fossil Energy Sources

It goes almost without saying that the CO_2-caused climate warming can be ameliorated by shifting to nonfossil energy sources. In the previous section we have seen that a shift to biomass would be at least CO_2 neutral; that is, as much CO_2 being emitted into the atmosphere from biomass use will be reabsorbed in future growth of trees and plants that produce the biomass. Shift to noncarbon energy sources eliminates CO_2 emissions completely, except perhaps for the fossil energy (mainly coal) used in smelting steel and other construction materials used in the nonfossil energy conversion devices.

In Section 2.3 we have seen that worldwide energy consumption is supplied by the following sources: 86% fossil, 6.5% nuclear, and 7.5% renewables, with the latter almost entirely hydroenergy. Because a shift to nuclear energy appears for the foreseeable future unacceptable to the public and body politic in most countries, the only recourse would be a shift to renewables.

Among the renewables, hydroenergy also appears to run into public opposition.[17] Thus, all expectations are turned toward solar, wind, geothermal, and ocean energy. These technologies are described in detail in Chapter 7. Their substitution for fossil energy is mostly dependent on economics. Their slice of the energy pie will only increase if the cost of renewable energy conversion devices becomes cheaper, or the price of fossil energy becomes greater.

10.5 CONCLUSION

Global warming is caused by increasing concentrations of far-infrared absorbing gases, called greenhouse gases (GHG), including CO_2, CH_4, CFCs, N_2O, and O_3. Most of these gases are of anthropogenic origin, primarily associated with fossil fuel usage. Water vapor is also a greenhouse gas, but of course it is of natural origin, although the global hydrogeologic cycle may be altered to some degree by anthropogenic activities.

The presence of GHG in the lower atmosphere causes the outgoing terrestrial far-infrared radiation to be absorbed and partially reradiated toward the earth's surface. This radiation increases the earth's surface temperature. There is substantial evidence that the average earth's surface temperature already has increased by about 0.5–1.0 °C since the middle of the nineteenth century, and by the end of the twenty-first century the surface temperature may rise by another 2–3 °C. This

---

[17]Note, for example, the national and international criticism that is aimed at the government of China for the construction of the 18-GW hydroelectric project on the Yangtze River at the Three Gorges Dam.

temperature rise may cause associated climatic effects, such as change in precipitation patterns, more frequent and extensive tropical storms, glacier and ice-sheet melting, ocean–atmosphere interaction variability, and a rise in sea level.

Global warming could be ameliorated by slowing, or even reversing, the trend of GHG concentration increases in the atmosphere by reducing their emissions. In this chapter we discussed the possible options for reducing CO_2 emissions, the gas that bears the bulk of the greenhouse effect. The options include the following:

- Demand-side conservation and efficiency improvements, including less space heating and better insulation, less air conditioning, fluorescent lighting, more energy-efficient appliances, process modification in industry, and, very importantly, more fuel-efficient automobiles. Such measures may even incur a negative cost (i.e., consumer savings by using less energy) or at least a rapid payback period for the investment in energy-saving devices.

- Supply-side efficiency measures. Here we mean primarily increasing the efficiency of coal-fired power plants. Coal gasification combined-cycle power plants have a thermal efficiency in the 45–50% range, compared with single-cycle pulverized-coal plants in the 35–40% range. However, coal gasification power plants are more expensive than single-cycle plants.

- Capture of CO_2 from the flue gas of power plants and sequestration in terrestrial or deep ocean reservoirs. This is an expensive option, and it will be exercised only if governments mandate or subsidize it.

- Utilization of the captured CO_2. The utilization for enhanced oil and natural gas recovery is economically attractive; the utilization of CO_2 as a raw material for the production of some fuels and chemicals requires extra energy input and does not appear to be economical.

- Shift to nonfossil energy sources. The choices here are agonizing, because the largest impact could be made by shifting to nuclear electricity and hydroelectricity, both presently very unpopular and fraught with environmental and health concerns. The shift to solar, wind, geothermal, and ocean energy are popular, but because of their limited availability and intermittency and because of their larger cost compared to fossil energy, a substantial shift to these energy sources can not be expected in the near future.

- Greater use of biomass, especially wood. The use of biomass is CO_2 neutral, because as much CO_2 that evolves in burning biomass is reabsorbed in the growth of the next generation of vegetation.

- Afforestation without using the trees for 100–200 years, during which period the CO_2 concentrations in the atmosphere will decline because of exhaustion of fossil fuel resources and shift to nonfossil energy sources.

- Stopping slash and burning practices of forests, especially tropical forests, which are prodigious absorbents of CO_2, and the burning of which releases CO_2.

None of these options can ameliorate global warming by itself. They have to be taken in combination and on an incremental basis, starting with the least expensive ones and progressing to the more expensive ones. Even if the predictions of global climate change were to turn out to be exaggerated, the fact that fossil fuel usage entails many other adverse environmental and health effects, and the certainty that fossil fuel resources are finite, makes it imperative that we curtail fossil energy usage as soon as possible.

PROBLEMS

Problem 10.1

Calculate the earth's radiative temperature (K) for albedos $\alpha = 0.27, 0.3$, and 0.33, assuming the solar constant is not changing.

Problem 10.2

Calculate the radiative temperatures (K) of the planets Mars and Venus, given their solar constants $S_0 = 589$ and 2613 W m^{-2}, respectively, and their albedos $\alpha = 0.15$ and 0.75, respectively.

Problem 10.3

Given the global surface temperature fluctuations shown in Figure 10.6, use a statistical program to plot a best-fit curve through the data. According to this best-fit curve, by how much did the temperature (°C) increase from 1860 to 2000?

Problem 10.4

The volume fraction of CO_2 in the atmosphere is 370 ppmV. What is the carbon content (Gt) of the atmosphere if CO_2 is the only carrier of carbon? (The radius of the earth is 6371 km, and the atmospheric mass per unit surface area is 1.033E(4) kg m^{-2}.)

Problem 10.5

Given the CO_2 concentrations as shown in Figure 10.8, calculate the rate of increase (%/y) of those concentrations in the years 1960–2000. Use an enlargement of the 1960–2000 segment of Figure 10.8, or preferably the data available on the internet from CDIAC, Oak Ridge National Laboratory. Use an exponential, not linear, growth.

Problem 10.6

The present concentrations of CO_2 and CH_4 are 370 and 1.7 ppmV, respectively. The former grows by 0.4%/y, the latter by 0.6%/y. What will be the concentrations (ppmV) of these gases in 2100? Use exponential, not linear, growth.

Problem 10.7

A 1000-MW(el) power plant working at 35% thermal efficiency 100% of the time (base load) uses coal with a formula C_1H_1 and a heating value of 30 MJ/kg. How much CO_2 does this plant emit (metric tons/y)?

Problem 10.8

This plant substitutes NG (formula CH_4) instead of coal, with a heating value of 50 MJ/kg, in a combined cycle mode having a thermal efficiency of 45%. (a) How much NG (m^3/y at STP) is consumed? (b) How much CO_2 (t/y) is emitted?

Problem 10.9

In order to ameliorate the CO_2 caused greenhouse effect, a power plant is built that enables the capture of CO_2 from the flue gas. The plant runs on pure methane and oxygen, so the flue gas consists only of CO_2 and H_2O:

$$CH_4 + 2O_2 \rightarrow CO_2 + 2H_2O \quad -0.244 \text{ kWh(th)/mole } CH_4$$

where the minus sign means that the process is exothermic; that is, energy is evolved. The plant must produce its own O_2. From oxygen suppliers we know that the energy requirement for air separation is 250 kWh (electric) per metric ton of O_2. An equal amount of electric energy is required for liquefaction of 1 metric ton of CO_2. The plant has a gross thermal efficiency $\eta_{gross} = 0.45$ (that is, thermal efficiency calculated before electricity is syphoned off for O_2 production and CO_2 liquefaction). (a) What percentage of the plant's electricity output is available for dispatch into the grid? (b) What is the net thermal efficiency (electricity output per heat input after O_2 production and CO_2 liquefaction)?

BIBLIOGRAPHY

Brown, M. A., M. D. Levine, J. P. Romm, A. H. Rosenfeld, and J. G. Koomey, 1998. Engineering–Economic Studies of Energy Technologies to Reduce Greenhouse Gas Emissions: Opportunities and Challenges. *Annu. Rev. Energy Environ.*, **23**, 287–387.

Eliasson, B., P. W. F. Riemer, and A. Wokaun, Eds., 1999. *Greenhouse Gas Control Technologies.* Amsterdam: Pergamon Press.

Herzog, H., D. Golomb, and S. Zemba, 1991. *Feasibility, Modeling and Economics of Sequestering Power Plant CO₂ Emissions in the Deep Ocean.* Environ. Prog., **10**, 64–74.

Houghton, J. T., 1997. *Global Warming.* Cambridge: Cambridge University Press.

Mahlman, J. D., 1998. Science and Nonscience Concerning Human-Caused Climate Warming. *Annu. Rev. Energy Environ.*, **23**, 83–107.

Matthai, C. V., and G. Stensland, Eds., 1994. *Global Climate Change: Science, Policy and Mitigation Strategies.* Pittsburgh: Air and Waste Management Association.

Mitchell, J. F. B., 1989. The Greenhouse Effect and Climate Change. *Rev. Geophys.,* **27**, 115–139.

Omerod, B., 1994. *The Disposal of CO₂ from Fossil Fired Power Stations.* Cheltenham: IEA Publications.

Ramanathan, V., and J. A. Coakley, 1978. Climate Modeling through Radiative-Convective Models. *Rev. Geophys. Space Phys.,* **16**, 465–490.

Riemer, P., 1993. *The Capture of CO₂ from Fossil Fired Power Stations.* Cheltenham: IEA Publications.

Schneider, S. H., 1989. The Greenhouse Effect: Science and Policy. *Science,* **243**, 771–781.

Concluding Remarks

11.1 ENERGY RESOURCES

At the beginning of this book we reviewed the energy resources available to mankind and the uses of these resources in the various countries of the world; in subsequent chapters we outlined the environmental effects associated with energy use. The supply and use of energy has other important consequences: economic and political. Energy is a necessary and significant factor of national economies; energy expenditures amount to 5–10% of the GDP in industrialized nations. The availability of adequate energy to enterprises and individuals is a national goal and is thereby affected by governmental policies.

Among fossil energy resources, coal appears to be available in abundance for at least two to three centuries, while the fluid fossil fuels, petroleum and natural gas, may last for less than a century. The availability of fluid fuel resources can be extended by manufacturing them from coal by coal gasification and liquefaction. (The manufactured fluid fuels are called synfuels.) Fluid fuels can also be obtained from unconventional resources, such as oil shale, tar sands, geopressurized methane, coal seam methane, and methane hydrates lying on the bottom of the oceans and under the icecaps. The manufacture of synfuels and the exploitation of unconventional fossil fuel resources will be more expensive than the exploitation of proven reserves, and the manufacturing and recovery processes will entail more severe environmental effects than those associated with exploitation of conventional reserves.

Electricity is an essential energy component of modern industrialized societies; its use is increasing worldwide. Electricity is a secondary form of energy; it has to be generated from primary energy sources. Presently about two-thirds of the world's electricity is generated from fossil fuels while the other third comes from hydroenergy and nuclear energy, with very minor contributions from wind, biomass, and geothermal sources.

In 1997, about 17% of the world's electricity and 6.3% of its energy was supplied by nuclear power plants. The global resources of the raw material for nuclear power plants—uranium and thorium—would last centuries, at current usage rates. These resources can be extended even further in the so-called breeder reactors, where artificial fissile isotopes can be generated from natural uranium and thorium. Nuclear power plants are much more complex and expensive to build and operate than fossil-fueled plants. Also, the real and perceived hazards of nuclear power plants, including the risks of the entire nuclear fuel cycle, from mining to refining to radioactive waste disposal, militates against building new nuclear power plants in many countries. However, with depleting fossil fuel resources and the associated environmental risks of fossil-fueled power plants, notably global warming due to CO_2 emissions, it is likely that in the future nuclear power will again assume a substantial share of the world's electricity generation.

Hydropower is a relatively clean source of electricity, but most of the high-gradient rivers and streams that are near population centers have already been dammed up for powering the turboelectric generators. Damming up more rivers is encountering increasing public resistance, because of the risk to the watershed ecology and because it may entail massive population displacement. While several new hydroelectric power generators are being built or planned, notably the 18-GW hydroelectric station on the Yangtze River, hydropower is not expected to increase its share substantially among other sources of electricity.

Other renewable energy sources, besides hydropower, hold the promise of occupying an increasing share among electricity generators. Renewable sources are biomass, geothermal, wind, solar thermal and thermal electric, photovoltaic, and ocean tidal energy. Biomass and geothermal plants are able to supply electric power dependably on a daily and annual basis. The other sources of renewable energy have diurnal and seasonal rhythms that do not necessarily match the demand for electric power. Because electricity cannot be directly stored, the renewable generators usually need to be backed up by conventional power sources. However, when producing electricity, the renewable sources can displace fossil fuel consumption and reduce air pollutant emissions. Renewable electricity generators require a greater capital investment than fossil power plants and are currently not economically competitive with these power plants.

11.2 REGULATING THE ENVIRONMENTAL EFFECTS OF ENERGY USE

Mitigating the adverse environmental effects of energy use has been a chronic problem afflicting nations worldwide, because national economies do not automatically respond to limit environmental degradation.

Economists label the release of pollutants into the environment as an externality, an activity that does not enter into the cost of production of a good or supply of a service. The capacity of the environment to tolerate the discharge of pollutants is considered a free good; the polluter pays no price for its use. But the environmental capacity to endure pollution is finite, and the cumulative effects of pollution from many sources degrades its quality, adversely affecting the interests of society as a whole to a much greater extent than the interests of one or even all the polluters. Although the total social cost of bearing the ill effects of pollution outweighs the cost of eliminating the pollution, the polluters share of these social costs are too small to offset his abatement cost, so that abatement is uneconomic for each polluter, both individually and collectively.

A common solution to this social and economic dilemma is government regulation of pollutant producing activities. Most often this takes the form of a performance requirement, such as a standard of maximum emissions per fuel input from energy using sources, or a requirement that certain pollution reduction technologies be employed. Less often, economic incentives to abatement are used, such as emission taxes, pollutant fines, or tax deductions and credits. Except for the case of deductions and credits, the cost of abatement is borne by the polluter, thereby internalizing the externality of pollution into the production process.

In almost all cases of pollution of air, water, or soil, the amounts of toxic pollutants emitted are only a tiny fraction of the fuel burned or material processed. In general, the cost of reducing ordinary pollutant emissions is only a small fraction of the economic value added to the production process, but the cost per unit of pollution removed is often very high, inevitably higher than any possible economic use of the sequestered pollutant. It is exceedingly rare that any pollutant-producing process pays for itself.

The goal of environmental regulation is to achieve social and environmental gains that accrue to society as a whole by regulating the activities of polluting enterprises without substantially vitiating their societal benefits. The role of technology in this effort is to provide the least costly way to achieve this goal. In the United States, for example, the responsibility for protecting the public health and welfare from harmful substances in the environment that are of anthropogenic origin is lodged with the Environmental Protection Agency (EPA). The regulatory power of the EPA is embodied in a series of legislative acts setting forth the activities to be regulated, the requirement for promulgation of regulations, and their enforcement in federal courts. Frequently, legislation is quite specific concerning how activities are to be regulated and also concerning timetables for achieving progress. In general, the legislation requires no balancing of costs and benefits; the economic costs of abating pollution are to be paid by the polluters, not the government, but funds have been appropriated to help local municipalities renovate municipal waste treatment plants and a superfund has been established to clean up abandoned toxic waste dumps. The EPA accomplishes its task by setting nationally uniform standards for air and water quality and also for the processes that lead to their contamination. For example, the EPA has promulgated National Ambient Air Quality Standards for several prominent air pollutants associated with fuel combustion, and it has also published emission or process standards for the sources of these pollutants. The EPA may require emission limits or the use of effective control technology for specific classes of sources, such as stationary sources (e.g., power plants) and mobile sources, or may regulate the properties of fuel, especially motor vehicle fuel. These means have been effective in reducing the environmental effects of energy use despite the steady growth in the consumption of energy.

Surprisingly, increasing the efficiency of energy use plays no direct role in environmental regulation of urban and regional pollutants because the degree of abatement needed is very much greater than can be garnered by the modest energy efficiency gains that are economical, while the cost of the requisite abatement technology is moderate. Nevertheless, there is some environmental benefit that accrues to energy efficiency improvement. Reducing electric power consumption by increasing the efficiency of its use would reduce the air pollutant emissions from power plants, given any level of control technology. Process modification could lead to lesser use of fossil fuels in manufacturing of industrial goods. In the commercial and residential sector, fossil energy use, and thereby pollution abatement, could be achieved by better insulation in buildings, replacing incandescent with fluorescent lighting, and using solar or geothermal space and water heating. In the transportation sector, great savings could be accomplished in fossil fuel usage and concomitant pollutant emissions by traveling in small, light vehicles or using either (a) hybrid internal combustion engines and electric motors for vehicle propulsion or (b) more efficient fuel-cell-powered electric motors.

11.3   GLOBAL WARMING

The accumulation in the atmosphere of greenhouse gases (mostly carbon dioxide, but with nonnegligible contributions from nitrous oxide, methane, and chlorofluorocarbons), which threatens to cause changes in the global climate and thus have adverse environmental effects, has generated international concern. Noncumulative urban and regional pollution, with but rare exceptions, has been seen as a national problem to be solved within the constraints of national economies. However, global climate change is clearly a global environmental problem requiring the coordinated action of many nations over long time periods, on the order of a century. The regulatory regime used by a nation to cope with environmental degradation within its borders does not exist in the international community. Therefore, it will be necessary to seek multilateral international agreements for GHG emission control by the major emitting nations if this growing problem is to be ameliorated.

Any effective program for limiting the amount of climate change to be experienced in the twenty-first and subsequent centuries must necessarily have significant impacts on the supply and consumption of energy by nations and on their economies. The magnitude and comprehensiveness of the needed adjustments to the energy economy will be much greater than what has been required for dealing with urban and regional pollution. The first steps toward seeking international agreement for climate change control were taken in the 1990s. These included seeking an international scientific consensus on the understanding of climate change and the prospective effects of remedial measures, along with a plan for allocation of national annual emission caps to be implemented within the first two decades of the twenty-first century. This consensus resulted in the Kyoto protocol of 1997, which aims to reduce annual global GHG emissions by an average of 5.5% below 1990 levels. However, as of the year 2000, none of the industrialized nations has ratified the Kyoto protocol. Nevertheless, individual countries are assessing (a) the measures they may need to secure some degree of control over their respective national GHG emissions and (b) the policies and procedures that will best meet their national objectives in dealing with climate change.

The emissions of non-CO_2 greenhouse gases can be reduced without resource to heroic measures. Anthropogenic methane emissions can be relatively easily reduced by preventing leakage of the gas from gas wells, pipes, tanks, tankers, coal mines, and landfills. However, nitrous oxide emission control is mired in uncertainty because we do not fully understand the sources of emission of this gas. The manufacture of CFC is being phased out worldwide as a consequence of an international treaty, the Montreal Convention of 1987. Unfortunately, because of their very long lifetimes in the atmosphere, CFC will contribute to global warming for many decades to come.

The reduction of emissions of carbon dioxide is a problem of a different kind and magnitude than reducing emissions of the other GHG pollutants. Eighty-eight percent of the world's primary energy sources and 63% of electricity generation comes from fossil fuels. Reducing CO_2 emissions to the atmosphere simply means lowering the rate of consumption of fossil fuels. Burning less fossil fuels or replacing them by other energy sources would involve a radical change in our energy supply structure. Traditionally, population and economic growth was always associated with increase of fossil fuel usage, not the inverse. Providing a burgeoning world population with energy and bridging the energy consumption inequalities that now exist between developed and less developed nations would normally require more fossil energy consumption, not less. There are no clear-cut solutions to this dilemma.

In this book we described some of the alternatives to increased fossil fuel consumption, such as replacing fossil-fueled electricity production by nuclear and renewable sources, energy efficiency improvements in fossil fuel usage, and sequestration methods for carbon dioxide. Most of these alternatives require much greater capital investments and/or higher prices for commodities (including electricity), that use primary energy for their production. These measures imply great changes in how energy is supplied and utilized and will undoubtedly require government intervention in the energy marketplace.

In conclusion, while urban and regional environmental pollution is still a major problem, especially in developing nations, the experience in industrialized nations has shown that it is technically and economically solvable and politically manageable on the time scale of several decades. It remains to be seen whether the much greater problem of global warming can be solved with comparable measures sustained over the next centuries.

Measuring Energy

Energy in materials, such as fossil or nuclear fuels, and electrical energy delivered by power lines are commodities in industrialized economies, being traded in the marketplace. There must be standards for measurement of the energy content and other pertinent properties of these commodities for this market to operate efficiently. These standards of measurement are derived from the development of modern science and technology, where agreement among researchers about how to quantify the results of their experiments is essential to continued scientific progress.

By common agreement among scientists of all nations, a system of units of measurement has been selected: the International System of Units, or SI for short. The SI system defines seven base units of measurement which are mutually independent of each other, it being impossible to measure one unit in terms of any other. Furthermore, all other physical quantities may be measured in terms of one or more of these units. The magnitudes of the base units have been arbitrarily chosen, but are clearly defined by agreement. The defined base units are the meter (length), kilogram (mass), second (time), ampere (electric current), kelvin (thermodynamic temperature), mole (amount of substance), and candela (luminous intensity). The first six of these are used in this text. They are listed in the first section of Table A.1 of this Appendix, "Base units," together with their abbreviated symbols.[1]

There are many physical quantities that arise in scientific studies for which it is useful to define a unit of measurement that is derived from the base units by a well-known physical law. For example, the SI unit of force, the newton, is defined as the magnitude of force which, when applied to a 1-kg mass, will cause the mass to experience an acceleration of 1 m/s^2. By Newton's law of motion (force = mass × acceleration), 1 newton must equal 1 kg m/s^2. For the physical and chemical quantities of interest in this book, Table A.1 lists these derived units in its second section, "Derived units."

Among the derived SI units, the unit of energy is the joule (J) and that of power, the time rate of energy use, is the watt (W), which equals one joule per sec (J/s). In terms of mechanical units, a joule equals one newton meter (Nm), or unit force times unit distance, and a watt is one newton meter per second (Nm/s), or unit force times unit velocity. In terms of common electrical units, a joule equals one volt ampere second (VAs), or unit charge times unit electric potential, and a watt equals one volt ampere (VA), or unit current times unit electric potential.

For various practical reasons, including the needs of commerce and historical usage that preceded modern science, additional SI units have been defined. The ones pertinent to this text are

---

[1]Units named after a scientist (newton, ampere, kelvin, etc.) are not capitalized when spelled out but their abbreviations (N, A, K, etc.) are capitalized.

TABLE A.1 SI Units

Measurement	Unit	Symbol	SI Base Unit Value[a]
Base units			
Length	meter	m	m
Mass	kilogram	kg	kg
Time	second	s	s
Electric current	ampere	A	A
Thermodynamic temperature	kelvin	K	K
Amount of substance	mole	mol	mol
Derived units			
Plane angle	radian	rad	1
Solid angle	stearadian	st	1
Frequency	hertz	Hz	$1/s$
Force	newton	N	$kg\ m/s^2$
Pressure	pascal	Pa	$N/m^2 = kg/m\ s^2$
Energy	joule	J	$Nm = kg\ m^2/s^2$
Power	watt	W	$J/s = kg\ m^2/s^3$
Electric charge	coulomb	C	$A\ s$
Electric potential	volt	V	$W/A = kg\ m^2/A\ s^3$
Electric capacitance	farad	F	$C/V = A^2\ s^4/kg\ m^2$
Electric resistance	ohm	Ω	$V/A = m^2\ kg/A^2\ s^3$
Magnetic flux	weber	Wb	$V\ s = kg\ m^2/A\ s^2$
Magnetic flux density	tesla	T	$Wb/m^2 = kg/A\ s^2$
Inductance	henry	H	$Wb/A = kg\ m^2/A^2\ s^2$
Activity (radionuclide)	becquerel	Bq	$1/s$
Absorbed dose	gray	Gy	$J/kg = m^2/s^2$
Dose equivalent	sievert	Sv	$J/kg = m^2/s^2$
Defined units[a]			
Plane angle	degree (angular)	°	$(\pi/180)$ **rad**
Length	nautical mile	nmile	**1852** m
Speed	knot (nmile/h)	kt	**0.51444** m/s
Area	hectare	ha	10^4 m²
Volume	liter	L	**1E(−3)** m³
Mass	ton (metric)	ton	**1E(3)** kg
Pressure	bar	bar	**1E(5)** Pa
Time	minute	min	**60** s
Time	hour	h	**60** min = **3600** s
Time	day	d	**24** h = **86400** s
Time	year (365 days)	y	**3.1536 E(7)** s
Temperature	centigrade	°C	K
Activity (radionuclide)	curie	Ci	**3.7E(10)** Bq
Absorbed dose	rad	rd	**1E(−2)** Gy
Dose equivalent	rem	rem	**1E(−2)** Sv
Energy	calorie (Int. table)	cal_{IT}	**4.1868** J

[a]Boldface values are exact.

TABLE A.2 U.S. Commercial Units

Quantity	Unit	Symbol	SI Unit Value[a]
Length	inch	in.	**2.54E(−2) m**
Length	foot	ft	**3.048E(−1) m**
Length	mile (statute)	mile	5280 ft = 1.609E(3) m
Area	acre		4.0469E(3) m^2
Volume	gallon (U.S.)	gal	3.7854E(−3) m^3
Volume	barrel	bbl	42 gal (US) = 1.5899E(−1) m^3
Force	pound (force)	lbf	4.448 N
Mass	pound (mass)	lbm	4.5359E(−1) kg
Mass	ton (short)		2000 lbm = 9.07185E(2) kg
Energy	British thermal unit	Btu	1.05506E(3) J
Energy	Quad (1E(15) Btu)	Q	1.05506E(18) J
Energy	therm	therm	1.05506E(8) J
Power	horsepower	hp	**7.46 E(2) W**
Pressure	pound (force)/square inch	psi	6.895E(3) Pa
Temperature	Fahrenheit	°F	**(5/9) K**

[a] Boldface values are exact.

listed in the third section of Table A.1, labeled "Defined units." Among these we note the nautical mile, which is the distance along the earth's surface corresponding to a minute of latitude, and the velocity of a knot, or one nautical mile per hour, the usual unit of wind speed. The unit for small geographic areas is the hectare, and the laboratory scale measure of volume is the liter (L). Large masses are usually measured in metric tons (ton). The chemist's unit of energy is the calorie, which equals the amount of heat required to warm a gram of water by one degree centigrade.

Unlike all the other industrialized nations, the United States has not adopted the metric system of measurement for domestic and commercial purposes, but continues with the system of units it inherited from England. To facilitate conversion to SI units, the values of pertinent U.S. commercial units are listed in Table A.2.

Many of these units are commonly used in the United States. For example, the gallon (gal) is the unit volume for retail sales of vehicle fuel, whereas the barrel is the preferred unit of international petroleum suppliers and refiners. The unit of force is the pound force (lbf), while the unit of mass is the pound mass (lbm). For large amounts of mass, the short ton is used, to be distinguished from the metric ton. The energy unit is the British thermal unit (Btu), the amount of heat required to warm a pound mass of water by one degree Fahrenheit. The unit of power is the horsepower (hp). A common pressure unit is the pound force per square inch (psi).

The conversion of quantities from one system of units is straightforward. For example, to convert x Btu/lbm to SI units of J/kg, multiply by the conversion factors from Table A.2,

$$x \text{ Btu/lbm} = x \left(\frac{\text{Btu}}{\text{lbm}}\right) \left(\frac{1.05506E(3) \text{ J}}{\text{Btu}}\right) \left(\frac{\text{lbm}}{4.5359E(-1) \text{ kg}}\right) = 2.326E(3) \, x \text{ J/kg}$$

In explaining the technology of energy systems in this text we have need to invoke the principles of physics, chemistry, and thermodynamics. The quantitative application of these sciences requires the measurement of certain properties of atoms and molecules that are used in defining important

TABLE A.3 Measured Quantities

Quantity	Symbol	SI Unit Value
Charge of electron		1.6030E($-$19) C
Electron volt	eV	1.6030E($-$19) J
Faraday	$\mathcal{F}$	9.6485E(4) C/mol
Unified atomic mass unit	amu	1.66054E($-$27) kg
Avogadro's number	N_0	6.0221E(23)/mol
Universal gas constant	R	8.3143E(3) J/kg K
Standard gravitational acceleration	g	9.80665 m/s^2
Standard atmospheric pressure		1.01325E(5) Pa
Melting point of ice (0 °C = 32 °F)		273.15 K
Stefan–Boltzmann constant	σ	5.6704E($-$8) W/m^2

TABLE A.4 SI Unit Prefixes

Factor	Prefix	Symbol	U.S. Word Modifier
1E(18)	exa	E	
1E(15)	peta	P	quadrillion
1E(12)	tera	T	trillion
1E(9)	giga	G	billion
1E(6)	mega	M	million
1E(3)	kilo	k	thousand
1E(2)	hecto	h	hundred
1E($-$1)	deci	d	
1E($-$2)	centi	c	percent
1E($-$3)	milli	m	
1E($-$6)	micro	μ	
1E($-$9)	nano	n	
1E($-$12)	pico	p	

physical constants. Some of these that we use are listed in Table A.3. Among these are the electron volt (eV), the Faraday ($\mathcal{F}$), Avogadro's number (N_0), and the universal gas constant (R), used in analyzing electrochemical processes.

The SI units of Table A.1 are often of inconvenient size. Just as paper currency comes in different denominations, physical quantities need to have different sizes to accommodate different uses. The SI system includes the use of prefixes to change the size of units by factors of 10, up or down—for example, kilometer (km), centimeter (cm), micrometer (μm). Table A.4 lists the SI unit prefixes that cover a range of 30 orders of magnitude, enough for most practical purposes. Occasionally we use practical, if not always logical, units such as the unit of electrical energy, the kilowatt hour (kWh), which equals 3.6 megajoules (MJ). The kilowatt hour, an amount of energy that will light a kilowatt bulb for one hour, is a better unit for commercial use than the megajoule.

INDEX